Lecture Notes in Applied and Computational Mechanics

Volume 13

Series Editors

Prof. Dr.-Ing. Friedrich Pfeiffer
Prof. Dr.-Ing. Peter Wriggers

Springer

Berlin
Heidelberg
New York
Hong Kong
London
Milan
Paris
Tokyo

Advanced Mathematical and Computational Geomechanics

Dimitrios Kolymbas (Editor)

Springer

Professor Dr. Dimitrios Kolymbas
Institute for Geotechnical and Tunnel Engineering
University of Innsbruck
Techniker Str. 13
A-6020 Innsbruck
Austria

e-mail: dimitrios.kolymbas@uibk.ac.at

With 156 Figures

Cataloging-in-Publication Data applied for
Bibliographic information published by Die Deutsche Bibliothek
Die Deutsche Bibliothek lists this publication in the Deutsche Nationalbibliografie;
detailed bibliographic data is available in the Internet at <http://dnb.de>

ISBN 3-540-40547-X Springer Verlag Berlin Heidelberg New York

Springer-Verlag Berlin Heidelberg New York
a member of BertelsmannSpringer Science+Business Media GmbH

http://www.springer.de

Printed in Germany

Cover design: design & production GmbH, Heidelberg
Typesetting: Digital data supplied by editor

Printed on acid-free paper 62/3020Rw -5 4 3 2 1 0

Preface

Geomechanics is the mechanics of geomaterials, i.e. soils and rocks, and deals with fascinating problems such as settlements (think of the Pisa tower), stability of excavations, tunnels and offshore platforms, landslides, earthquakes and liquefaction. In a long range it is conceivable that geomechanics will explain the deformation and fracture of the earth crust, as this is being analyzed by structural geology. Geomaterials, however, have an extremely complex mechanical behaviour: They can undergo very large deformations (cf. folding of large rock strata) exhibiting brittle or ductile behaviour, plastic flow, they can change their density upon shearing, they can interact with fluids entrapped within their pores, they can creep etc. The range of relevant scales in geometry and time is from nanometers to kilometers and from milliseconds to millions of years. On the top, geomaterials are rarely encountered in a virgin state, as they are the result of huge deformations, weathering and fracture. Research in Geomechanics is faced with problems which have — to an increasing degree — mathematical contents. To treat such problems engineers have to resort to their mathematical education which, as a rule, is rather limited and cannot be compared with the one of professional mathematicians. Recent advances of mathematics are, often, inaccessible to them.

In fact, the field of Geotechnical Engineering and Geomechanics is full of mathematical problems: We still do not possess a mathematical model that describes all aspects of soil behaviour, as revealed by experiments. The principal difficulties arise from the hysteretic behaviour. Even more difficult is the description of rock which, being a discontinuum cannot be described as a so-called simple material and must be endowed with a non-Euclidean internal geometric structure. On the top, the behaviour of rock depends on the time scale of observation. Whenever geomaterials have to be considered as mixtures of two or three phases (solid, liquid, gas), then the underlying balance and constitutive equations have to be formulated using complex averaging procedures, since experiments are often unfeasible. The underlying averaging procedures are mathematically very demanding, especially when we have to consider surface effects (as is the case for unsaturated soil). Wave propagation in such media poses additional mathematical difficulties. In view of the aforementioned problems the present mathematical models (i.e. constitutive models) are altogether simple approximations, although some of them exhibit a prohibitive complexity. Even though, the numerical solutions of boundary value problems of practical relevance encounter very soon limits, where the problems become ill-posed in such a way that many of our numerical tools break down. These mathematical problems are inherently related to peculiar pattern formations that are observed at the surface and in the interior of samples tested in the laboratory (see the article by J. Desrues in this volume).

Many mathematical applications in geotechnical problems can be drastically subsumed as follows: The problem is modeled by a system of linear equations that can be solved by computer. The encountered mathematical problems arise from several sources of non-linearity (geometric and/or material) and from the appearance of vanishing eigenvalues that signalizes non-uniqueness of solution. Engineers try to overcome such mathematical problems in a 'pragmatic way', which is in accordance to their education. Often, numerical tricks are applied that reside in an appropriate manipulation of numerical control parameters. As a result, the transparency and reliability of solutions is suffering. Therefore a deeper cooperation of mathematicians and engineers is urgently needed.

Another mathematical problem is how to express safety of geotechnical constructions? Engineers use the "factor of safety", which however is not a physical quantity (as e.g. temperature) and has many conceptual deficiencies. The probabilistic approach appears reasonable but is burdened by many difficulties.

In concluding I wish to mention that geotechnical engineering and geomechanics seeks to comply with LEONARDO DA VINCI saying that

> *Nessuna umana investigazione si può denominare vera scienza s' essa non passa per le matematiche dimonstrazioni*

i.e. research can only lead to true science if it is guided by mathematical methods.

Interdisciplinarity is a very fruitful but still a very cumbersome venture. When people having acquired different languages and approaches in a long series of years are brought together, they need a considerable amount of patience and enthusiasm, which is rarely encountered. All the more, I wish to cordially thank

- The participants and speakers of the three Euroconferences GeoMath (" Mathematical Methods in Geomechanics") that were held in 1-3 March 2000 (Innsbruck), 14-16 February 2001 (Innsbruck), 2-5 July 2002 (Horto/Greece), out of which the present collection of contributions emanated.
- The sponsors:
  - European Commission
Research Directorate-General
Directorate F: Human Potential and Mobility
Research Training Networks
High-Level Scientific Conferences, Contract No: HPCFCT-1999-00046
  - Federal Ministry for Education, Science and Culture
Minoritenplatz 5
Vienna

- All those persons who contributed to the success of the above mentioned events and the preparation of this volume. A special thank deserves Mrs. Christine Neuwirt, who tireless and with exemplary efficiency contributed to the successful organisation creating a pleasant and relaxed atmosphere and solved all financial problems. I also wish to thank cand. ing. Josef Wopfner, who was responsible for the re-formatting of the manuscripts. Despite all modern computer programmes (perhaps also: because of them) this is a very lengthy endeavor which he mastered with bravery rendering thus an attractive and homogeneous look to the present volume.

Innsbruck,
May 2003

Dimitrios Kolymbas

Contents

Authors (alphabetical listing)

Marta Castelli
Department of Structural and Geotechnical Engineering
Politecnico di Torino, Corso Duca degli Abruzzi 24, I 10129 Torino
Phone: 0039 011 564 4877
Fax: 0039 011 564.4899
e-mail: marta.castelli@polito.it

Jacques Desrues
CNRS - Universit de Grenoble, Laboratoire 3S - BP 5X, 38041 Grenoble Cedex 9, FRANCE
Phone: (33) 4 76 82 51 73
Fax: (33) 4 76 82 70 00
e-mail: jacques.desrues@hmg.inpg.fr
URL: http://www.3s.hmg.inpg.fr

Wolfgang Fellin
Institute of Geotechnical and Tunnel Engineering,
University of Innsbruck, Techniker Str. 13, A-6020 Innsbruck
Phone: 0043 (0)512 507 6672
Fax: 0043 (0)512 507 2996
e-mail: wolfgang.fellin@uibk.ac.at
URL: http://geotechnik.uibk.ac.at/staff/fellin-de.html

Ivo Herle
Institute of Geotechnical and Tunnel Engineering,
University of Innsbruck, Techniker Str. 13, A-6020 Innsbruck
Phone: + 43 512 507-6674
Fax: + 43 512 507-2996
e-mail: ivo.herle@uibk.ac.at
URL: http://geotechnik.uibk.ac.at/staff/ivo.html

Guy T. Houlsby
Department of Engineering Science
Oxford University, Parks Road, Oxford OX1 3PJ,UK
Phone: 0044 (0)1865 273162
Fax: 0044 (0)1865 283301
e-mail: guy.houlsby@eng.ox.ac.uk
URL: http://www-civil.eng.ox.ac.uk/people/gth.html

Antonio Huerta
Laboratori de Càlcul Numèric (LaCàN)
Departament de Matemàtica Aplicada III, E.T.S. de Ingenieros de Caminos
Edifici C2, Campus Nord, Universitat Politècnica de Catalunya,
E-08034 Barcelona, Spain.
Phone: (+34) 93 401-6916
Fax: (+34) 93 401-1825
e-mail: antonio.huerta@upc.es
URL: http://www-lacan.upc.es

Milan Jirásek
Laboratory of Structural and Continuum Mechanics, Swiss Federal Institute of Technology at Lausanne (EPFL), 1015 Lausanne, Switzerland
Phone: +41-21-6932413
Fax: +41-21-6936340
e-mail: milan.jirasek@epfl.ch
URL: http://lscwww.epfl.ch/jirasek.page

Dimitrios Kolymbas
Institute of Geotechnical and Tunnel Engineering,
University of Innsbruck, Techniker Str. 13, A-6020 Innsbruck
Phone: + 43 512 507 6670,
Fax: + 43 512 507 2996
e-mail: dimitrios.kolymbas@uibk.ac.at
URL: http://geotechnik.uibk.ac.at

Irene Morata
Laboratori de Càlcul Numèric (LaCàN)
Departament de Matemàtica Aplicada I, E.T.S. d'Enginyeria Industrial
Universitat Politècnica de Catalunya
E-08028 Barcelona, Spain.
Phone: (+34) 93 401-7077
Fax: (+34) 93 401-1713
e-mail: irene.morata@upc.es
URL: http://www-lacan.upc.es

Michael Oberguggenberger
Institut für Technische Mathematik, Geometrie und Bauinformatik
Universität Innsbruck, Technikerstr. 13, 6020 Innsbruck, Austria
Phone: 0043 (0)512 507 6824
Fax: 0043 (0)512 507 2941
e-mail: michael@mat1.uibk.ac.at
URL: http://techmath.uibk.ac.at/numbau/michael/

Alexander Ostermann
Institut für Technische Mathematik, Geometrie und Bauinformatik
Universität Innsbruck, Technikerstr. 13, 6020 Innsbruck, Austria
Phone: 0043 (0)512 507 682
Fax: 0043 (0)512 507 2941
e-mail: alexander.ostermann@uibk.ac.at
URL: http://techmath.uibk.ac.at/numbau/alex/

Antonio Rodríguez-Ferran
Laboratori de Càlcul Numèric (LaCàN)
Departament de Matemàtica Aplicada III, E.T.S. de Ingenieros de Caminos
Edifici C2, Campus Nord, Universitat Politècnica de Catalunya
E-08034 Barcelona, Spain.
Phone: +34 - 93 401 56 97
Fax: 34 - 93 401 18 25
e-mail: antonio.rodriguez-ferran@upc.es
URL: http://www-lacan.upc.es

Alexander M. Puzrin
Geosystems Group, Georgia Institute of Technology,
School of Civil and Environmental Engineering, 790 Atlantic Drive, Atlanta, Georgia 30332-0355, USA
Phone: (404) 894-7598
Fax: (404) 894-2281
e-mail: alexander.puzrin@ce.gatech.edu
URL: http://www.ce.gatech.edu/ apuzrin/

Simon Rolshoven
Laboratory of Structural and Continuum Mechanics, Swiss Federal Institute of Technology at Lausanne (EPFL), 1015 Lausanne, Switzerland
Phone: + 41 21 69 32418
Fax: +4121 693 63 40
e-mail: simon.rolshoven@epfl.ch
URL: http://dgcwww.epfl.ch/WWWLSC/lsc.html

Valeria Saetta
Department of Structural and Geotechnical Engineering
Politecnico di Torino, Corso Duca degli Abruzzi 24, I 10129 Torino
Phone: 0039 011 564 4911
Fax: 0039 011 564.4899

Claudio Scavia
Department of Structural and Geotechnical Engineering
Politecnico di Torino, Corso Duca degli Abruzzi 24, I 10129 Torino
Phone: 0039 011 564.4823
Fax: 0039 011 564.4899
e-mail: claudio.scavia@polito.it
URL: http://staff.polito.it/scavia/

Krzysztof Wilmanski
Weierstrass-Institut Institut für Angew. Analysis und Stochastik
Forschungsgruppe Kontinuumsmechanik
Mohrenstrasse 39, D 10117 Berlin, Deutschland
Phone: 0049 30 20372 545
Fax: 0049 30 2044975
e-mail: wilmansk@wias-berlin.de
URL: http://www.wias-berlin.de/research-groups/kontinua/wilmansk/

Part I

Material Modelling

On thermodynamic modeling and the role of the second law of thermodynamics in geophysics

Krzysztof Wilmanski

Weierstrass Institute for Applied Analysis and Stochastics, Berlin, Germany,
wilmansk@wias-berlin.de

Summary. The article contains a brief review of elements of thermodynamic modeling in theoretical geophysics. We motivate the existence of the second law of thermodynamics in macroscopic theoretical physics and demonstrate its evaluation. In particular we show its consequences in the construction of constitutive laws for a two-component poroelastic medium. This construction is also related to microstructural properties verified by means of the second law.

1 Introduction

The second law of thermodynamics as a fundamental principle of macroscopic physics has a rather untypical history. It has been discovered by an engineer for purely practical purposes and, as such, has not been questioned. However its justification within theoretical physics led to vehement discussions with, sometimes tragic consequences such as the suicide of Boltzmann. It was related to pessimistic philosophical considerations (the "entropic death" of the universe), enthusiastic expectations of some scientists that it may give them a tool to construct automatically models for complicated processes, etc. During the last thirty years the situation became normal – the second law of thermodynamics is now one of basic principles in the construction of macroscopic models but it is, of course, not a panacea for all our problems. Very often we can solely find some restrictions in form of inequalities but not prescriptions for determining parameters. A prominent example of such a thermodynamic result is the inequality restricting the coefficient of thermal expansion α

$$0 \leq \alpha^2 \leq \frac{1}{vT} c_p \kappa,$$

where v, T denote the specific volume and the absolute temperature, respectively, c_p is the specific heat under constant pressure and κ is the compressibility of the fluid. This inequality says that one cannot construct a model of a fluid by assuming incompressibility ($\kappa = 0$) and simultaneously admitting the thermal expansion ($\alpha^2 > 0$) which is, however, done in the so-called Boussinesq approximation [1] important for the stability analysis of fluids.

In this work we want to show some facets of the second law of thermodynamics within theoretical geophysics. We begin with a motivation of the existence of this law. To this aim we present a Gedankenexperiment whose origin goes back to Ehrenfest. This experiment was designed to reproduce considerations of Boltzmann in which he was justifying the existence of macroscopic irreversibility derived from a microscopic reversible model (the famous H-Theorem).

In the third section I present two simple examples of application of the second law of thermodynamics. The first example stems from soil mechanics and it is related to the evaluation of macroscopic material parameters by means of some microscopic data. This evaluation relies on certain assumptions concerning properties of the surface of contact between two different media. Using the second law of thermodynamics we demonstrate limitations on the practical applicability of classical results. The second example illustrates two different ways of exploitation of the second law. We show that they yield the same results. The most complicated technical part of the work is contained in section 4. We show some consequences of the second law for the construction of a nonlinear two-component (hence – fully saturated) model of poroelastic materials. The aim of this section is twofold. We present the technique which is used in the evaluation of the second law in continuum physics, and, simultaneously, we obtain the thermodynamic framework for the most popular model of porous materials used in acoustics of these media – the Biot model. Finally in the fifth section we discuss some problems of construction of macroscopic parameters of a linear model by means of the transition from the microscopic description. In this example we show the limitations imposed on such a transition by thermodynamic conditions applied to the microstructure and discussed in the example of section 2.

References to original works are rather scarce. I am quoting some more or less standard books and solely as examples some papers. There exists a vast literature on the subject and it would be irrational to account properly for all contributions in the article of this size.

2 Motivation

The second law of thermodynamics has a somewhat mysterious ring among people who do not work professionally with its applications. For this reason I begin this work with a brief motivation for the existence of this principle of macroscopic physics.

I proceed to describe one of the most impressive models or Gedankenexperiments demonstrating the source of irreversibility of macroscopic processes, and, consequently, the origin of the second law of thermodynamics. This model which I quoted in my book [2] goes back to Ehrenfest but it has been designed and discussed in many details in an almost unknown book of Mark Kac [3].

Let us consider a circle on which we locate n points in equal distance from each other. They are vertices of a polygon. A set S of m vertices is chosen, and these points have a special property which we define further. In each point n we locate a ball which may be either white or black. Now we define the dynamics of this model. We move all balls simultaneously in – say – anticlockwise direction in discrete time steps Δt. In each step balls which were located in points of the set S change the colour. All other balls do not change their colour. For simplicity we assume that at the initial instant of time $t = 0$ all balls are black. However this assumption is immaterial for properties of the dynamics of the system. It is intuitively clear that after sufficiently many steps and for the points of the set S randomly distributed on the circle approximately a half of the balls will be white and the other half black. Simultaneously if we rotate the system by the angle 4π it returns to its initial state – each ball changes its colour even times $2m$, and hence it will have again the initial colour, i.e. sufficiently long processes are periodic. Moreover processes are reversible, i.e. they are invariant with respect to the transformation $t \to -t$. Such a transformation means that we have to rotate the system in the clockwise direction. But this leads to the initial state again if we make sufficiently many steps. In addition the clockwise rotation describes exactly the same processes as the anticlockwise rotations – they become identical by renaming black balls to white and *vice versa*.

In order to describe this model analytically we denote by $N_b(t)$ the number of black balls, and by $N_w(t)$ the number of white balls. Initially $N_b(t = 0) = n$. Let $N_b(S,t)$, and $N_w(S,t)$ denote the number of black and white balls, respectively, occupying points of the set S. We have the following conservation laws

$$N_b(t) + N_w(t) = n, \quad N_b(S,t) + N_w(S,t) = m. \tag{1}$$

Simultaneously the following relations describe the discrete dynamics of the model. The number of black balls after the time step $t \to t + \Delta t$ is

$$N_b(t + \Delta t) = N_b(t) + N_w(S,t) - N_b(S,t), \tag{2}$$

because the balls occupying points of the set S in the instant of time t have changed the colour in this step. Similarly for the white balls we have

$$N_w(t + \Delta t) = N_w(t) + N_b(S,t) - N_w(S,t). \tag{3}$$

Now we use the **assumption** on random distribution of points of the set S among all points on the circle. After sufficiently long time lapse from the beginning of motion we expect that the fraction of black balls in the set S will be the same as the fraction of these balls in the set of all balls: $\frac{N_b(S,t)}{m} = \frac{N_b(t)}{n}$. The same concerns the white balls. Hence we have

$$N_b(S,t) = \frac{m}{n} N_b(t), \quad N_w(S,t) = \frac{m}{n} N_w(t). \tag{4}$$

The set of equations describing our model is now complete and we can proceed to its analysis.

If we subtract equations (2), (3) and substitute the assumption (4) the following equation for the increments follows

$$N_b(t+\Delta t) - N_w(t+\Delta t) = \left(1 - 2\frac{m}{n}\right)\left[N_b(t) - N_w(t)\right]. \tag{5}$$

This simple equation in finite differences can be immediately solved and we obtain

$$\frac{N_b(t) - N_w(t)}{n} = \frac{N_b(0) - N_w(0)}{n}\left(1 - 2\frac{m}{n}\right)^t. \tag{6}$$

The solution determines the surplus of black balls over the white balls at any instant of time t. As $N_w(0) = 0$ we can write it also in the following form

$$N_b(t) + N_w(t) = n \Longrightarrow N_b(t) = \frac{n}{2}\left[1 + \left(1 - 2\frac{m}{n}\right)^t\right]. \tag{7}$$

Reasonable values (existence of solutions under the assumption (4)!) follow under the assumption that the number m of points in the set S is smaller than $\frac{n}{2}$. Then the surplus of black balls over the white balls will decrease and in the limit $t \longrightarrow \infty$ it will be zero, i.e. a half of the balls will be black and another half – white. This agrees with our intuition. However it contradicts the periodicity of the system!

In the above considerations there is only one assumption. It is described by relations (4). This assumption is an analogon of the famous "Stoßzahlansatz" which Boltzmann made in his derivation of the kinetic theory of gases. The whole construction presented above is a very simplified version of the reasoning of Boltzmann and the conclusion following from the relation (6) corresponds to his H-theorem. This theorem is the prototype of the modern second law of thermodynamics. Simultaneously the assumption (4) is a stochastic element of the model and such a feature possess all our macroscopic models in physics.

In the situation of such a contradiction we either have to change the model or we have to apply a different interpretation of results. As both the assumption and the result seem to correspond well with our intuition we try to apply a different interpretation.

If the number of balls corresponded to the number of particles of a real substance (say a gas such as the air) then in typical conditions on the earth it would be of the order $\sim 10^{23}$ (Avogadro number). For the time step $\Delta t = 10^{-2}$ sec this would mean that the system would return to its initial state (reccurence time) after $2 \cdot 10^{21}$ seconds, i.e. after approximately $6 \cdot 10^{13}$ years. This is more than 1000 times longer than the lifetime of the universe since the big bang! Consequently we may consider the solution (6) to be a good approximation if the time lapse from the beginning of the process is neither too long nor too short.

Certainly the above very simple model cannot give even a hint whether real processes in nature are reversible or even periodic in a sense of very long observation times. However it justifies the statement that a stochastic element involved in any description of macroscopic processes in realistic times of observation yields **irreversibility**. This is also the reasonable approach in our macroscopic theories of geophysical systems.

The theory constructed by Boltzmann yields a criterion for the time behaviour described by such solutions as (6). Namely for isolated systems, i.e. systems which do not interact with the external world, one can introduce a function which depends on parameters of the system and which grows in time until it reaches a limit state – the thermodynamic equilibrium – and then remains constant. We denote this function by H and call it the **entropy function**. We have then

$$\frac{dH}{dt} \geq 0, \tag{8}$$

in each instant of time for real processes, i.e. for solutions of equations describing the system. In a different way without refering to stochastic arguments unknown at that time such a criterion for admissibility of models has been introduced by Rudolf Clausius in 1850 who invented the name entropy and by William Thomson Kelvin in 1851 who called it the degradation of energy. Both were motivated by the famous work of Nicolas Carnot on the efficiency of heat engines published in 1824.

The inequality (8) which is predesignated to lead to the second law of thermodynamics possesses already the most important feature of this law. It predicts the **relaxation** of isolated systems to the thermodynamic equilibrium in which $\frac{dH}{dt} = 0$.

If the system is not isolated from the external world there exists an exchange of mass, momentum, energy etc. between the system and the rest of the world which may contribute to changes of H. Then we have to correct the inequality (8) by a supply term Φ describing the amount of the entropy H carried into the system from the external world per unit time

$$\frac{dH}{dt} + \Phi \geq 0. \tag{9}$$

As the entropy itself, it depends on parameters of the system. In the next section we show some simple examples of such a supply.

The entropy function H and its supply Φ are defined for the whole system. However we can extend this statement requiring that this law must hold as well for some subsystems defined, for instance, on a part $\mathcal{P}$ of the domain $\mathcal{B}$ occupied by the system in the three-dimensional Euclidean space of motion. In such cases, most common in practical applications, it is also required that the entropy is an **additive** function, i.e for two separate subsystems $\mathcal{P}_1, \mathcal{P}_2$, with $\mathcal{P}_1 \cap \mathcal{P}_2 = \emptyset$, the entropy of the sum of these two systems is equal to the sum of entropies: $H(\mathcal{P}_1 \cup \mathcal{P}_2) = H(\mathcal{P}_1) + H(\mathcal{P}_2)$. Certain additional mathematical assumptions which are the same as in any continuum model (e.g. [4]) lead then to the existence of the entropy density η which satisfies the following relations

$$H(\mathcal{P}) = \int_{\mathcal{P}} \rho\eta dV, \quad \frac{d}{dt}\int_{\mathcal{P}} \rho\eta dV + \underbrace{\oint_{\partial\mathcal{P}} \mathbf{h}\cdot\mathbf{n} dS - \int_{\mathcal{P}} \rho s dV}_{=\Phi} \geq 0. \tag{10}$$

In these relations η is called the specific entropy, ρ is the mass density of the system, i.e. $M(\mathcal{P}) = \int_{\mathcal{P}} \rho dV$ is the mass of $\mathcal{P}$, $\mathbf{h}$ denotes the flux of entropy through the unit surface and in the unit time, s is the so-called entropy radiation to the external world, and $\mathbf{n}$ denotes a unit normal vector of the surface (boundary) $\partial\mathcal{P}$ of the system. This inequality can be written in the local form by means of the Stokes theorem on the transformation of surface integrals for closed surfaces into volume integrals

$$\frac{\partial(\rho\eta)}{\partial t} + \operatorname{div}(\rho\eta\mathbf{v} + \mathbf{h}) - \rho s \geq 0, \tag{11}$$

where $\mathbf{v}$ is the velocity of particles of the system. It must hold **for all real processes** in the system in almost all points of the set $\mathcal{B}$. The last condition means that the local form (11) can be violated in single points, on some lines or surfaces of the system for which, however, the global form (10) still should be true in the limit sense.

The relation (11) – the so-called entropy inequality – is the modern form of the **second law of thermodynamics for continua**.

Further in this work we show how this condition can be evaluated in the construction of various models in geophysics. In the next section we discuss first two very simple examples which have, however, a certain practical bearing.

3 Two simple examples of evaluation of the second law

We begin the demonstration of consequences of the second law of thermodynamics with an example appearing in procedures of transformation of data for real granular materials into a macroscopic multicomponent model. We return to this problem in its full generality further in this work. The purpose of the present analysis is solely to derive restrictions for the microscopic model. We describe it for the so-called **unjacketed** test of a two-component granular material (e.g. [5], [6])[1]. In this model we consider homogeneous, quasistatic

[1] I am grateful to Dr. W. Dreyer (WIAS, Berlin) for suggesting this problem.

processes in a heap of a compressible granular material immersed within a fluid. On the upper surface of the fluid the pressure p' is acting. We assume that the system preserves a constant temperature $T = T_0$ (isothermal processes).

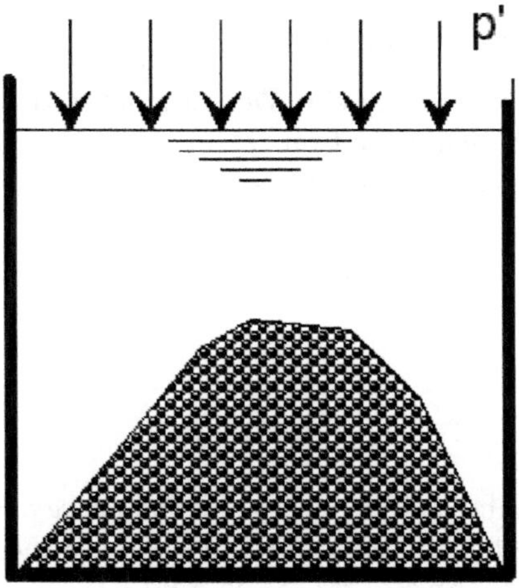

Fig. 1. Scheme of the unjacketed test

Processes in this system are described by means of the following variables: M^S – the mass of the solid granular material, M_i^F – the mass of the fluid inside the pores of the granular material, M_e^F – the mass of the fluid outside the heap, V^S – the volume of the solid granular material, V_i^F – the volume of the fluid inside the heap, V_e^F – the volume of the fluid outside the heap. We call the variables $\left\{M^S, M_i^F, M_e^F, V^S, V_i^F, V_e^F, T_0\right\}$ the process parameters. The purpose is to describe conditions between the heap and the exterior.

We proceed to exploit two laws of thermodynamics: the **conservation of energy** and the **second law of thermodynamics**. Let us denote by $\dot{Q}$ the amount of the energy which is carried into the system in a nonmechanical form (heat supply) from the external world in the unit time. Then the change of the energy E in the quasistatic process (the influence of the kinetic energy is neglected!) is described by the following first law of thermodynamics (conservation of energy)

$$\frac{dE}{dt} + \dot{Q} + \underbrace{p' \frac{d}{dt}\left(V^S + V_i^F + V_e^F\right)}_{\text{mechanical working}} = 0, \tag{12}$$

The heat supply $\dot{Q}$ is necessary for processes to pertain to isothermal conditions.

The **second law of thermodynamics** has the following form

$$\frac{dH}{dt} + \frac{\dot{Q}}{T_0} \geq 0, \tag{13}$$

where the form of the entropy supply $\Phi = \dfrac{\dot{Q}}{T_0}$ follows from the original considerations of Carnot and Kelvin. We show later that this is not always an appropriate form of such a relation.

Substitution of (12) in (13) yields

$$\frac{d\Psi}{dt} + p' \frac{d}{dt}\left(V^S + V_i^F + V_e^F\right) \leq 0, \tag{14}$$

where

$$\Psi := E - T_0 H, \tag{15}$$

is the so-called Helmholtz free energy. As the energy E and the entropy H it is assumed to be additive, i.e. it is in our example the sum of the free energy of the solid, and of the two parts of the fluid.

We make now the **constitutive assumption** which defines materials involved in the process. As the free energies possess densities this assumption shall be made for densities in a homogeneous process. Namely

$$\Psi = M^S \psi^S \left(T_0, \frac{V^S}{M^S}\right) + M_i^F \psi^F \left(T_0, \frac{V_i^F}{M_i^F}\right) + M_e^F \psi^F \left(T_0, \frac{V_e^F}{M_e^F}\right), \tag{16}$$

i.e. we assume that the densities of free energy depend on the temperature (a constant in our case), and on the specific volume $\frac{V}{M}$ for each component. The latter is, of course, identical with the inverse of the mass density, and the specific volumes have been chosen solely for historical reasons. Both fluid contributions contain the same function $\psi^F(T_0, \cdot)$ because it is **the same** fluid in the heap and outside of the heap. Now it follows from (14) by means of the chain rule of differentiation

$$\begin{aligned} &\left(\psi^S - \psi^{S\prime} \frac{V^S}{M^S}\right) \frac{dM^S}{dt} + \left(\psi_i^F - \psi_i^{F\prime} \frac{V_i^F}{M_i^F}\right) \frac{dM_i^F}{dt} + \\ &+ \left(\psi_e^F - \psi_e^{F\prime} \frac{V_e^F}{M_e^F}\right) \frac{dM_e^F}{dt} + \left(p' + \psi^{S\prime}\right) \frac{dV^S}{dt} + \\ &+ \left(p' + \psi_i^{F\prime}\right) \frac{dV_i^F}{dt} + \left(p' + \psi_e^{F\prime}\right) \frac{dV_e^F}{dt} \leq 0, \end{aligned} \tag{17}$$

where the indices i, e denote the evaluation at the corresponding values of the arguments, and the prime denotes the derivative with respect to the specific volume. This inequality must be satisfied for all real processes in our system. However real processes must satisfy the following constraints

$$M^S = \text{const.}, \quad M_i^F + M_e^F = \text{const.}, \tag{18}$$

which are, of course, the mass conservation laws. Otherwise the values of time derivatives in (17) are arbitrary. It means that the inequality can be satisfied solely in the case when the coefficients of these derivatives vanish identically. We obtain

$$p^S := -\frac{d\psi^S}{d\left(\frac{V^S}{M^S}\right)} = p', \quad p_i^F := -\left.\frac{d\psi^F}{d\left(\frac{V^F}{M^F}\right)}\right|_{\frac{V^F}{M^F}=\frac{V_i^F}{M_i^F}} = p', \tag{19}$$

$$p_e^F := -\left.\frac{d\psi^F}{d\left(\frac{V^F}{M^F}\right)}\right|_{\frac{V^F}{M^F}=\frac{V_e^F}{M_e^F}} = p',$$

$$\psi_i^F - \psi_i^{F\prime}\frac{V_i^F}{M_i^F} = \psi_e^F - \psi_e^{F\prime}\frac{V_e^F}{M_e^F} \Longrightarrow \underline{\psi_i^F + p_i^F\frac{V_i^F}{M_i^F} = \psi_e^F + p_e^F\frac{V_e^F}{M_e^F}},$$

where p^S is the pressure in the solid, p_i^F – pressure in the fluid inside the heap (**pore pressure**), p_e^F – pressure in the fluid outside the heap.

The last condition contains the combination of the Helmholtz free energy with the pressure contribution. This combination is called the **free enthalpy**. The condition means that this quantity must be the same in the fluid inside of the heap and outside of the heap (continuity across the interface between the two parts). Simultaneously the first three relations reduce the number of necessary constitutive relations from three to two for the two Helmholtz free energies. Let us note that the above relations reduce the second law to the identity which means that the system is in thermodynamic equilibrium.

According to the above relations we can make still a further simplification provided constitutive relations are monotone with respect to the specific volume[2]

$$p_i^F = p_e^F \equiv p' \quad \Longrightarrow \quad \frac{V_i^F}{M_i^F} = \frac{V_e^F}{M_e^F}. \tag{20}$$

Hence the mass density of the fluid is the same inside and outside of the heap and the condition for the enthalpy becomes trivial.

[2] Such a relation would not hold, for instance, in the case of hysteresis (i.e. materials with a phase transformation) in which at least two different mass densities correspond to the same pressure.

The above considerations yield rather trivial results anticipated in previous works on this subject. However they expose simplifications which must be made in order to obtain the condition (20). We list the most important of them.

- The solid granular material must react on volume changes in an elastic way, i.e. its pressure is dependent solely on changes of the mass density $\frac{M^S}{V^S}$. For instance we have not incorporated any viscoplastic changes of the volume;
- The fluid is ideal, i.e. its pressure depends also solely on changes of mass density. We have neglected, for instance, viscous properties or friction between the fluid in the heap and the grains.

Viscoplastic deformations of the solid and viscous properties of the fluid would require a considerable modification of the second law (14).

- There is no influence of capillary forces which would have to appear in the energy balance through an additional term defined on the contact surface (interface) between the fluid in the heap and outside of the heap;
- All processes are homogeneous, i.e we neglect, for example, the pattern formation (e.g. a heterogeneous deformation in the form of layered structure) in the heap;
- The heap is fully saturated by the fluid;
- There is no mass exchange between the solid and the fluid, and phase transformations (e.g. evaporation and condensation) are excluded.

Certainly one can add more items to this list.

The results for this simple example show the limitations which are imposed on the classical Gassmann relations between microscopic and macroscopic properties of granular materials (see: Sect. 5 for the presentation of this problem), and, simultaneously, they indicate the way in which these relations can be improved. Clearly, if any of the above listed issues will be corrected the continutity of the pressure in the fluid (20) must be replaced by the continuity of the enthalpy $(19)_3$.

We proceed to discuss the second simple example. We investigate thermodynamic limitations for the so-called **rigid heat conductor**. It is a solid in which solely the temperature field $T(\mathbf{x}, t)$ determines processes. The field equation for this field follows from the energy conservation law. In the local form it is as follows

$$\rho \frac{\partial \varepsilon}{\partial t} + \operatorname{div} \mathbf{q} = \rho r, \tag{21}$$

where ρ is the constant mass density ε denotes the specific internal energy per unit mass $\mathbf{q}$ is the so-called heat flux vector, and r – radiation density. In a further thermodynamic analysis we neglect the last contribution because it is controlled by external agents and, for this reason, can be switched off without any influence on properties of the material. In order to transform the equation (21) into the field equation for the temperature T we have to specify the dependence of ε and $\mathbf{q}$ on the temperature and its derivatives. In the classical case it is assumed that the following simplified caloric relation for the specific energy and the Fourier relation for the heat flux, respectively, hold true

$$\varepsilon = \varepsilon(T), \quad \mathbf{q} = -K(T)\,\mathbf{g}, \quad \mathbf{g} := \operatorname{grad} T. \tag{22}$$

Substitution in (21) yields the classical heat conduction equation

$$\rho c_v \frac{\partial T}{\partial t} = \operatorname{div}\left(K \operatorname{grad} T\right)\mathbf{q} + \rho r, \quad c_v := \frac{d\varepsilon}{dT}. \tag{23}$$

We evaluate the second law for a slightly more general case. Namely we assume that the specific energy ε, the heat flux $\mathbf{q}$, the specific entropy η, and the entropy flux $\mathbf{h}$, are functions of the temeperature T and its gradient $\operatorname{grad} T$. In addition we assume this system to be isotropic. We specify consequences of the latter assumption later. Then the energy balance (21) has the form

$$\rho\left(\frac{\partial \varepsilon}{\partial T}\frac{\partial T}{\partial t} + \frac{\partial \varepsilon}{\partial \mathbf{g}} \cdot \frac{\partial \mathbf{g}}{\partial t}\right) + \frac{\partial \mathbf{q}}{\partial T} \cdot \mathbf{g} + \frac{\partial \mathbf{q}}{\partial \mathbf{g}} \cdot (\operatorname{grad} \mathbf{g}) = 0, \tag{24}$$

and the entropy inequality (11) is as follows

$$\rho\left(\frac{\partial \eta}{\partial T}\frac{\partial T}{\partial t} + \frac{\partial \eta}{\partial \mathbf{g}} \cdot \frac{\partial \mathbf{g}}{\partial t}\right) + \frac{\partial \mathbf{h}}{\partial T} \cdot \mathbf{g} + \frac{\partial \mathbf{h}}{\partial \mathbf{g}} \cdot (\operatorname{grad} \mathbf{g}) \geq 0. \tag{25}$$

We have to evaluate the above inequality provided the field T satisfies its field equation, i.e. it must satisfy the above energy conservation law. This condition is a **constraint** on the class of admissible solutions of the inequality. Such constraints are eliminated in modern thermodynamics by means of Lagrange multipliers. We discuss this issue in some details further in this work. However, in order to show that the choice of a method for the evaluation of the second law of thermodynamics is solely a matter of convenience we proceed here by means of a direct method and then we show that the results are the same.

Namely among all possible local values of derivatives of the temperature the derivative, say, with respect to time can be calculated from the energy conservation provided the coefficient $c_v = \frac{\partial \varepsilon}{\partial T}$ is different from zero. This shall be so assumed. Then

$$\frac{\partial T}{\partial t} = -\left(\frac{\partial \varepsilon}{\partial T}\right)^{-1}\left\{\left[\frac{\partial \varepsilon}{\partial \mathbf{g}} \cdot \frac{\partial \mathbf{g}}{\partial t}\right] - \frac{1}{\rho}\left[\frac{\partial \mathbf{q}}{\partial T} \cdot \mathbf{g} + \frac{\partial \mathbf{q}}{\partial \mathbf{g}} \cdot (\operatorname{grad} \mathbf{g})\right]\right\}. \tag{26}$$

On the right hand side of this relation derivatives can be chosen arbitrarily because no other field equation appears in this model. We substitute this result in the entropy inequality (25). We obtain then

$$\rho\left[\Lambda\frac{\partial\varepsilon}{\partial\mathbf{g}}-\frac{\partial\eta}{\partial\mathbf{g}}\right]\cdot\frac{\partial\mathbf{g}}{\partial t}+\left[\Lambda\frac{\partial\mathbf{q}}{\partial T}-\frac{\partial\mathbf{h}}{\partial T}\right]\cdot\mathbf{g}+$$

$$+\left[\Lambda\frac{\partial\mathbf{q}}{\partial\mathbf{g}}-\frac{\partial\mathbf{h}}{\partial\mathbf{g}}\right]\cdot\operatorname{grad}\mathbf{g}\leq 0,\quad \Lambda:=\left(\frac{\partial\eta}{\partial T}\right)\left(\frac{\partial\varepsilon}{\partial T}\right)^{-1}. \tag{27}$$

As the derivatives $\frac{\partial\mathbf{g}}{\partial t}$, $\operatorname{grad}\mathbf{g}$ contribute in linear way to this inequality and they can be chosen arbitrary their coefficients must vanish. Hence we obtain

$$\Lambda\frac{\partial\varepsilon}{\partial\mathbf{g}}-\frac{\partial\eta}{\partial\mathbf{g}}=0,\quad \operatorname{sym}\left(\Lambda\frac{\partial\mathbf{q}}{\partial\mathbf{g}}-\frac{\partial\mathbf{h}}{\partial\mathbf{g}}\right)=0. \tag{28}$$

We proceed to use the assumption of isotropy. Then the vector functions $\mathbf{q},\mathbf{h}$ must depend on the vector argument in the following way

$$\mathbf{q}=-K\left(T,g^2\right)\mathbf{g},\quad \mathbf{h}=-H\left(T,g^2\right)\mathbf{g},\quad g^2:=\mathbf{g}\cdot\mathbf{g}. \tag{29}$$

Consequently the second relation (28) can be written in the form

$$-\left(\Lambda K-H\right)\mathbf{1}-2\left(\Lambda\frac{\partial K}{\partial g^2}-\frac{\partial H}{\partial g^2}\right)\mathbf{g}\otimes\mathbf{g}=0. \tag{30}$$

Separation of the deviatoric and spherical parts in this relation yields

$$\Lambda K-H=0,\quad \Lambda\frac{\partial K}{\partial g^2}-\frac{\partial H}{\partial g^2}=0. \tag{31}$$

Now the substitution of the first identity in the second one yields

$$\frac{\partial\Lambda}{\partial g^2}=0\quad\Longrightarrow\quad\frac{\partial\eta}{\partial T}=\Lambda\left(T\right)\frac{\partial\varepsilon}{\partial T}\quad\Longrightarrow\quad\Lambda=\frac{1}{T}, \tag{32}$$

where the classical Gibbs thermostatics (i.e. Gibbs equation in the thermodynamic equilibrium: $d\eta=\frac{1}{T}\left(d\varepsilon+pdV\right)$, $dV\equiv 0$, V - volume; e.g. see: [7]) has been used in the last implication. This yields

$$H=\frac{1}{T}K\quad\Longrightarrow\quad\mathbf{h}=\frac{1}{T}\mathbf{q}. \tag{33}$$

We have anticipated this result in the previous example.

It is convenient to introduce the Helmholtz free energy function

$$\psi:=\varepsilon-T\eta. \tag{34}$$

Then according to (28),(32)

$$\frac{\partial \psi}{\partial T} = \frac{\partial \varepsilon}{\partial T} - T\frac{\partial \eta}{\partial T} - \eta = -\eta \quad \Longrightarrow \quad \varepsilon = \psi + T\frac{\partial \psi}{\partial T},$$

$$\frac{\partial \varepsilon}{\partial \mathbf{g}} - T\frac{\partial \eta}{\partial \mathbf{g}} = 0 \quad \Longrightarrow \quad \psi = \psi(T) \quad \Longrightarrow \quad \varepsilon = \varepsilon(T). \tag{35}$$

Hence we have proven that the specific energy cannot depend on the temeperature gradient. This has been assumed in the simplified derivation of the heat conduction equation mentioned previously.

It remains to exploit the remaining part of the inequality (27) which contains a nonlinear dependence on the temperature gradient $\mathbf{g}$

$$\left[\Lambda\frac{\partial \mathbf{q}}{\partial T} - \frac{\partial \mathbf{h}}{\partial T}\right] \cdot \mathbf{g} \leq 0. \tag{36}$$

This is the so-called **residual inequality** which defines the **dissipation** in the system. If it is zero we have the state of thermodynamic equilibrium. Bearing the results (32), (33) in mind we can write this inequality in the following form

$$\frac{1}{T^2}Kg^2 \geq 0. \tag{37}$$

Hence the heat conductivity K cannot be negative. This means that the heat flux $\mathbf{q}$ is oriented from the hotter to the colder part of the body. This property is sometimes mistakenly identified with the second law of thermodynamics.

We have completed the exploitation of the second law of thermodynamics for our example. However we repeat these considerations in a different way to demonstrate the application of Lagrange multipliers in thermodynamics. This method has been designed in the PhD-Thesis of I-Shih Liu in 1973 (e.g. see [7, 2]). He has shown that instead of the direct elimination of some derivatives as we did above one can exploit an extended inequality in which all field equations are incorporated as constraints. In our simple example we have to replace the inequality (11) by the following one

$$\rho\frac{\partial \eta}{\partial t} + \operatorname{div}\mathbf{h} - \Lambda\left(\rho\frac{\partial \varepsilon}{\partial t} + \operatorname{div}\mathbf{q}\right) \geq 0, \tag{38}$$

where the Lagrange multiplier Λ, similarly to $\varepsilon, \mathbf{q}, \eta, \mathbf{h}$, can be a function of T and $\mathbf{g} = \operatorname{grad} T$. The existence of this multiplier and its dependence on the constitutive variables follows from Liu's theorem. If now the field T is chosen in such a way that it satisfies the field equation then the expression in the brackets vanishes and we have to satisfy the entropy inequality in the usual form. If it is not the case then the Lagrange multiplier must be chosen accordingly in order to correct the result for a non-zero value of the expression in the brackets. Hence this is indeed the same procedure as the one appearing in the classical mechanics with constraints.

The exploitation of the inequality (38) proceeds now in the same way as before. The chain rule of differentiation yields

$$\frac{\partial T}{\partial t}\left[\rho\frac{\partial \eta}{\partial T}-\Lambda\rho\frac{\partial \varepsilon}{\partial T}\right]+\frac{\partial \mathbf{g}}{\partial t}\cdot\left[\rho\frac{\partial \eta}{\partial \mathbf{g}}-\Lambda\rho\frac{\partial \varepsilon}{\partial \mathbf{g}}\right]+$$

$$+\mathbf{g}\cdot\left[\frac{\partial \mathbf{h}}{\partial T}-\Lambda\frac{\partial \mathbf{q}}{\partial T}\right]+\operatorname{grad}\mathbf{g}\cdot\left[\frac{\partial \mathbf{h}}{\partial \mathbf{g}}-\Lambda\frac{\partial \mathbf{q}}{\partial \mathbf{g}}\right]\geq 0. \qquad (39)$$

This inequality must hold for arbitrary derivatives $\frac{\partial T}{\partial t}, \frac{\partial \mathbf{g}}{\partial t}, \operatorname{grad}\mathbf{g}$ and, simultaneously, it is linear with respect to these derivatives. Hence their coefficients must vanish and we obtain

$$\frac{\partial \eta}{\partial T}-\Lambda\frac{\partial \varepsilon}{\partial T}=0, \quad \frac{\partial \eta}{\partial \mathbf{g}}-\Lambda\frac{\partial \varepsilon}{\partial \mathbf{g}}=0, \quad \operatorname{sym}\left(\frac{\partial \mathbf{h}}{\partial \mathbf{g}}-\Lambda\frac{\partial \mathbf{q}}{\partial \mathbf{g}}\right)=0, \qquad (40)$$

and there remains the residual inequality

$$\mathbf{g}\cdot\left[\frac{\partial \mathbf{h}}{\partial T}-\Lambda\frac{\partial \mathbf{q}}{\partial T}\right]\geq 0. \qquad (41)$$

These relations are identical with $(27)_2$, (28), (36). Hence together with the isotropy assumption we obtain exactly the same results as before.

Further in this work we use solely the method of multipliers. It seems to be simpler in the evaluation and it possesses certain additional properties convenient in the analysis of the so-called hyperbolic field equations which we will not discuss in this work.

4 Simple and Biot-type poroelastic nonlinear models

Macroscopic modeling of porous and granular materials belongs to the class of continuum theories. For many practical purposes it is sufficient to construct a single component model similar to this of elasticity or plasticity (e.g. hypoplasticity [8]), and, if necessary to extend such a model by a diffusion equation describing the relative motion of real components. Such procedures are commonly used in soil mechanics. However, dynamics and propagation of waves in porous materials as well as nonlinear effects such as large deformations, swelling, mass exchange processes, etc. require more sophisticated modeling. This follows usually the line proposed for mixtures of fluids by C. Truesdell in 1957 (see: [4], [9]) and extended by R. Bowen in 1982 ([10], [11]) to multicomponent porous materials. A linear multicomponent model of porous materials has been designed much earlier – in 1941– by Biot (see the collected papers of Biot on this subject [12]). This model and its various modifications are still successfully used in geophysical and biomechanical applications.

In this section we discuss some thermodynamic aspects of such a modeling. We begin with an introduction to a nonlinear continuum model in the so-called **Lagrangian description** and then present few important conclusions from the second law of thermodynamics for poroelastic materials. At the end of the section we present a linear version of poroelastic model which will be discussed in the next section of this work.

We consider a porous medium whose pores are filled with a mixture of A fluid components. The model is constructed on a chosen reference configuration $\mathcal{B}_0$ of the skeleton (the solid component), i.e. all fields are functions of a spatial variable $\mathbf{X} \in \mathcal{B}_0$, and time $t \in \mathcal{T}$. We consider a thermomechanical model in which the governing fields are as follows:

1. ρ^S – mass density of the skeleton in the reference configuration,
2. $\rho^\alpha, \alpha = 1, \ldots, A$ – partial mass densities of fluid components refering to the unit volume of the reference configuration of the skeleton,
3. $\acute{\mathbf{x}}^S$ – velocity field of the skeleton,
4. $\mathbf{F}^S$ – deformation gradient of the skeleton,
5. $\acute{\mathbf{x}}^\alpha, \alpha = 1, \ldots, A$ – velocity fields of fluid components,
6. T - absolute temperature of the skeleton,
7. n – porosity (the volume fraction of voids).

The mechanical part of these fields, i.e. fields 1. – 5., do not require any special justification. It should be solely stressed that the multiple velocity field $\{\acute{\mathbf{x}}^S, \acute{\mathbf{x}}^\alpha\}$ means that we include the diffusion in the system which is the main difference between this model and a model of composite materials. However, a single temperature requires already some explanation. In some physical systems such as ionized gases (plasma) different temperatures of components are quite natural. In plasma, for instance, electrons and ions possess different temperatures because they relax very slowly to a thermodynamic equilibrium with a common temperature. This is due to a big difference in mass of charged particles: ions – heavy and electrons – light. In soils different temperatures of components are primarily related to a different rate of heat transfer in a solid and in a fluid. Consequently, in many practical applications (for example in processes of freezing of water in soils) it would be justified to introduce more than only one temperature common for all components. This is not being done because a thermodynamics of such systems does not exist yet.

Another problem is connected with the choice of porosity as the only microstructural variable. First of all let us note that some properties of the microstructure are already appearing due to the fact that we are using a multicomponent model. One could work in such a model with a total mass density $\rho := \rho^S + \sum_{\alpha=1}^{A} \rho^\alpha$ and concentrations $c_\alpha := \dfrac{\rho^\alpha}{\rho}$. This would be the case for a mixture of fluids and concentrations would be microstructural variables of such a model. The presence of a solid requires an extension. For instance for empty pores all concentrations would be zero but still the microstructure would not be trivial. Hence the scalar field of porosity is a minimum extension which we have to make for porous materials. It does not seem to make sense to introduce any further volume fractions if there is solely one solid component. However, in many practical applications, for instance – for rocks, one may need two porosities (e.g. [13]) – one on the microscale (say – 10^{-2}m) and one on the macroscale of big macrocracks and clefts (say – 10m). There may be also the necessity to incorporate some additional parameters describing the geometry of channels such as tortuosity. We do not consider these generalizations in this work even though some of them can be treated by thermodynamic methods.

The above listed fields can be, of course, written in the Eulerian description characteristic for mixtures of fluids. We have then

$$\begin{aligned}
&\rho_t^S(\mathbf{x},t) := \rho^S\left(\mathbf{f}^{-1}(\mathbf{x},t),t\right) J^{S-1}\left(\mathbf{f}^{-1}(\mathbf{x},t),t\right), \quad J^S := \det \mathbf{F}^S, \\
&\rho_t^\alpha(\mathbf{x},t) := \rho^\alpha\left(\mathbf{f}^{-1}(\mathbf{x},t),t\right) J^{S-1}\left(\mathbf{f}^{-1}(\mathbf{x},t),t\right), \quad \alpha = 1,\dots,A, \\
&\mathbf{v}^S(\mathbf{x},t) := \acute{\mathbf{x}}^S\left(\mathbf{f}^{-1}(\mathbf{x},t),t\right), \qquad (42) \\
&\mathbf{v}^\alpha(\mathbf{x},t) := \acute{\mathbf{x}}^\alpha\left(\mathbf{f}^{-1}(\mathbf{x},t),t\right), \quad \alpha = 1,\dots,A, \\
&T(\mathbf{x},t) := T\left(\mathbf{f}^{-1}(\mathbf{x},t),t\right), \quad n(\mathbf{x},t) := n\left(\mathbf{f}^{-1}(\mathbf{x},t),t\right), \qquad (43)
\end{aligned}$$

where the function of motion of the skeleton

$$\mathbf{x} = \mathbf{f}(\mathbf{X},t), \tag{44}$$

is related to the velocity of the skeleton $\acute{\mathbf{x}}^S$ and to its deformation gradient $\mathbf{F}^S$ in the following way

$$\acute{\mathbf{x}}^S = \frac{\partial \mathbf{f}}{\partial t}, \quad \mathbf{F}^S = \operatorname{Grad}\mathbf{f}. \tag{45}$$

All fields listed above have purely macroscopic interpretation and, as usual in the continuum theory of mixtures, particles of all components are appearing simultaneously in each point of the domain. This means that processes which lead to a separation of components must be modelled on different domains separated by an interface. For instance the contact between a porous material saturated with water, and a porous material with empty pores requires the application of two models to two different domains and requires some contact conditions on the interface surrounding the saturated domain.

Macroscopic interpretation of fields yields frequently misunderstandings because some parts of the porous scientific community, particularly this stemming from the tradition of the pioneering work of Terzaghi, work with the so-called real (true) quantities, effective stresses, etc. Consequently in their interpretation the deformation of the skeleton is not describing changes of the macroscopic geometry of the system as it is the case in the present model but rather true deformations of grains on the microscopic level (real solid material), the partial mass densities are not refering to a common macroscopic volume of mixture but rather to partial volumes occupied by real components, etc. We return to this problem later in the work.

We proceed to define field equations for the above fields. Most of them follow as usual from partial balance equations.

In particular we have[3]

partial mass balance equations

$$
\begin{aligned}
\mathcal{E}^{\rho^S} &:= \frac{\partial \rho^S}{\partial t} - \hat{\rho}^S = 0, \\
\mathcal{E}^{\rho^\alpha} &:= \frac{\partial \rho^\alpha}{\partial t} + \operatorname{Div}\left(\rho^\alpha \acute{\mathbf{X}}^\alpha\right) - \hat{\rho}^\alpha = 0, \quad \alpha = 1, \ldots, A, \\
\acute{\mathbf{X}}^\alpha &:= \mathbf{F}^{S-1}\left(\acute{\mathbf{x}}^\alpha - \acute{\mathbf{x}}^S\right),
\end{aligned}
\tag{46}
$$

partial momentum balance equations

$$
\mathcal{E}^{v^S} := \frac{\partial\left(\rho^S \acute{\mathbf{x}}^S\right)}{\partial t} - \left[\operatorname{Div}\mathbf{P}^S + \hat{\mathbf{p}}^S + \rho^S \mathbf{b}^S\right] = \mathbf{0},
$$

$$
\begin{aligned}
\mathcal{E}^{v^\alpha} &:= \frac{\partial\left(\rho^\alpha \acute{\mathbf{x}}^\alpha\right)}{\partial t} + \operatorname{Div}\left(\rho^\alpha \acute{\mathbf{x}}^\alpha \otimes \acute{\mathbf{X}}^\alpha\right) - \\
&- \left[\operatorname{Div}\mathbf{P}^\alpha + \hat{\mathbf{p}}^\alpha + \rho^\alpha \mathbf{b}^\alpha\right] = \mathbf{0}, \quad \alpha = 1, \ldots, A,
\end{aligned}
\tag{47}
$$

partial energy balance equations

$$
\begin{aligned}
\mathcal{E}^{\varepsilon^S} &:= \frac{\partial}{\partial t}\left[\rho^S\left(\varepsilon^S + \tfrac{1}{2}\acute{x}^{S2}\right)\right] - \\
&- \left[\operatorname{Div}\left(\mathbf{Q}^S - \mathbf{P}^{ST}\acute{\mathbf{x}}^S\right) + \rho^S \mathbf{b}^S \cdot \acute{\mathbf{x}}^S + \rho^S r^S + \hat{r}^S\right] = 0,
\end{aligned}
$$

$$
\begin{aligned}
\mathcal{E}^{\varepsilon^\alpha} &:= \frac{\partial}{\partial t}\left[\rho^\alpha\left(\varepsilon^\alpha + \tfrac{1}{2}\acute{x}^{\alpha 2}\right)\right] + \operatorname{Div}\left[\rho^\alpha\left(\varepsilon^\alpha + \tfrac{1}{2}\acute{x}^{\alpha 2}\right)\acute{\mathbf{X}}^\alpha\right] - \\
&- \left[\operatorname{Div}\left(\mathbf{Q}^\alpha - \mathbf{P}^{\alpha T}\acute{\mathbf{x}}^\alpha\right) + \rho^\alpha \mathbf{b}^\alpha \cdot \acute{\mathbf{x}}^\alpha + \rho^\alpha r^\alpha + \hat{r}^\alpha\right] = 0, \quad \alpha = 1, \ldots, A,
\end{aligned}
\tag{48}
$$

balance equation of porosity

$$
\mathcal{E}^n := \frac{\partial n}{\partial t} + \operatorname{Div}\mathbf{J} - \hat{n} = 0. \tag{49}
$$

In these equations, all functions are defined on the reference configuration $\mathcal{B}_0$ of the skeleton. In this sense we may call it the **Lagrangian description** even though partial balance equations for fluid components contain convective parts with respect to the corresponding **Lagrangian velocities** $\acute{\mathbf{X}}^\alpha$.

[3] we define equations by various $\mathcal{E}$-symbols for typografical reasons – it simplifies the form of the entropy inequality appearing in further considerations.

The two-point tensors $\mathbf{P}^S, \mathbf{P}^\alpha$ denote the Piola-Kirchhoff partial stress tensors, $\mathbf{b}^S, \mathbf{b}^\alpha$ are partial body forces, $\varepsilon^S, \varepsilon^\alpha$ are partial densities of the internal energy, $\mathbf{Q}^S, \mathbf{Q}^\alpha$ – partial heat fluxes, r^S, r^α are partial energy radiations, $\mathbf{J}$ is the flux of porosity, and all quantities with a hat denote productions.

The balance equation of porosity (49) requires some explanation. We have argued in previous works on this subject (e.g. [14], [15]) that the balance equation for n follows from an averaging procedure for a representative elementary volume accounting for geometrical properties of the microstructure. However this argument is not needed if we make an extension of the continuous model of mixtures on the macroscopical phenomenological level. In such a case a new scalar field satisfies in the most general case a balance equation. Second order equations for microstructural variables appearing in some works on this subject (e.g. [16], [17]) indicate that most likely two variables rather than one additional microstructural variable should be introduced and one of them has to be eliminated from the model by substitution of one balance equation in another. The above balance equation for porosity specified for two-component poroelastic materials does not require additional boundary conditions – it possesses all properties of an evolution equation. Thermodynamic considerations indicate that the flux $\mathbf{J}$ results from the diffusion (relative motion of fluid components with respect to the skeleton), and the source $\hat{n}$ describes relaxation to the thermodynamic equilibrium.

Certainly we do not need all partial energy balance equations if we have to determine only one temperature field T. In such a case one relies on the bulk balance which follows from (48) by addition of equations for all components. Due to its relative complexity we will not present this equation in this work and refer the reader to the paper [18].

All bulk equations follow in the same way by adding partial equations. We make an assumption similar to this introduced by C. Truesdell for mixtures of fluids [9] that the bulk productions of mass, momentum, and energy vanish, i.e. the corresponding balance equations reduce to conservation laws. Hence

$$\hat{\rho}^S + \sum_{\alpha=1}^{A} \hat{\rho}^\alpha = 0, \quad \hat{\mathbf{p}}^S + \sum_{\alpha=1}^{A} \hat{\mathbf{p}}^\alpha = \mathbf{0}, \quad \hat{r}^S + \sum_{\alpha=1}^{A} \hat{r}^\alpha = 0. \tag{50}$$

Under these conditions we can introduce bulk quantities which correspond to those introduced by Truesdell for fluid mixtures which satisfy conservation laws of a single component continuum. Due to the fact that we have chosen one of the components – the skeleton – as a reference the form of these laws differs from the classical Lagrangian form of conservation equations of a single continuum.

As an example which we use later to transform the entropy inequaliy we present here two bulk quantities – the specific energy ε and the bulk heat flux vector $\mathbf{Q}$ – whose definitions follow by addition of partial energy equations $\mathcal{E}^{\varepsilon^S} + \mathcal{E}^{\varepsilon^\alpha}$. We quote here the results of the work [18]:

$$\rho\varepsilon := \sum_{\alpha=1}^{A} \rho^{\alpha}\varepsilon^{\alpha} + \rho^{S}\varepsilon^{S} + \tag{51}$$
$$+\tfrac{1}{2}\left[\rho^{S}\mathbf{C}^{S}\cdot\left(\dot{\mathbf{X}}\otimes\dot{\mathbf{X}}\right) + \sum_{\alpha=1}^{A}\rho^{\alpha}\mathbf{C}^{S}\cdot\left(\acute{\mathbf{X}}^{\alpha}-\dot{\mathbf{X}}\right)\otimes\left(\acute{\mathbf{X}}^{\alpha}-\dot{\mathbf{X}}\right)\right],$$

$$\mathbf{Q} := \sum_{\alpha=1}^{\mathbf{A}} \mathbf{Q}^{\alpha} + \mathbf{Q}^{S} + \sum_{\alpha=1}^{A}\rho^{\alpha}\varepsilon^{\alpha}\left(\acute{\mathbf{X}}^{\alpha}-\dot{\mathbf{X}}\right) - \rho^{S}\varepsilon^{S}\dot{\mathbf{X}} -$$
$$-\sum_{\alpha=1}^{A}\mathbf{P}^{\alpha T}\mathbf{F}^{S}\left(\acute{\mathbf{X}}^{\alpha}-\dot{\mathbf{X}}\right) + \mathbf{P}^{ST}\mathbf{F}^{S}\dot{\mathbf{X}} + \tag{52}$$
$$+\tfrac{1}{2}\left[\sum_{\alpha=1}^{A}\rho^{\alpha}\left(\acute{\mathbf{X}}^{\alpha}-\dot{\mathbf{X}}\right)\otimes\left(\acute{\mathbf{X}}^{\alpha}-\dot{\mathbf{X}}\right)\otimes\left(\acute{\mathbf{X}}^{\alpha}-\dot{\mathbf{X}}\right) -\right.$$
$$\left. -\rho^{S}\dot{\mathbf{X}}\otimes\dot{\mathbf{X}}\otimes\dot{\mathbf{X}}\right]\mathbf{C}^{S},$$

where

$$\mathbf{C}^{S} := \mathbf{F}^{ST}\mathbf{F}^{S}, \quad \rho\dot{\mathbf{X}} := \sum_{\alpha=1}^{A}\rho^{\alpha}\acute{\mathbf{X}}^{\alpha}, \quad \rho := \rho^{S} + \sum_{\alpha=1}^{A}\rho^{\alpha}. \tag{53}$$

Obviously the bulk flux contains the contribution of partial heat fluxes. However it is related as well to the transport of energy due to the relative motion of components as well as the working of stresses on these relative motions. Hence in contrast to the single component continuum it is different from zero even in the case of isothermal processes in which partial heat fluxes vanish.

The formal thermodynamic construction of a continuous model proceeds as follows. We need field equations for the following set $\mathcal{F}$ of fields

$$\mathcal{F} := \left\{\rho^{S}, \rho^{\alpha}, \mathbf{F}^{S}, \acute{\mathbf{x}}^{S}, \acute{\mathbf{x}}^{\alpha}, T, n\right\}, \quad \alpha = 1, \ldots, A. \tag{54}$$

They follow from the balance equations for mass and momentum (46), (47), added energy balance equations (48), porosity balance (49), and relations (45) which indicate the following integrability condition

$$\mathcal{E}^{F} := \frac{\partial \mathbf{F}^{S}}{\partial t} - \operatorname{Grad}\acute{\mathbf{x}}^{S} = \mathbf{0}. \tag{55}$$

This plays the role of balance equation for the deformation gradient $\mathbf{F}^{S}$.

However, in order to transform these equations into field equations we have to perform the so-called **closure**. Namely the set $\mathcal{R}$ of the following quantities

$$\mathcal{R} := \left\{\hat{\rho}^{\alpha}, \mathbf{P}^{S}, \mathbf{P}^{\alpha}, \hat{\mathbf{p}}^{\alpha}, \varepsilon, \mathbf{Q}, \mathbf{J}, \hat{n}\right\}, \quad \alpha = 1, \ldots, A, \tag{56}$$

must be specified in terms of fields and their derivatives in order to close the system. This is the **constitutive problem** defining materials contributing to the mixture. The mass and momentum sources for the skeleton do not appear in the above list because, according to (50), they are not independent. Let us remark that in many cases of practical bearing additional constitutive relations may have the form of evolution equations. For instance this is the case when the skeleton has some plastic properties. Then the effective stress must satisfy an evolution equation – the so-called Prandtl–Reuss equation for models of metals under small deformations or Kolymbas equation for hypoplastic models of soils [8]. When mass sources result from chemical reactions or adsorption/desorption processes their form is also given by an evolution equation, for instance by the Langmuir equation for adsorption processes.

We do not consider such problems in this work and limit further our attention to the so-called **poroelastic** materials. This assumption yields realistic models for biological tissues (e.g. lungs), woods, spongs, rocks etc. It is not very realistic for soils whose elastic properties are limited to some incremental processes. However even in this case such models are useful in description of waves of small amplitude, and, consequently, in a nondestructive testing of soils. Then the set of **constitutive variables** is as follows

$$\begin{aligned} \mathcal{C} &:= \left\{\rho^S, \rho^\alpha, \mathbf{F}^S, \acute{\mathbf{X}}^\alpha, T, \mathbf{G}, n, \mathbf{N}\right\}, \\ \mathbf{G} &:= \operatorname{Grad} T, \quad \mathbf{N} := \operatorname{Grad} n, \quad \alpha = 1, \ldots, A, \end{aligned} \tag{57}$$

Usually this set is still much too complicated for the full thermodynamic analysis and one considers simpler models. For example in the case of a **simple two-component isotropic isothermal model without mass exchange** scalar constitutive functions depend on the following set of constitutive variables

$$\mathcal{C}_{simple} := \left\{\rho^F, I, II, III, IV, V, VI, n\right\}, \tag{58}$$

where the six invariants $I, \ldots, VI$ are defined as follows

$$\begin{aligned} I &:= \operatorname{tr} \mathbf{C}^S, \quad II := \tfrac{1}{2}\left(I^2 - \operatorname{tr} \mathbf{C}^{S2}\right), \quad III := \det \mathbf{C}^S, \\ IV &:= \acute{\mathbf{X}}^F \cdot \acute{\mathbf{X}}^F, \quad V := \acute{\mathbf{X}}^F \cdot \mathbf{C}^S \acute{\mathbf{X}}^F, \quad VI := \acute{\mathbf{X}}^F \cdot \mathbf{C}^{S2} \acute{\mathbf{X}}^F, \end{aligned} \tag{59}$$

with $\acute{\mathbf{X}}^F$ being the Lagrangian velocity of the single fluid component: $\alpha = F$. We present some results for such a model further in this paper.

The fundamental assumption of a continuous modeling has the form of the following constitutive relation

$$\mathcal{R} = \mathcal{R}(\mathcal{C}), \tag{60}$$

where the mapping $\mathcal{R}$ is assumed to be at least once continuously differentiable. The above relation is a short-cut notation for constitutive relations for each quantity of the set (56) which should be functions of variables (57).

The constitutive functions (60) are said to be **thermodynamically admissible** if any solution of field equations satisfies identically the following **entropy inequality**

$$\frac{\partial(\rho\eta)}{\partial t} + \mathrm{Div}\left(\rho\eta\dot{\mathbf{X}} + \mathbf{H}\right) \geq 0, \quad \eta = \eta(\mathcal{C}), \quad \mathbf{H} = \mathbf{H}(\mathcal{C}). \tag{61}$$

This is the Lagrangian form of the second law of thermodynamics proposed by I. Müller [7] for mixtures. The bulk entropy and the bulk entropy flux **H** are related to partial quantities in the following way

$$\rho\eta := \sum_{\alpha=1}^{A} \rho^{\alpha}\eta^{\alpha} + \rho^{S}\eta^{S},$$

$$\mathbf{H} = \sum_{\alpha=1}^{\mathbf{A}} \mathbf{H}^{\alpha} + \mathbf{H}^{S} + \sum_{\alpha=1}^{A} \rho^{\alpha}\eta^{\alpha}\left(\dot{\mathbf{X}}^{\alpha} - \dot{\mathbf{X}}\right) - \rho^{S}\eta^{S}\dot{\mathbf{X}}. \tag{62}$$

The flux **H** is **not** proportional to the bulk heat flux **Q** as it would be the case for many single component models. This is due to the fact that both fluxes contain explicit contributions of relative velocities. However we may still assume that the Fourier relations hold for partial fluxes

$$\mathbf{H}^{S} = \frac{\mathbf{Q}^{S}}{T}, \quad \mathbf{H}^{\alpha} = \frac{\mathbf{Q}^{\alpha}}{T}. \tag{63}$$

As we have already mentioned the exploitation of the condition (61) is based on the elimination of constraints imposed on the inequality by field equations. This can be done either by direct substitution or by means of auxiliary functions – Lagrange multipliers. We have demonstrated these pocedures on the example of the rigid heat conductor. In the present case we use the method of Lagrange multipliers besides the relation between the heat flux and the entropy flux which follows from the Fourier relations and the explicit relations for bulk fluxes. These will be substituted during the evaluation of the second law. Bearing the balance equations and constitutive relations in mind we write the entropy inequality in the following form

$$\frac{\partial(\rho\eta)}{\partial t} + \mathrm{Div}\left(\rho\eta\dot{\mathbf{X}} + \mathbf{H}\right) - \Lambda^{\rho^S}\mathcal{E}^{\rho^S} - \sum_{\alpha=1}^{A}\Lambda^{\rho^\alpha}\mathcal{E}^{\rho^\alpha} - \tag{64}$$

$$-\boldsymbol{\Lambda}^{v^S}\cdot\mathcal{E}^{v^S} - \sum_{\alpha=1}^{A}\boldsymbol{\Lambda}^{v^\alpha}\cdot\mathcal{E}^{v^\alpha} - \Lambda^{\varepsilon}\left(\mathcal{E}^{\varepsilon^S} + \sum_{\alpha=1}^{A}\mathcal{E}^{\varepsilon^\alpha}\right) - \boldsymbol{\Lambda}^{n}\mathcal{E}^{n} - \boldsymbol{\Lambda}^{F}\cdot\mathcal{E}^{F} \geq 0,$$

where the Lagrange multipliers $\Lambda^{\rho^S}, \Lambda^{\rho^\alpha}, \boldsymbol{\Lambda}^{v^S}, \boldsymbol{\Lambda}^{v^\alpha}, \Lambda^{\varepsilon}, \Lambda^{n}, \boldsymbol{\Lambda}^{F}$ are functions of constitutive variables (57) (e.g. [2], [19]). The above inequality should hold not only for solutions of field equations but for **all** fields.

The evaluation of the above condition is tedious but straightforward. We have to apply the chain rule of differentiation and select linear contributions to the inequality. The requirement that this linear part is zero yields relations for multipliers and a set of thermodynamic identities restricting constitutive relations. There remains a residual, nonlinear inequality which defines the dissipation in the system. We quote here solely some results for a two–component system without presenting detailed derivations. These can be found in my earlier papers.

We proceed to present some results for isothermal processes in two-component systems: $\alpha = F$, where F denotes a single fluid component. It is convenient to introduce the specific partial Helmholtz free energies defined by the relations

$$\psi^S := \varepsilon^S - T\eta^S, \quad \psi^F := \varepsilon^F - T\eta^F. \tag{65}$$

Evaluation of the second law of thermodynamics has been made under the simplifying assumption that the constitutive dependence on two vectorial constitutive variables $\acute{\mathbf{X}}^F, \mathbf{N}$ is linear. This can be divided into two assumptions: partial Helmholtz free energies are independent of both variables, and in particular

$$\frac{\partial \psi^S}{\partial \acute{\mathbf{X}}^F} = \frac{\partial \psi^F}{\partial \acute{\mathbf{X}}^F} = 0, \tag{66}$$

and the constitutive dependence of the momentum source on the relative velocity $\acute{\mathbf{X}}^F$ as well as on the gradient of porosity $\mathbf{N}$ is linear, i.e.

$$\hat{\mathbf{p}} := \hat{\mathbf{p}}^S \equiv \hat{\mathbf{p}}^F = \pi \mathbf{F}^S \acute{\mathbf{X}}^F - \mathcal{N}\mathbf{N}. \tag{67}$$

Under these conditions thermodynamic identities imply the following relations.

- Partial stress tensors possess potentials - the partial Helmholtz free energies

$$\mathbf{P}^S = \rho^S \frac{\partial \psi^S}{\partial \mathbf{F}^S}, \quad \mathbf{P}^F = -\left(\rho^{F2} \frac{\partial \psi^F}{\partial \rho^F} + n\mathcal{N}\right) \mathbf{F}^{S-T}, \tag{68}$$

  i.e. the stress tensor in the fluid component reduces to its spherical part – a partial pressure;
- Partial Helmholtz free energies depend on the following constitutive variables

$$\psi^S = \psi^S\left(I, II, III, \rho^F\right), \quad \psi^F = \psi^F\left(III, \rho^F\right), \tag{69}$$

  and, in addition,

$$2III \frac{\partial \psi^F}{\partial III} + \rho^F \frac{\partial \psi^F}{\partial \rho^F} = -\frac{\mathcal{N}}{\rho_0^{FR}}, \tag{70}$$

  where ρ_0^{FR} is a constant (the initial true mass density of the fluid).

As we see, both partial free energies are dependent on the third invariant of deformation of the skeleton, III, and on the partial mass density of the fluid, ρ^F. This means that the partial stress tensors contain a mechanical coupling between components – volume changes of one of them yield changes of stresses in the other component. In order to obtain this result the set of constitutive variables must include the gradient of porosity ($\mathcal{N} \neq 0$). Otherwise processes in components would be solely coupled through the source of momentum describing an influence of the diffusion. Let us mention that an analogous property has been discovered by I. Müller for mixtures of fluids (e.g. [7]). He has proven in 70'ies that mixtures of ideal fluids in which constitutive relations do not contain a dependence on gradients of partial mass densities (or, equivalently, concentrations) yield the discoupling of partial pressures. Such mixtures are called **simple**.

This type of coupling is characteristic for the so-called linear Biot model of porous materials. We present some of its features in the next section.

There remains the residual inequality in the form

$$\pi \left(\mathbf{F}^{ST} \acute{\mathbf{X}}^F \right) \cdot \left(\mathbf{F}^{ST} \acute{\mathbf{X}}^F \right) \geq 0. \tag{71}$$

Hence the only source of dissipation in these materials is the diffusion with the diffusion velocity $\mathbf{F}^{ST} \acute{\mathbf{X}}^F = \acute{\mathbf{x}}^F - \acute{\mathbf{x}}^S$.

We complete this section with **linearized** equations for two-component isotropic poroelastic materials in isothermal conditions. These equations serve the purpose to describe acoustic waves in porous materials which, in turn, form the basis for nondestructive testing of soils and rocks. In such a case the Lagrangian description is identical with the Eulerian description and we can use fields transformed according to relations (42). The assumption on linearity means that we consider processes satisfying the following restrictions

$$\max \left\{ \left| \lambda^{(\alpha)} \right| \right\}_{\alpha=1,2,3} \ll 1, \quad |\epsilon| \ll 1, \quad \epsilon := -\frac{\rho_t^F - \rho_0^F}{\rho_0^F}, \tag{72}$$

where $\lambda^{(\alpha)}$ are eigenvalues of the Almansi-Hamel deformation tensor of the skeleton

$$\det \left(\mathbf{e}^S - \lambda \mathbf{1} \right) = 0, \quad \mathbf{e}^S := \frac{1}{2} \left(\mathbf{1} - \mathbf{F}^{ST-1} \mathbf{F}^{S-1} \right), \tag{73}$$

and ρ_0^F, ϵ denote a constant initial partial mass density of the fluid, and small macroscopic volume changes of the fluid, respectively.

Under these assumptions the constitutive relations for partial Cauchy stresses follow from the general relations (68) in the following form

$$\begin{aligned} \mathbf{T}^S &= \mathbf{T}_0^S + \lambda^S e \mathbf{1} + 2\mu^S \mathbf{e}^S + \beta \left(n - n_E \right) \mathbf{1} + n_0 \mathcal{N} \epsilon \mathbf{1}, \\ \mathbf{T}^S &:= (III)^{-\frac{1}{2}} \mathbf{P}^S \mathbf{F}^{ST}, \quad e := \operatorname{tr} \mathbf{e}^S, \end{aligned}$$

$$\mathbf{T}^F = -p_0^F \mathbf{1} + \left(\rho_0^F \kappa\epsilon + n_0 \mathcal{N} e\right) \mathbf{1} + \beta \left(n - n_E\right) \mathbf{1}, \tag{74}$$
$$\mathbf{T}^F := (III)^{-\frac{1}{2}} \mathbf{P}^F \mathbf{F}^{ST},$$

where the material parameters $\lambda^S, \mu^S, \kappa, \beta, \mathcal{N}$ depend solely on the initial constant porosity n_0. $\mathbf{T}_0^S, p_0^F$ are an initial partial stress in the skeleton, and an initial partial pressure in the fluid, respectively. The equilibrium value of the porosity n_E is given by the relation

$$n_E = n_0 \left(1 + e - \epsilon\right), \tag{75}$$

which follows from the evaluation of the porosity equation in the thermodynamic equilibrium: $\mathbf{v}^F = \mathbf{v}^S, \hat{n} = 0$. It is clear that changes of porosity described by the difference $n_E - n_0$ correspond to undrained conditions and there is no spontaneous relaxation of porosity in the thermodynamic equilibrium.

The momentum balance equations have in this case the following form

$$\rho_0^S \frac{\partial \mathbf{v}^S}{\partial t} = \operatorname{div} \mathbf{T}^S + \pi \left(\mathbf{v}^F - \mathbf{v}^S\right) - \mathcal{N} \operatorname{grad} n_E + \rho_0^S \mathbf{b}^S, \tag{76}$$
$$\rho_0^F \frac{\partial \mathbf{v}^F}{\partial t} = \operatorname{div} \mathbf{T}^F - \pi \left(\mathbf{v}^F - \mathbf{v}^S\right) + \mathcal{N} \operatorname{grad} n_E + \rho_0^F \mathbf{b}^F,$$

where ρ_0^S is the initial partial mass density of the skeleton, and $\rho_0^S \mathbf{b}^S, \rho_0^F \mathbf{b}^F$ are partial external body forces.

Simultaneously an influence of the difference $n - n_E$ on in the analysis of acoustic waves can be approximately neglected due to the smallness of parameter β. Substitution of relations (74) and (75) in the above set yields

$$\rho_0^S \frac{\partial \mathbf{v}^S}{\partial t} = \operatorname{div} \left(\mathbf{T}_0^S + K e \mathbf{1} + 2\mu^S \operatorname{dev} \mathbf{e}^S + Q\epsilon \mathbf{1}\right) + \pi \left(\mathbf{v}^F - \mathbf{v}^S\right) + \rho_0^S \mathbf{b}^S,$$
$$\rho_0^F \frac{\partial \mathbf{v}^F}{\partial t} = -\operatorname{grad} \left(p_0^F + R\epsilon + Qe\right) \operatorname{div} \mathbf{T}^F - \pi \left(\mathbf{v}^F - \mathbf{v}^S\right) + \rho_0^F \mathbf{b}^F, \tag{77}$$

where

$$K := \lambda^S + \frac{2}{3}\mu^S - n_0 \mathcal{N}, \quad Q := 2 n_0 \mathcal{N}, \quad R := \rho_0^F \kappa - n_0 \mathcal{N}, \tag{78}$$
$$\operatorname{dev} \mathbf{e}^S := \mathbf{e}^S - \frac{1}{3} \operatorname{tr} \mathbf{e}^S \mathbf{1}.$$

In the next section we discuss some aspects of the identification of the material parameters appearing in this set of equations.

5 Micro-macro-transitions for linear poroelastic materials

One of the main problems in applications of the linear poroelastic model is the dependence of material parameters K, μ^S, R, Q on the porosity n_0. In order to find this dependence one has to relate macroscopic and microscopic descriptions of processes. We should stress that the meaning of the microscopic description for geotechnical materials is different from this used, for instance, in materials science. The characteristic length of geotechnical microstructure is almost macroscopic (say – 1mm) and, therefore, we can describe microstructural processes by continuum theories as we do in the case of macroscopic description. The difference is solely related to very complicated domains in the microscopic description which make impossible a solution of real boundary value problems and require a construction of some averages.

In the case of the shear modulus μ^S for granular materials the problem cannot be solved on the basis of linear considerations. Properties of this modulus depend on interactions between grains and these in turn must be at least dependent on the pressure confining the system. Consequently this macroscopic parameter cannot be constant – it must be dependent on one of the macroscopic fields. Hence the model cannot be linear. There are some attempts to solve this problem by means of a kinetic theory of granular materials (e.g. [20]) but these are still rather far from engineering applications.

The situation is different in the case of the compressibility moduli. Some relations have been proposed already at the beginning of 20th century (e.g. [21]) and the results obtained by Gassmann 1951 [22] are used until today. We present here only the main features of this approach in order to point out the role of rational thermodynamics in this problem.

We consider a chosen point of the two-component continuum and assume that macroscopic constitutive properties of this medium are described by volume averages calculated over the so-called Representative Elementary Volume (REV) of the microstructure. This means that a real material behaviour is replaced by a certain amount of smoothed out data. In the case of mechanical properties related to the compressibility we assume that states of the system in REV are **homogeneous** and all processes are **quasistatic**. In addition we define the elementary representative volume as **material** with respect to the solid component.

On the macroscopic level the constitutive relations are given by (74). In the case of purely volumetric changes we can write them in the following form

$$p^S = p_0^S - Ke + Q\frac{\rho_t^F - \rho_0^F}{\rho_0^F}, \quad p^F = p_0^F + R\frac{\rho_t^F - \rho_0^F}{\rho_0^F} - Qe, \tag{79}$$

$$p^S := -\frac{1}{3}\,\mathrm{tr}\,\mathbf{T}^S, \quad p^F := -\frac{1}{3}\,\mathrm{tr}\,\mathbf{T}^F,$$

where the notation (78) has been used. In these relations p_0^S, p_0^F denote the reference partial pressures in the skeleton and in the fluid, respectively. We are using here again the mass densities of the fluid component ρ_t^F, ρ_0^F rather than volume changes ϵ because the partial mass density or the partial specific volume (compare section 3) are proper local constitutive variables for the fluid, and the volume change ϵ may be not for homogeneous processes. This

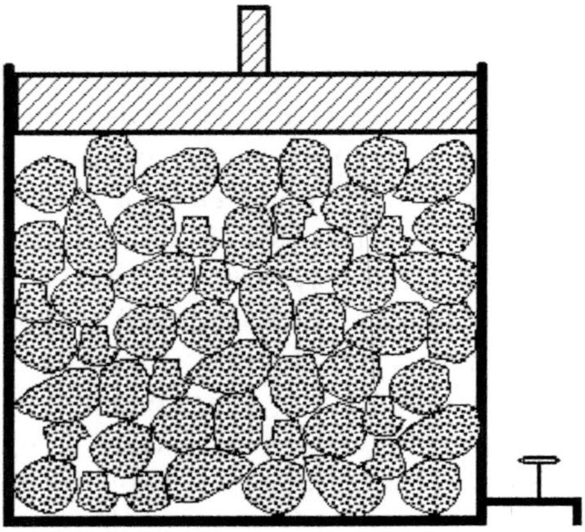

Fig. 2. Scheme of the jacketed test

appears, for instance, in simple tests considered in this section when the total mass of the fluid is not preserved (drainage!). In the local description this problem would not appear because the local mass conservation does hold. However it requires a local velocity of fluid to be different from zero which cannot be properly incorporated in global homogeneous models. The problem does not appear for the skeleton because the mass of skeleton is conserved also in homogeneous tests.

Let M^S denote the instantaneous mass of the solid component in REV, and M^F – the instantaneous mass of the fluid in REV. We assume that the constitutive relations on the microscopic level are of the form

$$p^{SR} = p_0^{SR} - K_s \frac{\rho_0^{SR} - \rho_t^{SR}}{\rho_0^{SR}}, \quad p^{FR} = p_0^{FR} - K_f \frac{\rho_0^{FR} - \rho_t^{FR}}{\rho_0^{FR}}, \tag{80}$$

where $\rho_t^{SR} = \dfrac{M^S}{V^S}$ is the true mass density of the skeleton (comp. section 3), $\rho_t^{FR} = \dfrac{M^F}{V^F}$ denotes the true mass density of the fluid, K_s is the so-called bulk modulus of the solid material composing the porous frame (compressibility modulus of grains), and K_f is the bulk (real) modulus of the fluid. The index zero refers to the initial state.

We want to find the relation between the macroscopic parameters K, R, Q, the microscopic parameters K_s, K_f, and the porosity n_0. To this aim Biot and Willis [5] designed two simple tests: the unjacketed test whose some elements we have already discussed earlier in this work, and the jacketed drained test shown schematically in Fig.2 (open tap). The third possibility appears for the jacketed undrained test (closed tap).

In all possible tests of this sort we have to satisfy the following relations

1. macro- and micro- constitutive relations given by (79), (80), respectively,

2. geometrical compatibility relations which follow from the assumption that REV is material with respect to the solid component,
3. dynamical compatibility relations between micro- and macro-pressures

$$p^S = (1 - n_0)\, p^{SR}, \quad p^F = n_0 p^{FR}, \tag{81}$$

which result from the assumption on randomness of microstructure (i.e. that the volume fraction of voids – porosity n_0 is identical with the fraction of the area of surface of voids to the total area in an arbitrary cross-section of REV),
4. equilibrium condition with an external excess pressure p' loading the system

$$p' = \left(p^S - p_0^S\right) + \left(p^F - p_0^F\right), \tag{82}$$

5. relations defining the tests.

We proceed to derive the geometrical compatibility relations. If we denote the volume of REV by V then in any homogeneous quasistatic process the mass $M^S = \rho_t^S V$ must be conserved

$$\frac{d\rho_t^S V}{dt} = 0 \quad \Longrightarrow \quad \frac{\rho_t^S}{\rho_0^S} = \frac{V_0}{V} = \frac{1}{1+e}, \tag{83}$$

where V_0 is the initial volume of REV. The formula means that we assume for the whole volume of REV the macroscopic rule of volume changes.

Simultaneously on the microscopic level we have $\rho_t^{SR} = \dfrac{M^S}{V^S}$ and due to the definition of porosity (volume fraction of voids) $1 - n = \dfrac{V^S}{V}$ we obtain

$$\rho_t^S = (1-n)\, \rho_t^{SR} \quad \Longrightarrow \quad \frac{1-n}{1-n_0} \frac{\rho_t^{SR}}{\rho_0^{SR}} = \frac{1}{1+e}. \tag{84}$$

On the other hand the mass conservation can be written in its microscopic form

$$\frac{d\rho_t^{SR} V^S}{dt} = 0 \quad \frac{\rho_t^{SR}}{\rho_0^{SR}} = \frac{V_0^S}{V^S} = \frac{1}{1+e^R}. \tag{85}$$

The combination of the results (84) and (85) yields the first geometrical compatibility relation

$$\frac{1-n}{1-n_0} = \frac{1+e^R}{1+e}. \tag{86}$$

This relation shows the difference between the notion of volume changes in the macroscopic model of porous materials and real volume changes of grains.

Even if the material of grains is assumed to be approximately incompressible ($e^R \equiv 0$) the macroscopic changes of volume of the skeleton e are different from zero due to changes of porosity.

The second relation follows from the transformation of fluid volume (voids)

$$V^F = V_0^F \left(1 + \epsilon^R\right) \quad \Longrightarrow \quad \frac{n}{n_0} = \frac{1 + \epsilon^R}{1 + e}. \tag{87}$$

This is the second geometrical compatibility relation.

Easy manipulations on those two compatibility relations yield the following relation between the macroscopic change of the volume and microscopic changes of volumes of both components.

$$e = (1 - n_0)\, e^R + n_0 \epsilon^R. \tag{88}$$

It remains to define the tests. It is easy to see that the following conditions must hold:

1. Undrained jacketed test: the mass of fluid is conserved, i.e.

$$\frac{d\rho_t^F V}{dt} = 0 \quad \Longrightarrow \quad \rho_t^F = \frac{\rho_0^F}{1 + e} \quad \Longrightarrow \quad \frac{n}{n_0}\frac{\rho_t^{FR}}{\rho_0^{FR}} = \frac{1}{1 + e},$$

   and

$$\frac{d\rho_t^{FR} V^F}{dt} = 0 \quad \Longrightarrow \quad \frac{\rho_t^{FR}}{\rho_0^{FR}} = \frac{1}{1 + \epsilon^R}.$$

   Combination of these two relations shows that we obtain again the relation (87). This means that this test does not give any additional information about micro-macrorelations.

2. Drained jacketed test: in this case the increment of the pore pressure in the fluid must be zero if the continuity of the pressure is preserved in the tap

$$p^{FR} - p_0^{FR} = 0. \tag{89}$$

   This seems to be a safe assumption for quasistatic processes because, in contrast to the unjacketed test discussed at the beginning of this work, there are no surface effects possible.This test is usually used in practice to define the so-called drained compressibility modulus of the skeleton K_d.

3. Unjacketed test: as we have shown in section 3 we can assume in this case that the pore pressure is in equilibrium with the excess external pressure

$$p^{FR} - p_0^{FR} = p'. \tag{90}$$

The above relations can be combined to give the desired relation between microscopic and macroscopic compressibilities.

This is not what is done in the literature on the subject. In those considerations the undrained jacketed test is fully ignored, the drained jacketed test is used, as already mentioned, to define the drained compressibility modulus K_d and the unjacketed test is considered under the assumption that the porosity remains constant (e.g. [6]). Then we obtain after easy calculations (e.g. see [23], [24])

$$K = K_d + \frac{\alpha^2}{\dfrac{\alpha - n_0}{K_s} + \dfrac{n_0}{K_f}}, \quad R = \frac{n_0}{\dfrac{\alpha - n_0}{K_s} + \dfrac{n_0}{K_f}}, \tag{91}$$

$$Q = \frac{n_0 \alpha}{\dfrac{\alpha - n_0}{K_s} + \dfrac{n_0}{K_f}}, \quad \alpha := 1 - \frac{K_d}{K_s}.$$

These are the so-called Gassmann relations. α is called the Biot-Willis parameter.

It seems to be a wonder that with all these assumptions the above relations give good results in practical applications, for instance – in the wave analysis. It is clear that the validity of the above relations is limited in many ways. The assumption on the constant porosity seems to be at least doubtful, the validity of the relation between pressures on the interface in the unjacketed experiment is limited to very simple systems, as discussed in Sect.3. If any of the effects listed in this section should be incorporated then we have to rely on the continuity of the free enthalpy rather than on the continuity of pressure. This should be, for instance, the case for viscous fluids even though the problem is ignored in the literature (compare the analysis of acoustic waves in porous materials in [6]).

Practical application in nondestructive testing follows now from measurements of speeds of propagation of bulk or surface waves in soils or rocks. These experimental data predict macroscopic elastic parameters combined with mass densities. Combination of those data with Gassmann relations (91) yields the relation for porosity. This procedure has been investigated in a recent paper of Foti, Lai and Lancellotta [25]. Results compare well with experimental findings.

6 Concluding remarks

Results presented in this article are typical for macroscopic thermodynamic modeling of continuous media. Consequently we can state that the results of evaluation of the second law of thermodynamics possess two features

- they reduce the extent of requirements concerning the construction of constitutive laws and frequently – this is the case for a rather broad class of poroelastic materials discussed in section 3 – they yield the existence of thermodynamic potentials. The latter not only simplify the construction of constitutive laws but enable an analysis of existence and stability of solutions which we did not discuss in this work;

- they clarify limitations of ad hoc models appearing in practical applications. The example of the Gassmann relations discussed in the paper shows that an extention of these relations on, for instance, viscous fluids may not be as straightforward as sometimes done in works on the subject.

Simultaneously the thermodynamic method and, in particular, the second law of thermodynamics do not give specific results which could replace experiments. In some cases of material properties such as the heat conductivity or permeability of porous media we obtain solely inequalities restricting the models but not even hints how to evaluate the material parameters without solving field equations and verifying solutions by experiments.

References

[1] Valentin Joseph Boussinesq; *Théorie Analytique de la Chaleur*, Gautier-Villars, Paris, 1903; also: D.D. Gray, A. Giorgini; The validity of the Boussinesq Approximation for Liquid and Gases, *Int. J. Heat Mass Transfer*, **19**, 545-551, 1976.

[2] Krzysztof Wilmanski; *Thermomechanics of Continua*, Springer, Berlin, 1998.

[3] Mark Kac; *Some Stochastic Problems in Physics and Mathematics*, Magnolia Petroleum Company, Dallas, 1957.

[4] Clifford A. Truesdell; *A First Course in Rational Continuum Mechanics*, The Johns Hopkins University, Baltimore, Maryland, 1972.

[5] Maurice A. Biot, D. G. Willis,; The Elastic Coefficients of the Theory of Consolidation, *Jour. Appl. Mech.*, **24**, 594-601, 1957.

[6] Robert D. Stoll; *Sediment Acoustics*, Lecture Notes in Earth Sciences, vol. 26, Springer-Verlag, Berlin, 1989.

[7] Ingo Müller; *Thermodynamics*, Pitman, Boston, 1985.

[8] Dimitrios Kolymbas; *Introduction to Hypoplasticity*, Balkema-Verlag, 2000.

[9] Clifford A. Truesdell; Sulle basi della termomeccanica, *Acad. Naz. dei Lincei, Rend. della Classe di Scienze Fisiche, Matematiche e Naturali*, **22**(8), 33-38, 158-166, 1957.

[10] Ray M. Bowen; Compressible Porous Media Models by Use of the Theory of Mixtures, *Int. J. Engn. Sci.*, **20**(6), 697-735, 1982.

[11] Clifford A. Truesdell; *Rational Thermodynamics*, second edition, Springer, N. Y., 1985.

[12] I. Tolstoy; *Acoustics, Elasticity, and Thermodynamics of Porous Media: Twenty-one Papers by M. A. Biot*, Acoustical Society of America, 1991.

[13] Hans-Jürgen Lang, Jachen Huder, Peter Amann; *Bodenmechanik und Grundbau*, Springer, Berlin, 1996.

[14] Krzysztof Wilmanski; Porous Media at Finite Strains - the New Model With the Balance Equation of Porosity, *Arch. Mech.*, **48**(4), 591-628, 1996.

[15] Krzysztof Wilmanski; A Thermodynamic Model of Compressible Porous Materials with the Balance Equation of Porosity, *Transport in Porous Media*, **32**, 21-47, 1998.

[16] M. A. Goodman, S. C. Cowin; A Continuum Theory For Granular Materials, *Arch. Rat. Mech. Anal.*, **48**, 249-266, 1972.

[17] Bob Svendsen, Kolumban Hutter; On the Thermodynamics of a Mixture of Isotropic Materials With Constraints, *Int. J. Engn. Sci.*, **33** 2021-2054, 1995.

[18] Krzysztof Wilmanski; On Thermodynamics of Poroelastic Materials, *J. Elasticity*, 2003 (to appear).

[19] I-Shih Liu; *Continuum Mechanics*, Springer, Berlin, 2002.

[20] James T. Jenkins, Luigi La Ragione; Induced Anisotropy, Particle Spin, and Effective Moduli in Granular Materials, in: P. Podio-Guidugli, M. Brocato (eds.), *Rational Continua, Classical and New*, Springer, Milano, 111-118, 2003.

[21] A. B. Wood; *Textbook of Sound*, G. Bell and Sons, London, 1911.

[22] F. Gassmann, Über die Elastizität poröser Medien, Vierteljahrsschrift der Naturforschenden Gesellschaft in Zürich, 96, 1-23, 1951.

[23] James G. Berryman, Graeme W. Milton; Exact results for generalized Gassmann's equations in composite porous media with two constituents, *Geophysics*, **56**, 1950-1960, 1991.

[24] James G. Berryman, David E. Lumley; Inverting Ultrasonic Data on Solid/Fluid Mixtures for Biot-Gassmann Parameters, in: *Proc. of the Second Int. Conf. on Mathematical and Numerical Aspects of Wave Propagation*, SIAM, 57-68, 1993.

[25] Sebastiano Foti, Carlo G. Lai, Renato Lancellotta; Porosity of Fluid-Saturated Porous Media from Measured Seismic Wave Velocities, *Géotechnique*, **52**, 5, 359-373, 2002.

Some mathematics for the constitutive modelling of soils

Guy T. Houlsby

Department of Engineering Science, Oxford University, Parks Road, Oxford, OX1 3PJ, UK, guy.houlsby@eng.ox.ac.uk

Summary. The purpose of this paper is to introduce some mathematical techniques which prove to be valuable in the constitutive modelling of soils. All the developments are related to an approach to constitutive modelling called hyperplasticity, in which strong emphasis is placed on the derivation of the entire behaviour of a material from two scalar potentials. Hyperplasticity includes all sufficient conditions to satisfy the laws of thermodynamics, but some conditions are not strictly necessary: they embody a slightly stricter statement than the second law. In this paper this issue is not addressed, but instead some simple models are developed to illustrate the hyperplastic approach. It is left for the reader to judge whether these models are valuable in representing real material behaviour or whether they are too restrictive. The hyperplasticity approach has its roots in the work of Ziegler, and also has much in common with much of the French work in plasticity theory, where the concept of "standard materials" is employed. Much of what is presented here is not new, but represents application of existing mathematical techniques in areas of geotechnics where they are not currently employed.

1 Potentials and the Legendre Transform

The Legendre transform is a simple and powerful technique with countless applications in theoretical mechanics. It is frequently used, implicitly if not explicitly. We introduce the transform here without proof, but further details can be found in the Appendix to Collins and Houlsby (1997). Suppose a function $X = X(x_i)$ is defined, and this function acts as a potential, so that $y_i = \partial X / \partial x_i$. The Legendre transform of X is a function $Y = Y(y_i)$ defined by $X + Y = x_i y_i$ (where the summation convention is used, so that right hand side is an inner product). The transform has the property $x_i = \partial Y / \partial y_i$. There is an obvious symmetry that X is also the Legendre transform of Y.

A marginally more complicated case occurs when $X = X(x_i, a_i)$ and the partial Legendre transform (again defined by $X + Y = x_i y_i$) is made to $Y = Y(y_i, a_i)$, with the a_i regarded as passive variables. In this case it is straightforward to show that the transform has the additional property $\partial Y/\partial a_i = -\partial X/\partial a_i$. If $b_i = \partial X/\partial a_i$ then a further partial transform is possible in which the a_i, b_i variables are active and the x_i, y_i variables passive. This procedure leads to a closed chain of four transformations, again with useful symmetry properties, see Collins and Houlsby (1997). Such chains of transformations play a very important role in thermodynamics.

For simplicity in the following we restrict ourselves to a small strain formulation in Cartesian coordinates. This allows use of specific energies per unit volume rather than per unit mass (as is strictly necessary in thermodynamics), thus avoiding a ubiquitous factor of the density ρ.

1.1 Stress-based and strain-based formulations

Probably the best known Legendre transformation in constitutive modelling is the relationship in elasticity theory between the strain energy $E = E(\varepsilon_{ij})$, and the complementary energy $C = C(\sigma_{ij})$. In hyperelasticity these both serve as potentials, such that $\sigma_{ij} = \partial E/\partial \varepsilon_{ij}$ and $\varepsilon_{ij} = \partial C/\partial \sigma_{ij}$, and are related by $E + C = \sigma_{ij}\varepsilon_{ij}$. Thus the two energies are Legendre transforms of each other. In thermodynamic terminology (if we exclude dependence on temperature) the strain energy can be identified with either the internal energy u or the Helmholtz free energy f, thus $E = u = f$. The complementary energy can be identified with either the enthalpy h or the Gibbs free energy g, but by usual convention with a change of sign, thus $C = -h = -g$. A fuller discussion of these relationships is given by Houlsby and Puzrin (2000).

In linear isotropic elasticity, and using a prime to indicate the deviator of a tensor, we can simply write either

$$f = 3K\frac{\varepsilon_{ii}\varepsilon_{jj}}{6} + 2G\frac{\varepsilon'_{ij}\varepsilon'_{ij}}{2} \quad \text{or} \quad g = -\frac{1}{3K}\frac{\sigma_{ii}\sigma_{jj}}{6} - \frac{1}{2G}\frac{\sigma'_{ij}\sigma'_{ij}}{2},$$

illustrating the case that if a function is a homogeneous function of degree 2 in its arguments, then so is its Legendre transform. In this case, given one of the functions it is straightforward to derive the other by applying the appropriate definitions. In more complicated cases though it can be difficult (or in practice even impossible) to perform the necessary eliminations to obtain the transform in the form of an explicit function $Y = Y(y_i)$.

As an illustration of a slightly more complex example consider the case of non-linear elasticity, with both the bulk and shear moduli proportional to pressure, expressed in triaxial stress and strain variables (p, q) and (v, ε). This may be expressed either by

$$f = p_o\kappa \exp\left(\frac{v}{\kappa} + \frac{3g\varepsilon^2}{2\kappa}\right) \quad \text{or} \quad g = -p\kappa\left(\log\left(\frac{p}{p_o}\right) - 1\right) - \frac{q^2}{6gp},$$

where p_o is a reference pressure, and the bulk and shear moduli on the isotropic axis are given by p/κ and gp respectively. In this case the transform relationship is less obvious between the two functions, but it can be verified easily.

Taking the Gibbs free energy as the starting point it is straightforward to derive

$$v = -\frac{\partial g}{\partial p} = \kappa \log\left(\frac{p}{p_o}\right) - \frac{q^2}{6gp^2} \quad \text{and} \quad \varepsilon = -\frac{\partial g}{\partial q} = \frac{q}{3gp} .$$

It further follows that the incremental relationship is:

$$\begin{bmatrix} \dot{v} \\ \dot{\varepsilon} \end{bmatrix} = \begin{bmatrix} \frac{\kappa}{p} + \frac{q^2}{3gp^3} & -\frac{q}{3gp^2} \\ -\frac{q}{3gp^2} & \frac{1}{3gp} \end{bmatrix} \begin{bmatrix} \dot{p} \\ \dot{q} \end{bmatrix} \tag{1}$$

Noting that for an isotropic elastic material this relationship could be written:

$$\begin{bmatrix} \dot{v} \\ \dot{\varepsilon} \end{bmatrix} = \begin{bmatrix} 1/K & 0 \\ 0 & 1/3G \end{bmatrix} \begin{bmatrix} \dot{p} \\ \dot{q} \end{bmatrix} \tag{2}$$

it follows that if $q \neq 0$ the material does not behave in an isotropic elastic manner. This stress-induced anisotropy is an inevitable consequence of hyperelasticity in the case where the moduli depend on the pressure. The occurrence of such anisotropy could not be predicted without recourse to the formulation based on a potential.

1.2 Hyperplasticity

Hyperelastic materials are defined by *one* potential function (or equivalently its Legendre transform). Hyperplastic materials (Houlsby and Puzrin, 2000) are defined by *two* potentials: one specifying the stored energy (and closely analogous to the strain or complementary energy in elasticity theory) and one specifying the dissipated power. In the formulation defined by Houlsby and Puzrin (2000), the entire constitutive behaviour follows from these two potentials.

The hyperplastic approach is firmly rooted in the science of thermodynamics with internal variables (see the review by Maugin, 1999), and we make use of internal variables α_{ij}, which will be seen in most cases to play exactly the same role as the plastic strains. We express the Helmholtz free energy per unit volume in the form $f = f(\varepsilon_{ij}, \alpha_{ij})$ or the Gibbs free energy per unit volume $g = g(\sigma_{ij}, \alpha_{ij})$, with the relationships $f - g = \sigma_{ij}\varepsilon_{ij}$, $\sigma_{ij} = \partial f / \partial \varepsilon_{ij}$ and $\varepsilon_{ij} = -\partial g / \partial \sigma_{ij}$. Note that, for consistency with terminology used elsewhere, it is sometimes appropriate to change the sign of the Legendre transform.

The second potential defined for a rate-independent material is the dissipation function d, expressed in the form $d(\varepsilon_{ij}, \alpha_{ij}, \dot{\alpha}_{ij})$ or $d(\sigma_{ij}, \alpha_{ij}, \dot{\alpha}_{ij})$, depending on whether the strain-based of stress-based formulation is used. Quantities called generalised stresses are then defined as $\bar{\chi}_{ij} = -\partial f / \partial \alpha_{ij} = -\partial g / \partial \alpha_{ij}$ and $\chi_{ij} = \partial d / \partial \dot{\alpha}_{ij}$. One can then show that, from considerations of thermodynamics $(\bar{\chi}_{ij} - \chi_{ij})\,\dot{\alpha}_{ij} = 0$. If, however, one adopts the somewhat stronger assumption that $\bar{\chi}_{ij} - \chi_{ij} = 0$, then it follows that the entire constitutive behaviour of the material can be derived once the form of the free energy and dissipation expressions have been defined, and no separate "evolution equations" need be specified. This allows a particularly compact and consistent expression of constitutive models. The strong assumption $\bar{\chi}_{ij} - \chi_{ij} = 0$ has a variety of interpretations. It is essentially the same as the orthogonality postulate of Ziegler (1977, 1983). It is a stronger statement than the second law of thermodynamics, which is a statement that dissipation is always positive. The orthogonality postulate is effectively an assumption that not only is energy dissipated, but that it is dissipated at a maximal rate, subject to any relevant constraints. This is not only an appealing physical concept, but also has far-reaching theoretical consequences. The appropriateness or otherwise of the stronger statement can only be measured by the consequences for models that embody it, but the author knows of no clear counterexamples to the postulate.

1.3 The relationship between dissipation and yield

Those familiar with plasticity theory will be more used to specifying a yield function as part of a plasticity model, so it is important to examine the relationship between dissipation and yield. In fact the dissipation function and the yield function prove to be related by a Legendre transform, but in this case a rather important special case. It is straightforward to show that if a material is to exhibit rate-independent behaviour, then the dissipation function must be a homogeneous first order (although not necessarily linear) function of the internal variable rates $\dot{\alpha}_{ij}$. When we form the Legendre transform $w = w(\sigma_{ij}, \alpha_{ij}, \chi_{ij}) = \chi_{ij}\dot{\alpha}_{ij} - d$ in this case, it follows from Euler's theorem that the right hand side is identically zero (see the appendix to Collins and Houlsby (1997)). In this degenerate special case it is in fact more convenient to introduce an arbitrary multiplier λ and define $w = \lambda y = \lambda y(\sigma_{ij}, \alpha_{ij}, \chi_{ij}) = \chi_{ij}\dot{\alpha}_{ij} - d = 0$. The condition $y(\sigma_{ij}, \alpha_{ij}, \chi_{ij}) = 0$ turns out to be none other than the yield condition. It then immediately follows that $\dot{\alpha}_{ij} = \lambda \partial y / \partial \chi_{ij}$, which plays the same role as the flow rule in conventional plasticity theory. Two cases emerge, either $\lambda = 0$ and $y \leq 0$, which corresponds to elastic behaviour, or $y = 0$ and $\lambda \geq 0$, corresponding to plasticity.

These concepts are best illustrated by an example. We will consider a free energy in the form $g = g_1(\sigma_{ij}) + g_2(\alpha_{ij}) - \sigma_{ij}\alpha_{ij}$, which immediately leads to the result $\varepsilon_{ij} = -\partial g_1 / \partial \sigma_{ij} + \alpha_{ij}$, so that the internal variable α_{ij} plays the role of the plastic strain, as it simply serves as an additive term to the elastic strain (which is just a function of the stress state) $-\partial g_1 / \partial \sigma_{ij}$. We also obtain $\bar{\chi}_{ij} = -\partial g_2 / \partial \alpha_{ij} + \sigma_{ij}$. In the case where $g_2 = 0$, as considered below, the generalised stress $\bar{\chi}_{ij}$ is simply equal to the stress σ_{ij}. In more general cases

$\sigma_{ij} = \partial g_2/\partial \alpha_{ij} + \bar{\chi}_{ij} = \rho_{ij} + \bar{\chi}_{ij}$, where $\rho_{ij} = \rho_{ij}(\alpha_{ij})$ is the "back stress" in kinematic hardening, which is a function of the plastic strains.

Consider the functions $g = -\frac{1}{3K}\frac{\sigma_{ii}\sigma_{jj}}{6} - \frac{1}{2G}\frac{\sigma'_{ij}\sigma'_{ij}}{2} - \sigma_{ij}\alpha_{ij}$ and $d = k\sqrt{2\dot{\alpha}_{ij}\dot{\alpha}_{ij}}$. We supplement these by a constraint specifying zero volumetric plastic strain rate $c = \dot{\alpha}_{kk} = 0$. The constraint is included in the formulation by considering the augmented dissipation function $d' = d + \Lambda c$, where Λ is a Lagrangian multiplier. The result is :

$$\chi_{ij} = \frac{\partial d'}{\partial \dot{\alpha}_{ij}} = k\sqrt{2}\frac{\dot{\alpha}_{ij}}{\sqrt{\dot{\alpha}_{kl}\dot{\alpha}_{kl}}} + \Lambda\delta_{ij} \tag{3}$$

The trace of this expression gives $\sigma_{kk} = \chi_{kk} = 3\Lambda$, so that the undetermined Lagrangian multiplier is none other than the mean stress. The deviator gives $\chi'_{ij} = k\sqrt{2}\frac{\dot{\alpha}'_{ij}}{\sqrt{\dot{\alpha}_{kl}\dot{\alpha}_{kl}}}$, from which we can derive the expression $\chi'_{ij}\chi'_{ij} - 2k^2 = 0$, which is in fact the required yield surface in generalised stress space. Since in this case $\sigma_{ij} = \chi_{ij}$, this yield surface also represents a surface in stress space, which can immediately be recognised as the von Mises yield surface.

Now consider the same model, but starting from the yield surface $y = \chi'_{ij}\chi'_{ij} - 2k^2 = 0$ instead of the dissipation function. We would then derive $\dot{\alpha}_{ij} = \lambda\partial y/\partial \chi_{ij} = 2\lambda\chi'_{ij} = 2\lambda\sigma'_{ij}$, which is of course the flow rule for the von Mises material.

In a conventional plasticity model for a soil, which might employ non-associated flow, one would specify both the yield surface and the plastic potential, but here we use only a single expression for the yield surface (since the free energy expression is principally related to the elastic behaviour and the hardening of the yield surface). How is non-associated flow therefore accommodated? Note that the plastic strain is given by the differential of the yield surface with respect to the *generalised* stress. However, the generalised stress and real stress are equal (at least in this model), so are essentially interchangeable in the expression for the yield surface. Substituting the stress for the generalised stress does not change the yield surface, but it does change the flow rule. Thus the equivalent of the difference between the yield surface and plastic potential is achieved by a partition of the terms in the yield function between the stress and generalised stress.

It is in this area where the mathematics of the Legendre transform leads to a powerful insight into plasticity models for soils. If only the generalised stress appears in the yield surface then the flow is "associated" in the conventional sense. Whenever the real stress is substituted in the yield expression non-associated flow occurs. A property of the partial Legendre transformation between yield and flow is, however, that $\partial d/\partial \sigma_{ij} = -\lambda\partial y/\partial \sigma_{ij}$. Thus whenever the yield surface is a function of the stress, then so is the dissipation (and *vice versa*). What is the meaning of the dissipation being dependent on the stress? An obvious physical interpretation is that this would serve as a definition of frictional behaviour – if the dissipation increases with the stress level, then that is precisely what is meant by friction. Thus we see that frictional behaviour is irretrievably linked to non-associated flow. It is of course universally observed that it is those materials that exhibit frictional behaviour that also exhibit non-association, but it is the mathematics of the Legendre transform that reveals the fundamental nature of this link. As a simple example (using triaxial variables), consider the yield surface expressed as $y = |\chi_q| - Mp - \beta\chi_p = 0$, where (χ_p, χ_q) are the generalized stresses corresponding to (p, q) and an energy function is specified such that $\chi_p = p$ (in practice this corresponds to no kinematic hardening of the yield surface). In stress space the yield condition simply becomes $|q| = (M + \beta)\, p$, so that the friction angle corresponds to the stress ratio $M + \beta$. The plastic strain rates are calculated as:

$$\dot{\alpha}_p = \lambda \frac{\partial y}{\partial \chi_p} = -\lambda\beta, \quad \dot{\alpha}_q = \lambda \frac{\partial y}{\partial \chi_q} = -\lambda \mathrm{sgn}\,(\chi_q) \tag{4}$$

So that $\dot{\alpha}_p / \dot{\alpha}_q = \beta sgn\,(\chi_q) = -\beta \mathrm{sgn}\,(q)$, and the angle of dilation is just related to the parameter β. Thus the model expressed by equation (4) describes the familiar pattern that the observed angle of friction is made up from two components: an angle of friction at constant volume (related to M) and a dilational component (related to β). The value of β would of course be expected to be a function of the relative density. Note, however, that this behaviour is captured by a single equation, and does not require a separate yield function and plastic potential.

Collins and Houlsby (1997) and Houlsby and Puzrin (2000) discuss friction in more detail.

1.4 Rate effects, pseudopotentials and how to avoid them

In the above the behaviour of a rate-independent material was specified by the dissipation function or yield function. If rate-dependent behaviour is to be specified, then the yield function is no longer a homogeneous first order function, and in this case the expression for the generalised stress becomes:

$$\chi_{ij} = \frac{d}{\left(\frac{\partial d}{\partial \dot{\alpha}_{kl}} \dot{\alpha}_{kl}\right)} \frac{\partial d}{\partial \dot{\alpha}_{ij}} \tag{5}$$

where the scaling factor $d \Big/ \left(\dfrac{\partial d}{\partial \dot{\alpha}_{kl}} \dot{\alpha}_{kl}\right)$, which is unity in the case of a rate-independent material, has to be introduced so that the equation $\chi_{ij}\dot{\alpha}_{ij} = d$ is satisfied (as it must be from thermodynamic considerations). In this case d no longer serves as a potential for the generalised stress, but is said to be a pseudopotential. The presence of the multiplying factor complicates the formulation considerably, and can be avoided if, instead of specifying the dissipation function directly, one makes the hypothesis that a force potential

$z(\varepsilon_{ij}, \alpha_{ij}, \dot{\alpha}_{ij})$ or $z(\sigma_{ij}, \alpha_{ij}, \dot{\alpha}_{ij})$ exists, such that $\chi_{ij} = \partial z / \partial \dot{\alpha}_{ij}$. The Legendre transform of the force potential is the flow potential $w(\varepsilon_{ij}, \alpha_{ij}, \chi_{ij})$ or $w(\sigma_{ij}, \alpha_{ij}, \chi_{ij})$ such that $\dot{\alpha}_{ij} = \partial w / \partial \chi_{ij}$ and $z + w = \chi_{ij}\dot{\alpha}_{ij} = d$. In the special case of rate-independent behaviour $z = d$ and $w = \lambda y = 0$. Rate-dependent materials using this formulation are explored by Houlsby and Puzrin (2002). An advantage of the rate-dependent formulation based on specification of the flow potential w is that it leads to a much more compact form of the incremental stress-strain relationship, which is very easy to implement numerically:

$$d\varepsilon_{ij} = -\frac{\partial^2 g}{\partial \sigma_{ij} \partial \sigma_{kl}} d\sigma_{kl} - \frac{\partial^2 g}{\partial \sigma_{ij} \partial \alpha_{kl}} \frac{\partial w}{\partial \chi_{kl}} dt \tag{6}$$

2 Functionals and Frechet derivatives

The formulation above places strong emphasis on the use of potential functions, and this proves to be a useful starting point for more complex models. A model with a single (tensorial) internal variable involves a single yield surface. To describe models with multiple yield surfaces, as are often used to describe the effects of load history for geotechnical materials, then multiple internal variables can be used. This process can be taken to its logical conclusion and the multiple internal variables replaced by a continuously varying "internal function". The potential functions are now replaced by potential functionals (loosely defined as "functions of functions"), see for example Puzrin and Houlsby (2001a). Thus for instance we replace $g(\sigma_{ij}, \alpha_{ij})$ by $g[\sigma_{ij}, \hat{\alpha}_{ij}(\eta)]$, where the square bracket denotes a functional, and $\hat{\alpha}_{ij}$ is in turn a function of some variable η which we call an *internal coordinate.*

In order to proceed we first need to generalise the concept of differentiation of a function, which is done as follows. The Frechet derivative $f'[u]$ of a functional $f[u]$ is a linear operator defined (for sufficiently well-behaved functionals) as satisfying the expression:

$$\lim_{\|\delta u\| \to 0} \frac{\|f[u + \delta u] - f[u] - f'[u]\,\delta u\|}{\|\delta u\|} = 0 \tag{7}$$

Where $\|\|$ indicates an appropriate norm, δu represents the variation of the argument function and $f'[u]\,\delta u$ indicates the action of the linear operator on the function δu.

Here we are particularly interested in functionals which can be written in the form $f[u] = \int_{\Upsilon} \hat{f}(u(\eta), \eta)\, d\eta$ where Υ is the domain of η, and for this case the Frechet derivative is given by:

$$f'[u]\,\delta u = \int_{\Upsilon} \frac{\partial \hat{f}(u, \eta)}{\partial u} \delta u(\eta)\, d\eta \tag{8}$$

The development of models using functionals is treated in more detail by Puzrin and Houlsby (2001a). Simply as an example of part of the development, we note that if $g(\sigma_{ij}, \alpha_{ij})$ in the original formulation is replaced by $g[\sigma_{ij}, \hat{\alpha}_{ij}] = \int\limits_{\Upsilon} \hat{g}(\varepsilon_{ij}, \alpha_{ij}(\eta), \eta)\, d\eta$, and $d(\dot{\alpha}_{ij})$ by $d\left[\dot{\hat{\alpha}}_{ij}\right] = \int\limits_{\Upsilon} \hat{d}(\dot{\alpha}_{ij}(\eta), \eta)\, d\eta$ then the equivalent of the condition $\bar{\chi}_{ij} = \chi_{ij}$ simply becomes $\hat{\bar{\chi}}_{ij} = \hat{\chi}_{ij}$, where $\bar{\chi}_{ij}(\eta) = -\partial\hat{g}(\sigma_{ij}, \hat{\alpha}_{ij}(\eta), \eta)/\partial\hat{\alpha}_{ij}(\eta)$ and $\chi_{ij}(\eta) = \partial\hat{d}\left(\dot{\hat{\alpha}}_{ij}(\eta), \eta\right)\Big/\partial\dot{\hat{\alpha}}_{ij}(\eta)$. This generalization of the orthogonality condition allows once more a complete specification of the constitutive model once the two potential functionals have been defined.

Applications of this approach to kinematic hardening plasticity are discussed by Puzrin and Houlsby (2001b).

2.1 Non-linearity at small strain

One of the most important motivations in the above development is to be able to model the observed non-linearity of soil at small strain in a realistic way. A model employing the method to describe the small-strain non-linearity of soils is given by Puzrin, Houlsby and Burland (2001), but a simple example, is given here. Consider the case of a one-dimensional model where

$$g[\sigma, \hat{\alpha}] = -\frac{\sigma^2}{2E} + \int_0^1 \frac{H(\eta)}{2} (\hat{\alpha}(\eta))^2\, d\eta - \sigma \int_0^1 \hat{\alpha}(\eta)\, d\eta \qquad \text{and}$$

$d\left[\dot{\hat{\alpha}}\right] = \int\limits_0^1 k\eta \left|\dot{\hat{\alpha}}(\eta)\right| d\eta$. This allows a rather general form of kinematic hardening with Masing-type hysteresis on unloading and reloading. Application of the Frechet derivative, and further manipulation (Puzrin and Houlsby, 2001b) allows the kernel hardening function $H(\eta)$ to be related the shape of the initial loading curve through $H(\sigma/k) = 1\big/\left(kd^2\varepsilon/d\sigma^2\right)$. Thus, for instance the expression $H(\eta) = E(1-\eta)^3\big/2$ gives rise to the hyperbolic stress-strain curve $\varepsilon = \dfrac{k}{E}\dfrac{\sigma}{k-\sigma}$. The result that the kernel function can be derived from the stress-strain curve is important in that it facilitates the derivation of soil models from material data.

3 Convex Analysis

The terminology of convex analysis allows a number of the issues relating to hyperplastic materials to be expressed in a succinct manner. In particular, through the definition of the subdifferential, it allows a more rigorous treatment of functions with singularities of various sorts. These arise, for instance, in the treatment of the yield function. A very brief summary of some basic concepts of convex analysis is given here, followed by some illustrations of the advantages in modelling soils. The terminology is based chiefly on that of Han and Reddy (1999). A more detailed introduction to the subject is given by Rockafellar (1970). No attempt is made to provide rigorous, comprehensive definitions here, and for a fuller treatment reference should be made to the above texts. Although it is currently used by only a minority of those studying plasticity, it seems likely that in time convex analysis will be come the standard paradigm for plasticity theory.

In the following C is a subset in a normed vector space V, usually with the dimension of R^n, but possibly infinite dimensional. The notation $\langle\, , \rangle$ is used for an inner product, or more generally the action of a linear operator on a function. The topological dual space of V (the space of linear functionals on V) is V'.

The set containing a range of numbers is denoted by $[\,,]$, thus $[a, b] = \{x \mid a \leq x \leq b\}$, where the meaning of the contents of the final bracket is "x, such that $a \leq x \leq b$".

3.1 Convex sets and functions

A set C is *convex* if and only if $(1 - \lambda)\, x + \lambda y \in C$, $\forall x, y \in C$, $\forall 0 < \lambda < 1$. A function f whose domain is a convex subset C of V and range is real or $\pm\infty$ is *convex* if and only if $f\left((1 - \lambda)\, x + \lambda y\right) \leq (1 - \lambda)\, f(x) + \lambda f(y)$, $\forall x, y \in C$, $\forall 0 < \lambda < 1$. This is illustrated for a function of a single variable in Figure 1. Convexity requires that NP $\leq$ NQ for all N between X and Y. This property has to be true for all pairs of X,Y within the domain of the function. A function is *strictly convex* if $\leq$ can be replaced by $<$ in the above expression for all $x \neq y$.

3.2 Subdifferentials and subgradients

The concept of the subdifferential of a convex function is a generalisation of the concept of a differentiation. It allows the process of differentiation to be extended to convex functions that are not smooth (*i.e.* continuous and differentiable in the conventional sense to any required degree). If V is a vector space and V' is its dual under the inner product $\langle\, , \rangle$, then $z \in V'$ is said to be a subgradient of the function $f(x)$, $x \in V$, if and only if $f(y) - f(x) \geq \langle (y - x), z \rangle$, $\forall y$.

The subdifferential, denoted by $\partial f(x)$ is the subset of V' consisting of all the vectors z satisfying the definition of the subgradient. For a function of one variable it is the set of the slopes of lines passing through a point on the graph of the function, but lying entirely on or below the graph. The concept is illustrated in Figure 2.

The concept of the subdifferential allows us to define "derivatives" of non-differentiable functions. For example the subdifferential of $w = |x|$ is:

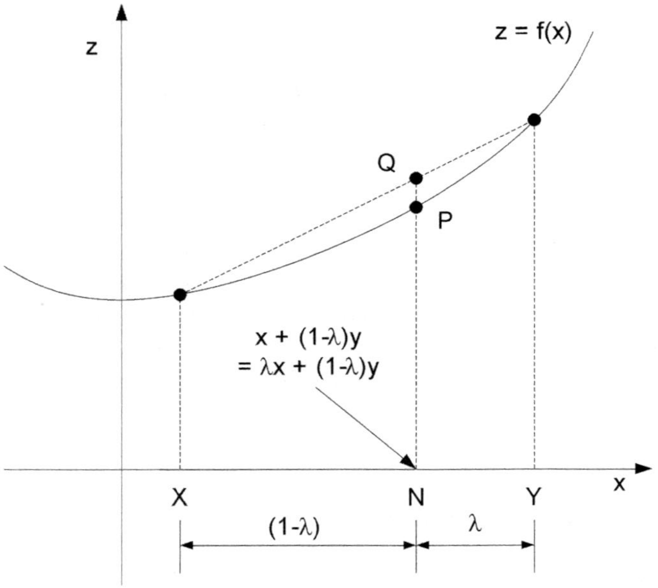

Fig. 1. Graph of a convex function of one variable

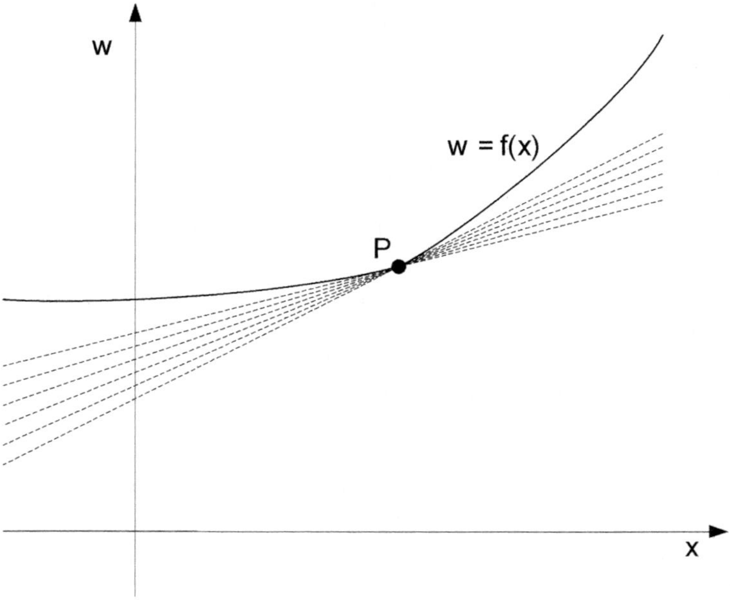

Fig. 2. Subgradients of a function at a non-smooth point

$$\partial w(x) = \begin{cases} \{-1\}, & x < 0 \\ [-1,+1], & x = 0 \\ \{+1\}, & x > 0 \end{cases} \tag{9}$$

Thus at a point x, $\partial f(x)$ may be a set consisting of a single number equal to $\partial f/\partial x$, or a set of numbers, or (in the case of a non-convex function) be empty.

3.3 Functions defined for convex sets

The *indicator function* of a set C is defined:

$$I_C(x) = \begin{cases} 0, & x \in C \\ +\infty, & x \notin C \end{cases} \tag{10}$$

So that the indicator function is simply zero for any x that is a member the set, and $+\infty$ elsewhere. Although this appears at first sight to be a rather curious function, it proves to have many applications.

The *normal cone* $N_C(x)$ of a convex set C, is the set defined by:

$$N_C(x) = \left\{ z \in V' \mid \langle z, y - x \rangle \leq 0,\ \forall y \in C \right\} \tag{11}$$

It is straightforward to show that $N_C(x) = \{0\}$ if $x \in \text{int}C$ (the point is in the interior of the set), and that $N_C(x)$ can be identified with the cone of normals at x if $x \in \text{bdy}C$ (the point is on the boundary of the set), and further that $N_C(x)$ is empty if $x \notin C$ (the point is outside the set). Furthermore, the subdifferential of an indicator function of any convex set is the normal cone of that set: $\partial I_C(x) = N_C(x)$.

Another important function defined for a convex set is the *gauge function* or *Minkowski function*, defined for a set C as:

$$\gamma_C(x) = \inf \left\{ \mu \geq 0 \mid x \in \mu C \right\} \tag{12}$$

where $\inf\{x\}$ denotes the *infimum*, or lowest value of a set.

In other words $\gamma_C(x)$ is the smallest positive factor by which the set can be scaled and x will be a member of the scaled set. The meaning is most easily understood for sets which contain the origin (which proves to be the case for all sets of interest in hyperplasticity). It is straightforward to see in this case that $\gamma_C(x) = 1$ for any point on the boundary of the set, is less than unity for a point inside the set and greater than unity for a point outside the set.

In the context of (hyper)plasticity, it is immediately obvious that the gauge may be related to the conventional yield function. If the set C is the set of (generalised) stresses that are accessible for any given state of the internal variables, then the yield function is a function conventionally taken as zero at the boundary of this set (the yield surface), negative within and positive without. One possible expression for the yield function would therefore be $y(\chi) = \gamma_C(\chi) - 1$. Other functions could of course be chosen as the yield function, but this is perhaps the most rational choice, so we follow Han and Reddy (1999) in calling this the *canonical yield function*. To emphasise when the yield surface is written in this way we shall give it the special notation $\bar{y}(\chi) = \gamma_C(\chi) - 1$.

The gauge function is always homogeneous of order one in its argument x. (In the language of convex analysis such functions are simply referred to as "positively homogeneous".) The canonical yield function is therefore conveniently written in the form of a positively homogeneous function of the (generalised) stresses minus unity.

It is straightforward to see that the definition (12) can be inverted. Given a positively homogeneous function $\gamma(x)$ one can define a set C, such that $\gamma(x)$ is the gauge function of C:

$$C = \{x \mid \gamma(x) \leq 1\} \tag{13}$$

It is worth noting too that the indicator function of a set containing the origin can always be expressed in the following way, which will prove to be useful in the application of this approach to (hyper)plasticity:

$$I_C(x) = I_{[-\infty,0]}(\gamma_C(x) - 1) \tag{14}$$

3.4 Legendre-Fenchel transformation

If $f(x)$ is a convex function defined for all $x \in V$, its Legendre-Fenchel transformation (or Fenchel dual, or conjugate function) is $f^*(x^*)$, where $x^* \in V'$, defined by:

$$f^*(x^*) = \sup_{x \in V}\{\langle x^*, x\rangle - f(x)\} \tag{15}$$

where $\sup\limits_{x \in V}$ means the *supremum*, or highest value for any $x \in V$.

It is straightforward to show the Fenchel dual (or conjugate function) is the generalisation of the Legendre transform. We use the notation that if $x^* \in \partial f(x)$ and $f^*(x^*)$ is the Fenchel dual of $f(x)$ then $x \in \partial f^*(x^*)$.

The support function

The final function that we use here for a convex set C in V is the *support function*. If $x^* \in V'$, then the support function is defined by:

$$\sigma_C(x^*) = \sup\{\langle x, x^*\rangle \mid x \in C\} \tag{16}$$

Note that although C is a set of values of the variable x, the argument of the support function is the variable x^* conjugate to x.

It can be shown that the support function is the Fenchel dual of the indicator function. The support function is always homogeneous of order one in x^*, *i.e.* it is "positively homogeneous".

It follows that any homogeneous order one function defines a set in the dual space. In (hyper)plasticity one can observe that the dissipation function is indeed homogeneous and order one in the internal variable rates. It can thus be interpreted as a support function, and the set it defines in the dual space of (generalised) stresses is the set of accessible (generalised) stress states. The Fenchel dual of the dissipation function is the indicator function for this set of accessible states, which is of course zero throughout the set. We can identify this indicator function with the Legendre transform $w = \lambda y$ of the dissipation function introduced earlier.

Equation 16 can be inverted to obtain the set C from the support function. If $f(x^*)$ is a homogeneous first order function in x^*, then the corresponding set can be found by solving the system of inequalities:

$$C = \{x \mid \langle x, x^* \rangle \leq f(x^*), \forall x^*\} \tag{17}$$

Both the gauge and support functions are positively homogeneous. If can be shown (see Han and Reddy, 1999) that:

$$\gamma_C(x) = \sup_{0 \neq x^* \in \mathrm{dom}\sigma_C} \frac{\langle x^*, x \rangle}{\sigma_C(x^*)} \tag{18}$$

and $\gamma_C(x)$ is called the *polar* of $\sigma_C(x^*)$, written $\gamma_C = \sigma_C^\circ$. The process is symmetric so that we have $\sigma_C = \gamma_C^\circ$ and:

$$\sigma_C(x^*) = \sup_{0 \neq x \in C} \frac{\langle x^*, x \rangle}{\gamma_C(x)} \tag{19}$$

Further we have the following inequality:

$$\sigma_C(x^*)\gamma_C(x) \geq \langle x^*, x \rangle, \ \forall x \in C, \ \forall x^* \in \mathrm{dom}\sigma_C \tag{20}$$

In summary we therefore have the following objects of interest:

- A convex set C in V.
- The indicator function $I_C(x)$ of the set.
- The gauge function $\gamma_C(x)$ of the set.
- The support function $\sigma_C(x^*)$ which is the Fenchel dual of the indicator, and is also the polar of the gauge function.
- The normal cone $N_C(x) = \partial I_C(x)$ which is a set in V' which is the subdifferential of the indicator.

3.5 The indicator function and constraints

For cases where the potentials are not differentiable in the conventional sense, convex analysis serves as the framework for expressing the constitutive behaviour, subject only to the limitation that the potentials must of course be convex. This does not prove too restrictive for our purposes. A complete exposition of hyperplasticity in convex analysis terminology would be lengthy, but suffice it to say (at least for simple examples) that each occurrence of a differential becomes a subdifferential. Thus instead of $\sigma = \partial f / \partial \varepsilon$ we have $\sigma \in \partial f(\varepsilon)$.

As an example of how convex analysis can be used to express constraints, consider now some simple variants on elasticity. Linear elasticity is given by either of the expressions $f = \dfrac{E\varepsilon^2}{2}$ or $-g = \dfrac{\sigma^2}{2E}$.

Using derivations based on the subdifferential (which in this case includes simply the derivative, because both the above are smooth strictly convex functions) $\sigma \in \partial f(\varepsilon)$ hence $\sigma = E\varepsilon$, or $\varepsilon \in \partial(-g(\sigma))$ hence $\varepsilon = \sigma/E$.

Now consider a rigid material, which can be considered as the limit as $E \to \infty$. The resulting f can be written in terms of the indicator function $f = I_{\{0\}}(\varepsilon)$, which has the Fenchel dual $-g = 0$.

The subdifferential of f gives $\sigma \in N_{\{0\}}(\varepsilon)$, which gives $\sigma \in [-\infty, +\infty]$ for $\varepsilon = 0$, and is otherwise empty, so that there is zero strain for any finite stress. Conversely the subdifferential of $-g$ (in this case just consisting of the derivative) gives $\varepsilon = 0$ directly. In a comparable way, the limit $E \to 0$, *i.e.* an infinitely flexible material, is obtained from either of $f = 0$ or $-g = I_{\{0\}}(\sigma)$.

The above considerations become of more practical application as one moves to two and three dimensional cases. For instance triaxial linear elasticity is given by $f = \dfrac{Kv^2}{2} + \dfrac{3G\varepsilon^2}{2}$ or $-g = \dfrac{p^2}{2K} + \dfrac{q^2}{6G}$. Incompressible elasticity ($K \to \infty$) is simply given by $f = I_{\{0\}}(v) + \dfrac{3G\varepsilon^2}{2}$ or $-g = \dfrac{q^2}{6G}$, without the need to introduce a separate constraint. Note that whenever it is required to constrain a variable x which is an argument of a function to zero, one simply adds the indicator function $I_{\{0\}}(x)$. In the dual form, the Fenchel dual does not depend on the conjugate variable to x.

The above results can of course very simply be extended to full continuum models.

Unilateral constraints can also be treated using convex analysis. A one-dimensional material with zero stiffness in tension (*i.e.* a "cracking" material) can obtained from $f = \dfrac{E_c \langle -\varepsilon \rangle^2}{2}$ or $-g = I_{[-\infty,0]}(\sigma) + \dfrac{\langle -\sigma \rangle^2}{2E_c}$, where the Macaulay bracket is defined such that $\langle x \rangle = x$ if $x \geq 0$ and $\langle x \rangle = 0$ if $x < 0$ (where we use a tensile positive convention). Such a model might for instance be the starting point for the modelling of masonry materials, concrete or soft rocks.

Another case, rigid in tension and with zero stiffness in compression (in other words the "light inextensible string" that is found in many elementary textbooks) is given by $f = I_{[-\infty,0]}(\varepsilon)$ or $-g = I_{[0,\infty]}(\sigma)$.

In each of the above cases elementary application of the subdifferential formulae gives the required constitutive behaviour, effectively applying the "constraints" (unilateral or bilateral) as required.

3.6 The yield surface revisited

The dissipation function, or force potential $d = z = d(\dot{\alpha})$ is a first order function of $\dot{\alpha}$, and the conjugate generalised stress is defined by $\chi \in \partial d(\dot{\alpha})$, which is the generalisation of $\chi = \partial d / \partial \dot{\alpha}$.

The set X (capital χ) of accessible stress states can be found by identifying the dissipation function as the support function of a convex set of χ, hence applying equation (17):

$$\mathrm{X} = \{\chi \mid \langle \chi, \dot{\alpha} \rangle \leq d(\dot{\alpha}), \forall \dot{\alpha}\} \tag{21}$$

The indicator and gauge functions of X can be determined in the usual way. Note that the indicator is of course the dual of the support function, so it is the flow potential:

$$I_{\mathrm{X}}(\chi) = \begin{cases} 0, & \chi \in \mathrm{X} \\ +\infty, & \chi \notin \mathrm{X} \end{cases} = w(\chi) \tag{22}$$

where $\dot{\alpha} \in \partial w(\chi) = N_{\mathrm{X}}(\chi)$, which is the generalisation of $\dot{\alpha} = \partial w / \partial \chi$. It is useful at this stage to obtain the gauge function:

$$\gamma_{\mathrm{X}}(\chi) = \inf \{ \mu \geq 0 \mid \chi \in \mu \mathrm{X} \} \tag{23}$$

The gauge may also be obtained directly as the polar of the dissipation:

$$\gamma_{\mathrm{X}}(\chi) = \sup_{0 \neq \dot{\alpha} \in \mathrm{dom} d} \frac{\langle \chi, \dot{\alpha} \rangle}{d(\dot{\alpha})} \tag{24}$$

And further we define the canonical yield function (in the usual sense adopted in hyperplasticity) as $\bar{y}(\chi) = \gamma_{\mathrm{X}}(\chi) - 1$. Applying then equation (14)

$$I_{\mathrm{X}}(\chi) = w(\chi) = I_{[-\infty,0]}(\bar{y}(\chi)) \tag{25}$$

So that applying the usual approach we obtain any of the following:

$$\dot{\alpha} \in \partial w(\chi) = \partial I_{\mathrm{X}}(\chi) = N_{\mathrm{X}}(\chi) = \lambda \partial \gamma(\chi) = \lambda \partial \bar{y}(\chi) \tag{26}$$

Where $\lambda \geq 0$ (see Lemma 4.5 of Han and Reddy (1999)). The above is the equivalent of the usual $\dot{\alpha} = \lambda \partial y / \partial \chi$. Clearly $\partial \bar{y}(\chi)$ plays the role of $\partial y / \partial \chi$, and λ has its usual meaning. In particular $\lambda = 0$ for a point within the yield surface (interior of X) and takes any value in the range $[0, +\infty]$ for a point on the yield surface (boundary of X).

It can be seen, however, that the assumption made in developments in earlier papers that, since $\dot{\alpha} = \partial w / \partial \chi = \lambda \partial y / \partial \chi$ with λ an arbitrary multiplier, one could therefore deduce that $w = \lambda y$ was slightly too simplistic a step.

We are now in a position to address the process of obtaining either a yield surface from a dissipation function or *vice versa*. If we start with $d = z = d(\dot{\alpha})$

then we apply (17) to find the set of admissible states X, and then use (12), together with the definition of the canonical yield function:

$$\begin{aligned}\bar{y}(\chi) &= \inf\left\{\mu \geq 0 \mid \chi \in \mu \mathrm{X}\right\} - 1 \\ &= \inf\left\{\mu \geq 0 \mid \chi \in \mu\left\{\chi' \left\langle \chi', \dot{\alpha}\right\rangle \leq d(\dot{\alpha}), \forall \dot{\alpha} \in \operatorname{dom} d\right\}\right\} - 1 \\ &= \inf\left\{\mu \geq 0 \mid \langle \chi, \dot{\alpha}\rangle \leq \mu d(\dot{\alpha}), \forall \dot{\alpha} \in \operatorname{dom} d\right\} - 1\end{aligned} \tag{27}$$

So that $\bar{y}(\chi)$ can in principle be determined directly from $d(\dot{\alpha})$. This is an important result. We then have $\dot{\alpha} \in \lambda \partial \bar{y}(\chi)$.

Conversely, if we first specify the yield surface $y(\chi)$ in the normal way, then X is easily obtained from $\mathrm{X} = \{\chi \mid y(\chi) \leq 0\}$, and the dissipation function is then the support function of this set:

$$d(\dot{\alpha}) = \sigma_{\mathrm{X}}(\dot{\alpha}) = \sup\left\{\langle \chi, \dot{\alpha}\rangle \mid \chi \in \mathrm{X}\right\} = \sup\left\{\langle \chi, \dot{\alpha}\rangle \mid y(\chi) \leq 0\right\} \tag{28}$$

So that $d(\dot{\alpha})$ can in principle be determined directly from $y(\chi)$. This too is an important result, although it is one that is rather more obvious than the transformation from dissipation to yield.

It is not essential for (28), but there is a clear preference for expressing the yield surface in canonical form such that $\gamma_{\mathrm{X}}(\chi) = \bar{y}(\chi) + 1$ is a homogeneous first order function of χ, so that it can be interpreted as the gauge function of the set X. Note that the yield function is not itself positively homogeneous, but it is, however, expressible as a positively homogeneous function minus unity. If it is chosen this way then $\bar{y}$ is dimensionless, so that λ has the dimension of stress times strain rate.

If y is expressed in canonical form then the dissipation function can be expressed directly as the polar:

$$d(\dot{\alpha}) = \sup_{0 \neq \chi \in \mathrm{X}} \frac{\langle \chi, \dot{\alpha}\rangle}{(\bar{y}(\chi) + 1)} \tag{29}$$

The results are summarised as follows:

Option 1: start from specified dissipation function $d = z = d(\dot{\alpha})$

$$\chi \in \partial d(\dot{\alpha}) \tag{30}$$

$$\bar{y}(\chi) = \inf\left\{\lambda \geq 0 \mid \langle \chi, \dot{\alpha}\rangle \leq \lambda d(\dot{\alpha}), \forall \dot{\alpha}\right\} - 1 = \sup_{0 \neq \dot{\alpha} \in \operatorname{dom} d} \frac{\langle \chi, \dot{\alpha}\rangle}{d(\dot{\alpha})} - 1 \tag{31}$$

Option 2: start from specified $y(\chi)$

$$w(\chi) = I_{[-\infty, 0]}(y(\chi)) \tag{32}$$

$$\dot{\alpha} \in \partial w(\chi) = \lambda \partial y(\chi) \tag{33}$$

$$d(\dot{\alpha}) = \sup\left\{ \langle \chi, \dot{\alpha} \rangle \mid y(\chi) \leq 0 \right\} \tag{34}$$

Note that if y is not expressed in canonical form it cannot be readily converted to the gauge, and so the dissipation function cannot simply be obtained as the polar of the gauge.

The function w (the flow potential) is the indicator of the set of admissible generalised stress states.

If $y(\chi)$ is in canonical form such that $\gamma(\chi) = \bar{y}(\chi) + 1$ is homogeneous of order one, then applying option 2 to obtain d, and then applying option 1 to obtain y will return the original function. If this condition is not satisfied then applying this procedure will give a different functional form of the yield function (in fact the canonical form), but specifying the same yield surface.

3.7 Examples in plasticity theory

A plastically incompressible cohesive material in triaxial space can be defined, for instance, by:

$$-g = \frac{p^2}{2K} + \frac{q^2}{6G} + q\alpha_q \tag{35}$$

$$d = 2c\left|\dot{\alpha}_q\right| \tag{36}$$

In which only a plastic shear strain is introduced. The canonical yield function can be obtained as:

$$\bar{y} = \frac{|\chi_q|}{2c} - 1 \tag{37}$$

Alternatively both the plastic strain components are introduced, but the volumetric component is constrained to zero. This approach proves to be more fruitful for further development. In the past this has been achieved by imposing a separate constraint, but now we do so by introducing an indicator function into the dissipation:

$$-g = \frac{p^2}{2K} + \frac{q^2}{6G} + p\alpha_p + q\alpha_q \tag{38}$$

$$d = 2c\left|\dot{\alpha}_q\right| + I_{\{0\}}\left(\dot{\alpha}_p\right) \tag{39}$$

The yield function is unchanged for this case.

This model is readily altered to frictional non-dilative plasticity by changing the dissipation to:

$$d = M\langle p \rangle \left|\dot{\alpha}_q\right| + I_{\{0\}}\left(\dot{\alpha}_p\right) \tag{40}$$

Where note that we have introduced a Macaulay bracket on p which we did not use before, but strictly is necessary. The corresponding canonical yield function is:

$$\bar{y} = \frac{|\chi_q|}{M\langle p\rangle} - 1 \tag{41}$$

The virtue of introducing the plastic volumetric strain is now seen in that the model can now be further modified to include dilation by changing d to:

$$d = M\langle p\rangle |\dot{\alpha}_q| + I_{\{0\}}(\dot{\alpha}_p + \beta|\dot{\alpha}_q|) \tag{42}$$

The canonical yield function for this case becomes:

$$\bar{y} = \frac{|\chi_q| - \beta\chi_p}{M\langle p\rangle} - 1 \tag{43}$$

Which can be compared with the yield locus $y = |\chi_q| - Mp - \beta\chi_p = 0$ used in the earlier example.

The above are some simple examples of the way by which expressions making use of convex analysis terminology can provide a succinct description of plasticity models for geotechnical materials. They may provide the starting point for using this approach in more sophisticated modelling.

4 Conclusions

The purpose of this paper has been to set out some mathematical results which are useful in the constitutive modelling of geotechnical materials. Emphasis has been placed on mathematics appropriate the "hyperplasticity", which is an approach that has certain benefits in this modelling, principally related to the strong use of potentials.

Legendre transforms are used to interchange between different energy potentials and also between the dissipation and yield functions.

The technique of Frechet differentiation of a functional is introduced to allow models with, in effect, an infinite number of yield surfaces to be described.

Finally concepts of convex analysis are introduced, following Han and Reddy's approach to plasticity, and it is shown how this terminology can be successfully used (a) in the treatment of constraints and (b) in a more rigorous formulation of the relationships between yield and dissipation.

Acknowledgement

The author acknowledges gratefully the continuing input to this work from Assoc. Prof. A.M. Puzrin. Useful discussions at the Horton conference with Prof. M. Brokate are also acknowledged.

References

[1] Collins, I.F. and Houlsby, G.T. (1997) "Application of Thermomechanical Principles to the Modelling of Geotechnical Materials", Proc. Royal Society of London, Series A, Vol. 453, 1975-2001

[2] Han, W. and Reddy, B.D. (1999) Plasticity: Mathematical Theory and Numerical Analysis, Springer-Verlag, New York

[3] Houlsby, G.T. and Puzrin, A.M. (2000) "A Thermomechanical Framework for Constitutive Models for Rate-Independent Dissipative Materials", Int. Jour. of Plasticity, Vol. 16 No. 9, 1017-1047

[4] Houlsby, G.T. and Puzrin, A.M. (2002) "Rate-Dependent Plasticity Models Derived from Potential Functions", Jour. of Rheology, Vol. 46, No. 1, January/February, 113-126

[5] Maugin, G. (1999) The Thermomechanics of Nonlinear Irreversible Behaviours, World Scientific, Singapore

[6] Puzrin, A.M., Houlsby, G.T. and Burland, J.B. (2001) "Thermomechanical Formulation of a Small Strain Model for Overconsolidated Clays", Proceedings of the Royal Society of London, Series A, Vol. 457, No. 2006, February, ISSN 1364-5021, 425-440

[7] Puzrin, A.M. and Houlsby, G.T. (2001b) "Fundamentals of Kinematic Hardening Hyperplasticity", Int. Jour. Solids and Structures, Vol. 38, No. 21, May, 3771-3794

[8] Puzrin, A.M. and Houlsby, G.T. (2001a) "A Thermomechanical Framework for Rate-Independent Dissipative Materials with Internal Functions", Int. Jour. of Plasticity, Vol. 17, 1147-1165

[9] Rockafellar, R.T. (1970) Convex Analysis, Princeton University Press

[10] Ziegler, H (1977, 2nd edition 1983) An Introduction to Thermomechanics, North Holland, Amsterdam

Elasticity in constitutive modeling of soils

Alexander M. Puzrin

School of CEE, Georgia Tech, Atlanta GA 30332-0355, USA,
alexander.puzrin@ce.gatech.edu

1 Introduction

In June 2002 I met Dr. Jacques Desrues at a conference in Minnesota. I overheard Jacques in a discussion with a colleague saying: "I do not believe in elastic strains in soils". I told him that this is exactly the topic Prof. Kolymbas asked me to present in July in Horton. He said that Prof. Kolymbas approached him with this topic a while ago, but he decided that it was too complicated to get involved. That was when I realized that I was in trouble, but it was already too late to back off.

Are elastic strains in soils really a matter of belief? This paper is an attempt to review the problem of elastic strains in a light of the recent advances in experimental and theoretical studies of the constitutive behavior of soils. Specifically, we shall try to find the answers to the following three questions:

- Does the linear elastic region (LER) of stresses/strains in soils exist?
- Can elastic strains exist in spite of the lack of the elastic region?
- Is there a more objective criterion for existence of the elastic strains?

We shall start with a brief review of applications of elasticity in Geotechnical Engineering and try to understand why there is a need for more accurate modeling of small strain behavior of soils. Then we shall try to summarize how different plasticity theories approach the issue of elasticity. It will be shown that neither lab tests nor plasticity theories agree on existence of the LER and of the elastic component of total strain. This urges us to look for more objective criteria for their existence, which brings us to thermo- and micromechanical approaches to elasticity. The criterion which these approaches suggest is the elasto-plastic coupling. In conclusion we shall attempt to answer the three above questions and look into the possible implications of these answers for constitutive modeling of soils.

2 Elasticity in Geotechnical Engineering

2.1 Definition

Before we move any further, let us agree on the terms. The general definition of elasticity requires a unique relationship between stresses and strains:

$$\sigma_{ij} = F\left(\varepsilon_{ij}\right) \quad \varepsilon_{ij} = F^{-1}\left(\sigma_{ij}\right) \tag{1}$$

The most popular form of elastic relationship is Hooke's Law:

$$\begin{aligned} \sigma_{ij} &= K\varepsilon_{mm}\delta_{ij} + 2G\varepsilon'_{ij} \\ \varepsilon_{ij} &= \frac{1}{9K}\sigma_{mm}\delta_{ij} + \frac{1}{2G}\sigma'_{ij} \end{aligned} \tag{2}$$

where K and G are the elastic bulk and shear moduli, prime stands for deviatoric tensors. If only soils were elastic materials, the life of practicing Geotechnical Engineers would be a paradise. However in this case most of the researchers working in the area of constitutive modeling would find themselves out of their job. Let us explore this controversy a little bit further in a brief review of applications of elasticity in Geotechnical Engineering

2.2 Soil behavior

Let us take a soil sample and perform a standard triaxial compression test with external strain measurements. A typical experimental stress-strain diagram for this test is shown in Figure 1a, where the deviatoric stress is plotted versus the axial strain. From this diagram it follows that the soil behavior is: (1) non-linear; (2) inelastic; and (3) has memory. Elastic theory on its own cannot handle this behavior, but it can still be used as a part of an elasto-plastic formulation (Figure 1b). This formulation assumes existence of linear or non-linear elastic region. When the stress state reaches the boundary of

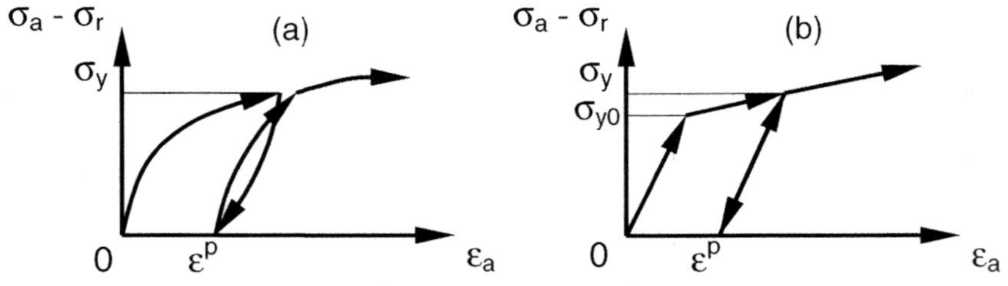

Fig. 1. Triaxial compression test: a) stress-strain curve; b) elasto-plastic simulation.

this region, known as a yield surface, the total strain is calculated as a sum of the elastic and plastic components. The yield surface may undergo hardening, but unloading and reloading inside this surface are elastic again. Within this context, many geotechnical problems have been treated using the theory of elasticity.

2.3 History and applications

Elastic solutions utilized in geotechnical engineering are associated with some great names in the Theory of Elasticity. Babinet, Saint-Venant, Lord Rayleigh studied waves in elastic bodies starting from the early 19th century. In 1852, Lame published his work on cavity expansion. Famous solution for the force on a boundary of a semi-infinite space is attributed to the 1885 work of Boussinesq while the problem for the force on a boundary of a semi-infinite plane was solved in 1892 by Flamant. In 1936 Mindlin extended these solutions to a force at an internal point of a semi-infinite space.

The first systematic overview of applications of elasticity in Geotechnical Engineering was given in 1943 by Terzaghi in his classical book "Fundamentals of Soil Mechanics". Section D "Elastic problems in Soil Mechanics" in this book is entirely devoted to these applications. Among the problems treated in this section we find: stress distribution in soils, displacements of shallow and deep foundations, displacements around underground excavations, wave propagation, etc.. The most extensive summary of these solutions has been presented by Poulos and Davis in their 1974 monograph "Elastic Solutions for Soil and Rock Mechanics". Advantage of these solutions is that most of them are given in a closed form or in the form of plots and tables. However, approximately by the same time as these solutions were published, development of computers and numerical methods made this advantage less meaningful, and more realistic constitutive models have been developed and implemented ever since. Still, most of the recent advances in Dynamic Pile Analysis and Geophysics have been based on elastic or visco-elastic analysis.

2.4 Limitations of elastic solutions

As convenient and tempting as the elastic solutions are, their application to geotechnical problems may be rather misleading. The general observation, made in 2002 by Prof. Harry Poulos in his 5th George Sowers Lecture "Critical role of ground characterization for foundation deformation prediction", is that elastic models may work for "in-line" problems, like calculation of vertical ground displacements due to vertical loads of foundations. Then even the linear elastic theory, with some adjustment of parameters, may produce a decent prediction. For "out-of-line" problems, e.g. vertical displacements due to horizontal loads or horizontal displacements due to vertical loads, simple models do not work. The striking example of an "out-of-line" problem is the problem of ground displacements, where the linear elastic theory predicts a heave of the ground surface around the borehole which contradicts the reality and cannot be corrected by adjustment of elastic parameters.

Generally, in large strain problems displacements are governed by the plastic strain component and elastic solution does not contribute a lot. Though this phenomenon should not affect the small strain problems, displacements in these problems are still miscalculated, when elastic stiffness is determined from standard lab tests. This discrepancy initiated 20 years ago a study into small strain behavior of soils which we are going to discuss in the next section.

2.5 Summary

Elasticity plays the historically important role in Geotechnical Engineering. It is used to solve both small and large strain problems – in the latter case as a part of elasto-plastic constitutive model. However, inability of elasticity to provide accurate quantitative and sometimes even qualitative predictions requires a better understanding and modeling of pre-failure behavior of soils as discussed in the following section.

3 Elasticity in Small Strain Behavior of Soils

3.1 Local deformation measurements

Over the last 20 years ago, researchers in England (e.g. Jardine et al [1]; Burland [2]), Italy (e.g. Jamiolkowski et al [3]), Japan (e.g. Tatsuoka&Kohata [4]) and USA (e.g. Stokoe et al [5]) started measuring both axial and radial deformations locally on the sample inside the triaxial chamber. This allowed for the stress-strain diagram to be corrected as shown in Figure 2.

The difference between the lower curve obtained by external measurements and the upper one, obtained from the local measurements, might seem not worthy of all the efforts. But if we zoom into the area of small strains, say up to 0.1%, this difference will be much more impressive (Figure 3).

3.2 Non-linearity and irreversibility at small strains

The curve obtained from external measurements (Figure 3) has almost a constant slope suggesting a linear behaviour at small strains, while local measurements indicate much higher initial stiffness and a high degree of non-linearity. However, after the strain exceeds a value of about 0.1%, there is no visible difference in stiffness. So does this phenomenon really affect solutions of geotechnical problems?

Jardine and Potts [6] demonstrated the practical significance of this finding by finite element analysis of the distribution of shear strains in ground caused by the pulling force applied to a tension pile. When the pulling force reaches a value providing a FOS=2, it causes a pile head vertical displacement of

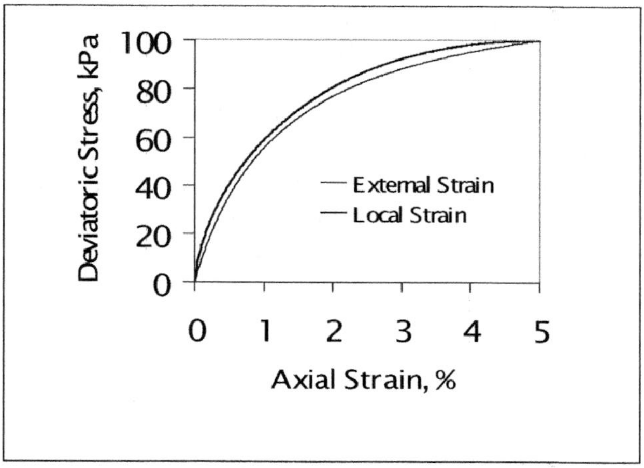

Fig. 2. Triaxial compression test: external and local strain measurements.

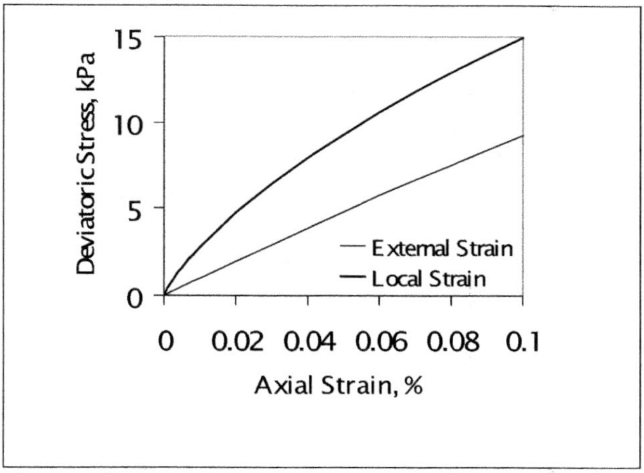

Fig. 3. External and local strain measurements at small strains.

1cm, which is very close to the working conditions. Surprisingly, only a tiny little area in a close vicinity of the pile appears to be strained beyond 0.1%, the rest of the soil media experiences very small strains. In order to obtain displacements, the strains have to be integrated over the area. Therefore, the small strains integrated over the vast area make a sound contribution into the displacement values, and miscalculation of stiffness at small strains may affect these values considerably.

This explains why it is difficult to obtain accurate predictions when the stiffness is taken from external strain measurements, suggesting linear behavior at small strains. But if we do take this non-linearity into account can we still use elasticity? Unfortunately, not in the problems where at least a certain part of the media is subsequently loading and unloading, because another important finding was that the soil behavior becomes irreversible at very small strains as well (Figure 4).

The strain separating between linear elastic and non-linear inelastic behavior was found to be very close to the resolution of measurements (e.g. Jardine [7]), which suggested that the linear elastic region (LER) may not exist at all. However, probably at that time it required too much courage to abandon the idea of the linear elastic region completely.

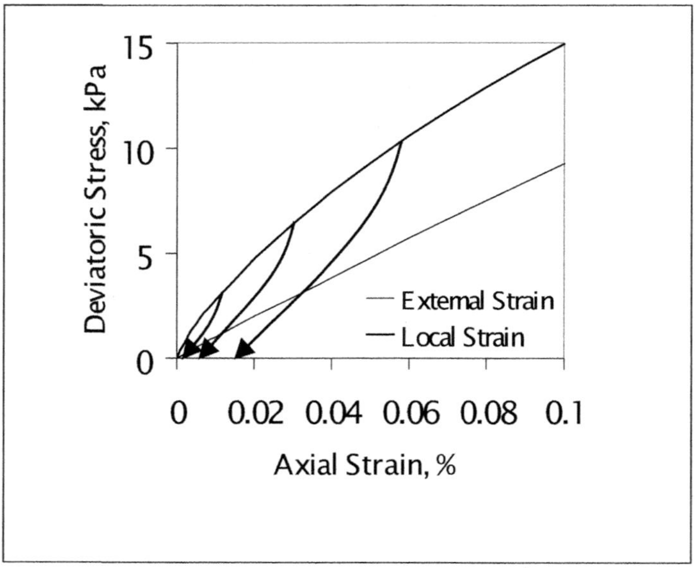

Fig. 4. Triaxial compression test: stress reversals at small strains.

3.3 The linear elastic region (LER)

We can see that, for example, in the concept of kinematic regions of high stiffness presented by Jardine [7] (Figure 5). The LER is floating within the non-linear SSR. Both of them are floating within the large-scale yield surface which in the past was considered to be itself the boundary of elastic region. We see that in the modern understanding of the small strain behavior of soils, the LER has shrunk tremendously but it is still there.

3.4 Threshold strains

Generally, the existing experimental data is very contradictory when it comes to the existence of the LER. The boundary of the LER in the strain space

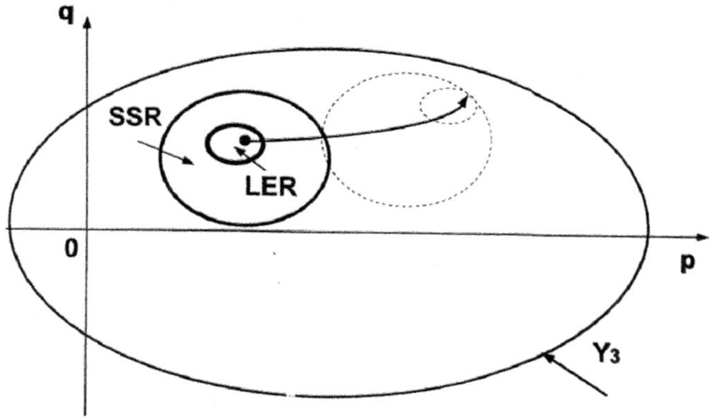

Fig. 5. Kinematic regions of high stiffness.

is called the Linear Threshold Strain (LTS). An excellent example of this contradiction is presented by Vinale et al [8], who performed RC cyclic tests on Metramo sand for different levels of controlled suction. When the normalized shear modulus is plotted versus the shear strain, the LTS, beyond which the behavior becomes non-linear, is clearly observed. However, when the same tests are performed at the constant shear rate, the behavior becomes non-linear at the smallest strain measured. Examples like this divide experimentalists into two camps – those who believe in existence of the LER and those who do not believe in it. Those who do not believe in LER, also do not believe in existence of the elastic strain component of the total strain, though we this may be quite a separate issue as will be shown in the next section.

3.5 Summary

Let us summarize what we know about the small strain behavior of soils. Local deformation measurements show that the soil behavior becomes non-linear and irreversible at very small strains. Experimentalists are divided with respect to existence of the linear elastic region and threshold strain. But even if purely elastic strain does not exist, can it still exist as a component of total strain? Experimentalists cannot answer this question, because the only thing they can measure is the reversible part of strain. But is this reversible strain elastic? To answer this question we need to have a constitutive model in mind. In the following section we are going to find out what do advanced plasticity theories have to say about the LER and elastic strains.

4 Elasticity in Advanced Plasticity Theories

Due to the lack of space this review of the plasticity theories is very brief, with references made only to the pioneering works from which the theories evolved. The conclusions, however, encompass a much broader body of literature on each particular theory discussed below.

4.1 Multiple Surfaces Kinematic Hardening Plasticity

Multiple surface kinematic hardening plasticity (Figure 6), introduced by Iwan [9] and Mroz [10], has a linear elastic region but it has no problem if it vanishes. In this case it will still use elastic strain as a component of total strain, but the reversible strain is not purely elastic.

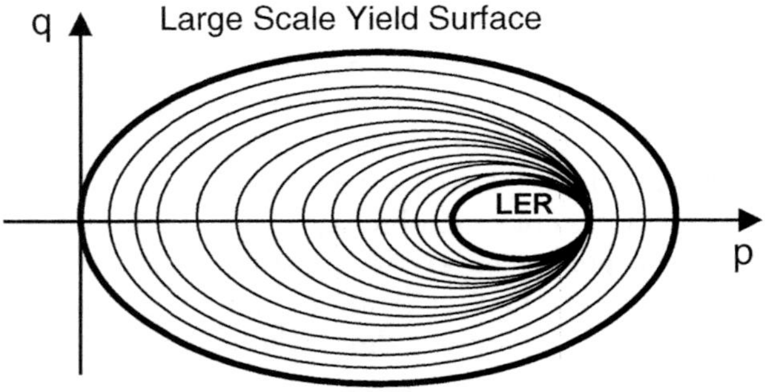

Fig. 6. Schematic layout of a multiple surfaces kinematic hardening plasticity model.

4.2 Bounding Surface Plasticity

Classical bounding surface plasticity (Figure 7) proposed by Krieg [11] and Dafalias & Popov [12] has no elastic region in loading and reloading but the behavior is elastic in unloading. In some versions there is no elastic region at all. In any case it also uses elastic strain as a component of total strain, but the reversible strain may not be purely elastic.

4.3 Hypoplasticity

Hypoplasticity formulated by Kolymbas [13] assumes a direct incrementally non-linear stress-strain relationship:

$$\dot{\sigma}_{ij} = L_{ijkl}\dot{\varepsilon}_{kl} + N_{ij}\sqrt{\dot{\varepsilon}_{kl}\dot{\varepsilon}_{kl}} \tag{3}$$

Therefore it has neither the elastic strain nor the elastic region.

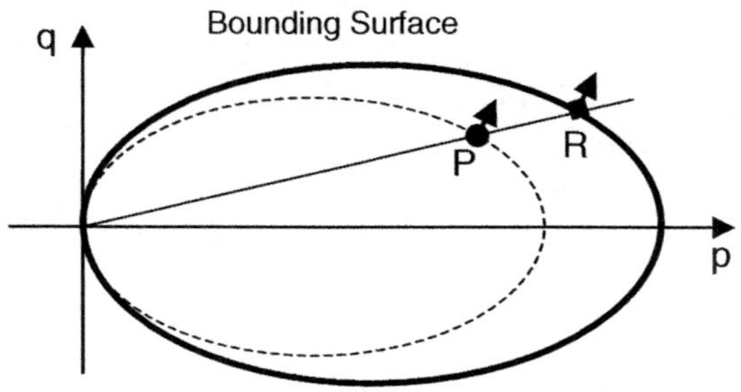

Fig. 7. Schematic layout of a bounding surface plasticity model.

4.4 Continuous Hyperplasticity

Continuous hyperplasticity, introduced by Puzrin&Houlsby [14], is based on thermo-mechanical approach to plasticity. Any model within this framework is specified by two thermodynamic potentials - Gibbs free energy and dissipation functionals:

$$G = G_1(\sigma_{ij}) - \sigma_{ij} \int_0^1 \varepsilon_{ij}^p(\eta)\, d\eta + \int_0^1 G_2\left(\varepsilon_{ij}^p(\eta), \eta\right) d\eta \tag{4}$$

$$D^G = \int_0^1 D^G\left(\sigma_{ij}, \dot{\varepsilon}_{ij}^p(\eta), \eta\right) d\eta \geq 0 \tag{5}$$

The model behavior is similar to that of the kinematic hardening model with infinite number of surfaces. The difference is that the continuous hyperplasticity formulation ensures that the 1st and the 2nd laws of Thermodynamics are satisfied. This framework is flexible with respect to LER and elastic strains, i.e. it may or may not accommodate them. However, in a lack of the LER, the elastic strain component is not equivalent to the reversible strain.

4.5 Summary

Table 1 summarizes assumptions of various plasticity theories with respect to existence of the LER and elastic strains. It looks like that, similar to the experimental data, plasticity theories do not agree on existence of the LER and the elastic component of total strain. Furthermore, if the LER assumption is not valid, none of these theories implies equivalence between the reversible and elastic strains, which is not of a great help to experimentalists. This urges us to look for more objective criteria for the LER and elastic strains existence, which brings us to thermo- and micro-mechanical approaches to elasticity and plasticity.

Theory	LER Existence	Elastic Strain
Multiple Surface Plasticity	+ / -	+
Bounding Surface Plasticity	+ / -	+
Hypoplasticity	-	-
Continuous Hyperplasticity	+ / -	+ / -

Table 1. Plasticity theories treatment of the LER and elastic strains.

5 Elasticity in Thermomechanics

Let us start with thermomechanics. In order to avoid misunderstanding, we are not interested here in problems related to heat transfer. Our concern is energy conservation and dissipation, governed by the First and the Second Laws of Thermodynamics, respectively.

5.1 Hyperelasticity

The First Law of Thermodynamics states that the total incremental work performed on the unit volume of soil is partially stored as an increment of the free energy and partially dissipated into heat:

$$\sigma_{ij}\dot{\varepsilon}_{ij} = \dot{f} + d \tag{6}$$

Because elastic processes are reversible, the rate of dissipation $d = 0$. It follows that in other to satisfy the First Law, specific Helmholtz free energy $f(\varepsilon_{ij})$ should be a potential for stresses:

$$\sigma_{ij} = \frac{\partial f}{\partial \varepsilon_{ij}} \tag{7}$$

This expression defines the hyperelastic constitutive law. The strain in this expression is entirely elastic and defines both the stress and amount of stored energy uniquely.

5.2 Hyperelasticity within perfect and isotropic hardening plasticity

For decoupled elasto-plastic behavior, the total strain can be decomposed into elastic and plastic components

$$\varepsilon_{ij} = \varepsilon_{ij}^{e} + \varepsilon_{ij}^{p} \tag{8}$$

Then for perfectly plastic and isotropic hardening/softening materials it can be shown that the specific Helmholtz free energy depends solely on the elastic component ε_{ij}^e. In this case the total work increment can also be decomposed:

$$\sigma_{ij}\dot{\varepsilon}_{ij} = \sigma_{ij}\dot{\varepsilon}_{ij}^e + \sigma_{ij}\dot{\varepsilon}_{ij}^p \tag{9}$$

where the first term is the full differential of specific Helmholtz free energy, $\sigma_{ij}\dot{\varepsilon}_{ij}^e = \dot{f}\left(\varepsilon_{ij}^e\right)$,while the second term gives the rate of dissipation $\sigma_{ij}\dot{\varepsilon}_{ij}^p = d$. Therefore, the stress has to be defined as:

$$\sigma_{ij} = \frac{\partial f\left(\varepsilon_{ij}^e\right)}{\partial \varepsilon_{ij}^e} \tag{10}$$

In this formulation, the elastic strain ε_{ij}^e is only a part of total strain, but because it defines both the stress and amount of stored energy uniquely, it can still be defined as elastic in thermomechanical sense.

5.3 Hyperelasticity within kinematic hardening plasticity

For decoupled elasto-plastic behavior $\varepsilon_{ij} = \varepsilon_{ij}^e + \varepsilon_{ij}^p$ in kinematic hardening materials it can be shown that the specific Helmholtz free energy takes the form of

$$f\left(\varepsilon_{ij}^e, \varepsilon_{ij}^p\right) = f_1\left(\varepsilon_{ij}^e\right) + f_2\left(\varepsilon_{ij}^p\right) \tag{11}$$

The total work increment can again be decomposed $\sigma_{ij}\dot{\varepsilon}_{ij} = \sigma_{ij}\dot{\varepsilon}_{ij}^e + \sigma_{ij}\dot{\varepsilon}_{ij}^p$, while the First Law takes the form:

$$\sigma_{ij}\dot{\varepsilon}_{ij} = \dot{f} + d = \frac{\partial f_1}{\partial \varepsilon_{ij}^e}\dot{\varepsilon}_{ij}^e + \frac{\partial f_2}{\partial \varepsilon_{ij}^p}\dot{\varepsilon}_{ij}^p + d \tag{12}$$

Comparing these two equations we conclude that the dissipation rate (which should be homogeneous of order one in $\dot{\varepsilon}_{ij}^p$) is

$$\left(\sigma_{ij} - \frac{\partial f_2}{\partial \varepsilon_{ij}^p}\right)\dot{\varepsilon}_{ij}^p = d \tag{13}$$

so that the stress is defined as

$$\sigma_{ij} = \frac{\partial f_1\left(\varepsilon_{ij}^e\right)}{\partial \varepsilon_{ij}^e} \tag{14}$$

In this formulation ε_{ij}^e defines the stress uniquely, and therefore formally can still be called elastic in a general sense. However it is not elastic in thermomechanical sense, because it cannot define the amount of stored energy uniquely – this will also depend on the plastic strain.

5.4 Elasto-plastic coupling

In real soils the so called elastic properties depend on plastic strains, therefore decomposition $f\left(\varepsilon_{ij}^e, \varepsilon_{ij}^p\right) = f_1\left(\varepsilon_{ij}^e\right) + f_2\left(\varepsilon_{ij}^p\right)$ is not possible and the First Law becomes:

$$\sigma_{ij}\dot{\varepsilon}_{ij} = \dot{f} + d = \frac{\partial f\left(\dot{\varepsilon}_{ij}^e, \dot{\varepsilon}_{ij}^p\right)}{\partial \varepsilon_{ij}^e}\dot{\varepsilon}_{ij}^e + \frac{\partial f\left(\dot{\varepsilon}_{ij}^e, \dot{\varepsilon}_{ij}^p\right)}{\partial \varepsilon_{ij}^p}\dot{\varepsilon}_{ij}^p + d \tag{15}$$

It is still possible to define the stress as

$$\sigma_{ij} = \frac{\partial f\left(\dot{\varepsilon}_{ij}^e, \dot{\varepsilon}_{ij}^p\right)}{\partial \varepsilon_{ij}^e} \tag{16}$$

but in this case neither stress nor stored energy are defined by ε_{ij}^e uniquely, therefore this is not an elastic strain in any sense. Furthermore, decomposition $\varepsilon_{ij} = \varepsilon_{ij}^e + \varepsilon_{ij}^p$ in this case is not valid. Therefore, there is a coupling between elastic and plastic components of the stress-strain behavior, and the strain component ε_{ij}^e in does not make much sense. It is much more logical to formulate the free energy for coupled materials in terms of total strains ε_{ij} and kinematic internal variables α_{ij}. Their rates $\dot{\alpha}_{ij}$ will also define the rate of dissipation. As will be shown in the following, formulations which use "elastic" strain for coupled materials have a higher risk to go wrong.

5.5 Summary

Within the thermomechanical approach to elasticity, elastic strains are required to define uniquely not only stresses but also the stored energy. In isotropic hardening elasto-plasticity both the stresses and the stored energy are defined uniquely by elastic strains. In kinematic hardening case, only stresses are defined uniquely by elastic strains. When elastic constants depend on plastic strains, elastic strains do not exist. Therefore, it looks like that the elasto-plastic coupling can be considered as a thermo-mechanical criterion for existence of elastic strains. In the following section we shall attempt to find a micro-mechanical criterion for existence of elastic strains.

6 Elasticity in Micromechanics

6.1 The Hertz Theory

As we know from the Hertz theory, two elastic spheres pushed together (Figure 8) exhibit non-linear behavior, which is, however, still elastic (Santamarina [15]):

$$\frac{r_c}{R} = \sqrt[3]{\frac{3-3\nu_g}{2}\frac{\sigma}{G_g}} \quad \frac{\delta}{2R} = \left(\sqrt[3]{\frac{3-3\nu_g}{2}\frac{\sigma}{G_g}}\right)^2 \tag{17}$$

where G_g and ν_g are the shear modulus and Posson's ratio of the particles; $\sigma = N/\pi r_c^2$.

A similar conclusion can be reached from analysis of the electric forces theory [15].

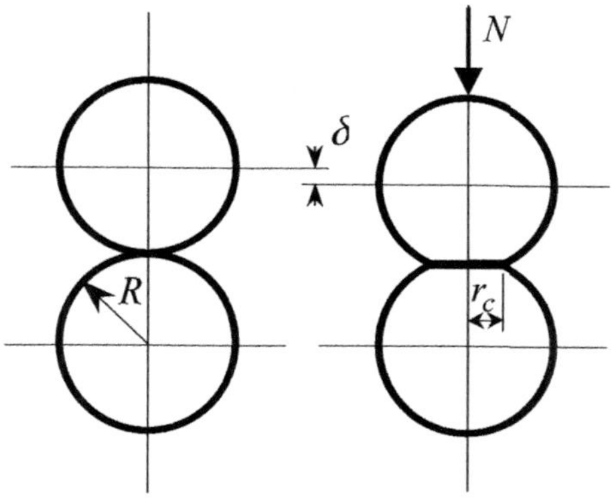

Fig. 8. The Hertz Theory

6.2 The Mindlin Theory

In the Mindlin theory, when a shear force is applied to the same two spheres with a frictional contact (Figure 9), the annular slippage region appears at infinitesimal load (Santamarina [15]). Therefore the behavior is not only non-linear:

$$\delta_{\tan} = \left[1 - \left(1 - \frac{T}{\mu N}\right)^{2/3}\right]\delta^* \quad \delta^* = \frac{3}{8}(2-\nu_g)\frac{\mu N}{G_g r_c} \tag{18}$$

(μ is the inter-particle friction coefficient) but also inelastic and, therefore, the LER does not exist even at infinitesimal strains. But what about the linear threshold strains observed in some experiments?

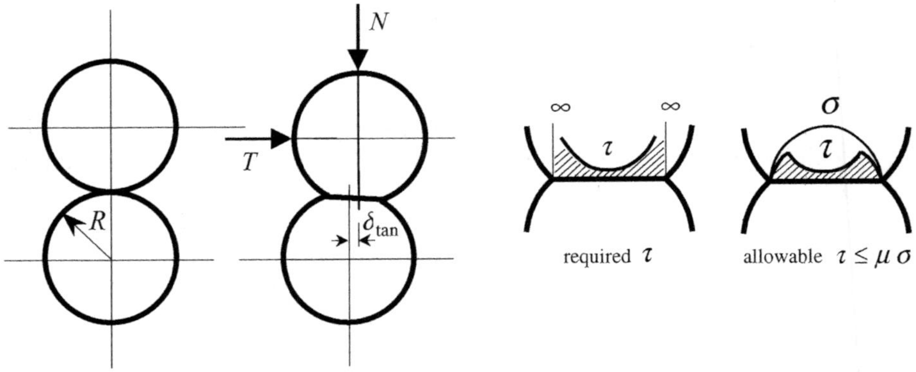

Fig. 9. The Mindlin Theory

6.3 Threshold Strains

One possible explanation of the observed linear threshold strain is that changes in stiffness, though starting almost instantaneously, become noticeable only when changes in fabric take place [15]. In the Hertz theory this happens when particles 1 and 3 loose contact (Figure 10), which takes place at

$$\gamma_{tl}^{fabric} = 1.3 \left(\frac{\sigma}{G_g} \right)^{2/3} \tag{19}$$

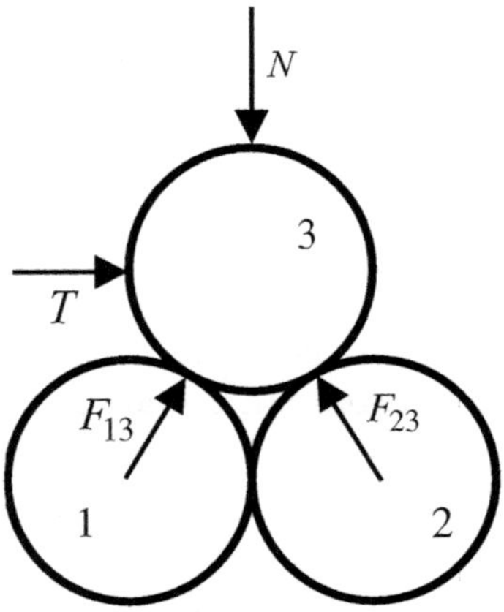

Fig. 10. Threshold strains in the Hertz Theory

In the Mindlin theory this happens when the two particles slip along the contact at

$$\gamma_{tl}^{slip} = 1.26\mu \left(\frac{\sigma}{G_g}\right)^{2/3} \tag{20}$$

The resulting threshold strains are surprisingly similar, in spite of the difference in the assumptions. In the Mindlin case, however, this threshold strain indicates existence of the apparent elastic region only, because the local slippage takes place at infinitesimal strain. The true elastic region in this case may exist only at atomic scale.

6.4 Thermomechanics of the Mindlin Model

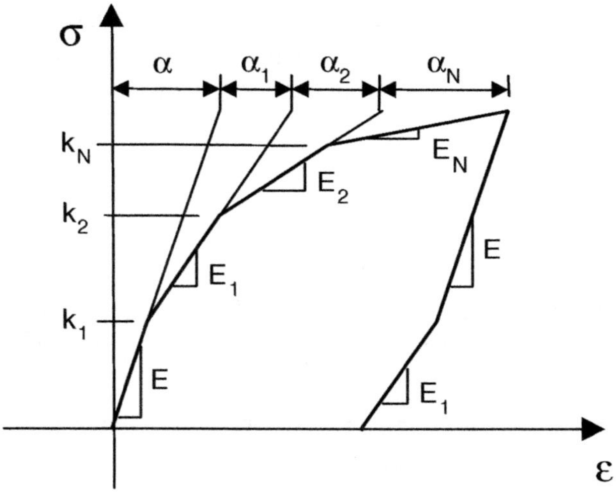

Fig. 11. Piecewise-linear approximation of the non-linear behavior of Mindlin's model.

The fact that the true elastic region in the Mindlin model does not exist, still does not answer the question about existence of the elastic component of total strain. Let us try to assess the behavior of this model from the thermomechanical point of view. Yielding starts immediately at the edges of the contact and propagates gradually towards the center with increase of the shearing force. Total work is dissipated at the part of the contact where the slippage takes place but it is also stored within the particles which keep deforming elastically. If we approximate the non-linear behavior of this model by a piecewise-linear function (Figure 11) we can analyze the mechanical model in Figure 12, which has a similar piecewise-linear stress-strain behavior.

This model is known as Iwan's model [9] and it consists of springs where energy is stored and slip elements where energy is dissipated:

$$f\left(\alpha_0 \ldots \alpha_{N+1}\right) = \frac{E}{2}\alpha_0^2 + \frac{1}{2}\sum_{n=1}^{N+1} H_n \alpha_n^2 \tag{21}$$

$$d\left(\dot{\alpha}_0 \ldots \dot{\alpha}_{N+1}\right) = k_0 \left|\dot{\alpha}_0\right| + \sum_{n=1}^{N+1} k_n \left|\dot{\alpha}_n\right| \tag{22}$$

Because all springs are in parallel with the slip elements, no elastic region and elastic strain are assumed in advance in accordance with the Mindlin model. If the weakest slip element had a slip stress larger then zero, then the behavior would be rigid at infinitesimal load. This is not possible because both spheres are elastic. Therefore, the smallest slip stress k_0 has to be equal to zero, but in this case, according to equation (22) there is no dissipation in this element, so that the strain component α_0 depends uniquely on stress and can be considered elastic in the general sense: $\alpha_0 = \varepsilon^e$.

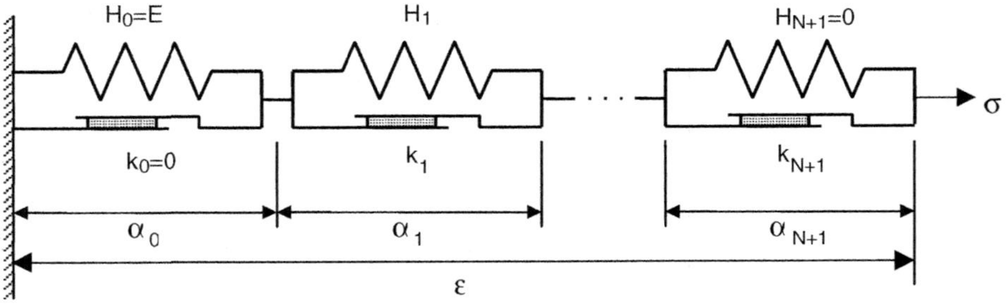

Fig. 12. Iwan's model [9].

6.5 Summary

Let us summarize our micromechanical study. The Hertz, Electric Forces and the Mindlin Theories do not justify small-strain linearity. The Mindlin theory does not justify existence of the linear or non-linear elastic region. In spite of this, the observed threshold strain phenomena can be explained by these theories. Thermomechanical analysis of the Mindlin model does not exclude existence of elastic strain, which is not really surprising, because elastic and plastic properties of the Mindlin model are not coupled. It looks like that the only consistent argument against existence of elastic component of strain is based on the elasto-plastic coupling.

7 Thermomechanics of Elasto-Plastic Coupling

7.1 Consistent Formulation: Collins & Houlsby [16]

In coupled materials the specific Helmholtz free energy $f(\varepsilon_{ij}, \alpha_{ij})$ and the specific Gibbs free energy $g(\sigma_{ij}, \alpha_{ij})$ depend on the kinematic internal variable tensor α_{ij}. These two potentials are related through Legendre Transformation: $g(\sigma_{ij}, \alpha_{ij}) = f(\varepsilon_{ij}, \alpha_{ij}) - \sigma_{ij}\varepsilon_{ij}$, where α_{ij} plays a role of a passive variable. The incremental elastic stress-stain response may now be affected by the plastic strain increment as well:

$$\dot{\sigma}_{ij} = \frac{\partial^2 f}{\partial \varepsilon_{ij} \partial \varepsilon_{kl}} \dot{\varepsilon}_{kl} + \frac{\partial^2 f}{\partial \varepsilon_{ij} \partial \alpha_{kl}} \dot{\alpha}_{kl} \tag{23}$$

$$\dot{\varepsilon}_{ij} = -\frac{\partial^2 g}{\partial \sigma_{ij} \partial \sigma_{kl}} \dot{\sigma}_{kl} - \frac{\partial^2 g}{\partial \sigma_{ij} \partial \alpha_{kl}} \dot{\alpha}_{kl} \tag{24}$$

The effect of plastic strain increments on elastic component of soil behavior is avoided only if the corresponding energy potentials can be decomposed in the following way:

$$f(\varepsilon_{ij}, \alpha_{ij}) = f_1(\varepsilon_{ij} - \alpha_{ij}) + f_2(\alpha_{ij}) \tag{25}$$

$$g(\sigma_{ij}, \alpha_{ij}) = g_1(\sigma_{ij}) - \sigma_{ij}\alpha_{ij} + g_2(\alpha_{ij})$$

where $f_2(\alpha_{ij}) \equiv g_2(\alpha_{ij})$. In this case only,

$$\frac{\partial^2 g}{\partial \sigma_{ij} \partial \alpha_{kl}} = -\delta_{ij}\delta_{kl} \text{ and } \frac{\partial^2 f}{\partial \varepsilon_{ij} \partial \alpha_{kl}} = -\frac{\partial^2 f}{\partial \varepsilon_{ij} \partial \varepsilon_{kl}} = -\frac{\partial^2 f}{\partial \varepsilon^e_{ij} \partial \varepsilon^e_{ij}} \tag{26}$$

so that the total strain can be decomposed into elastic and plastic components $\varepsilon_{ij} = \varepsilon^e_{ij} + \alpha_{ij}$ and the incremental elastic stress-strain response is given by:

$$\dot{\sigma}_{ij} = \frac{\partial^2 f}{\partial \varepsilon^e_{ij} \partial \varepsilon^e_{kl}} \dot{\varepsilon}^e_{kl} \quad \dot{\varepsilon}^e_{ij} = -\frac{\partial^2 g}{\partial \sigma_{ij} \partial \sigma_{kl}} \dot{\sigma}_{kl} \tag{27}$$

Any other form of the energy potentials would lead to coupling between elastic and plastic components of the model behavior. If these coupling terms are ignored anyway, expressions for the rate of change of Gibbs or Helmholtz free energy may not be full differentials and the First Law of Thermodynamics will be violated. Unfortunately, this complication has been overlooked in some recent hyperelastic formulations.

7.2 Elasto-plastic coupling and violation of the First Law

Puzrin&Tatsuoka [17] demonstrated the existence of the strain energy potential for a cross-anisotropic hypoelastic model for sands proposed by Tatsuoka&Kohata [4]. In this model, cross-anisotropic tangent elastic moduli depend on the soil density through the void ratio. This implies dependency of the energy functions on volumetric plastic strain, which cannot be decomposed. The incremental elastic stress-strain response of the model can be derived from the proposed energy potential only if the coupling terms are taken equal to zero, which is not true.

Hashiguchi&Collins [18] utilize Gibbs free energy potential dependent on plastic volumetric strain. In spite of the fact that this potential cannot be decomposed, they ignored the resulting coupling term $\frac{\partial^2 g}{\partial \sigma_{ij} \partial \alpha_{kl}} \neq -\delta_{ij}\delta_{kl}$, wich lead to substantial errors in calculation of elastic strains (see discussion by Einav&Puzrin [19]). In his closure to this discussion, Hashiguchi [20] have corrected his formulation by incorporating missing coupling terms.

Borja&Regueiro [21] utilize Helmholtz free energy potential dependent on strain-like vector of plastic variables. In their general formulation, long before a particular form of the potential function was specified, they assumed the elastic incremental response. Fortunately, the potential function utilized in that work can be decomposed, so that in this particular case the elastic and plastic strains are not coupled anyway. However, when more general forms of the Helmholtz free energy function are utilized (Borja et al [22]), the elastic component will be affected by plastic strains. Unfortunately, this function is not presented explicitly. Nevertheless, from the incremental equation given for the generalized stress, it may be concluded that $\frac{\partial^2 f}{\partial \alpha_{ij} \partial \alpha_{kl}} = F(\varepsilon_{ij}, \alpha_{ij})$. Clearly, this equation implies that the Helmholtz free energy function cannot be decomposed, so that the coupling term $\frac{\partial^2 f}{\partial \varepsilon_{ij} \partial \alpha_{kl}} \neq -\frac{\partial^2 f}{\partial \varepsilon_{ij} \partial \varepsilon_{kl}}$. Therefore, it is inappropriate to use the incremental elastic strains to describe the incremental stress-strain response of the coupled material.

7.3 Elasto-plastic coupling and violation of the Second Law

Elasto-plastic coupling may lead to violation of the Second Law as well. Collins [23] showed that for non-frictional materials when attempt is made to satisfy the both Laws, coupling inevitably leads to non-associated flow rule for internal variable. Puzrin&Einav [24] showed that for frictional materials associated flow rule though in principle possible, may lead to negative dissipation and violate the Second Law.

7.4 Summary

Elastic strains exist only for particular form of decoupled energy potentials. For real soils – with elasto-plastic coupling – energy potentials cannot be decoupled. Assumption of elastic strains in this case will lead to violation of energy conservation.

8 Conclusions

Let us see now if we can answer the three questions formulated in the introduction:

- Does the linear elastic region of stresses/strains in soils exist?
  - Micromechanics gives the negative answer to this question.
- Can elastic strains exist in spite of the lack of elastic region?
  - Both micro- and thermomechanics give the positive answer.
- Is there a more objective criterion for existence of elastic strains?
  - Thermomechanics suggests that elasto-plastic coupling, can serve a criterion.

What are the implications of these conclusions for constitutive modeling of soils? First of all, it looks like that there is no theoretical justification for LER. When using elastic strains within any conventional elasto-plastic model, it is important to make sure that there is no elasto-plastic coupling. If the behavior is coupled, it is recommended to use hypo- or hyperplasticity. Advantage of hypoplasticity is that there is no need in internal variables responsible for dissipation. Advantage of hyperplasticity is that the Laws of Thermodynamics are satisfied automatically.

References

[1] Jardine, R.J., Potts, D.M., Fourie, A.B. & Burland, J.B. (1986). Studies of the influence of non-linear stress-strain characteristics in soil-structure interaction. *Geotechnique*, **36**, 377-396.

[2] Burland, J.B. (1989). Small is beautiful - the stiffness of soil at small strains. Ninth Laurits Bjerrum Memorial Lecture. *Can. Geotech. J.* , **16**(4), 499-516.

[3] Jamiolkowski, M., Leroueil, S. & Lo Presti, D.C.F. (1991). Design parameters from theory to practice. *Geo-Coast '91*, Yokohama, Japan.

[4] Tatsuoka, F. & Kohata, Y. (1995). Stiffness of hard soils and soft rocks in engineering applications. *Pre-failure deformation of geomaterials*, Shibuya, Mitachi & Mura (eds). Rotterdam, Balkema, **2**, 947-1066.

[5] Stokoe II, K.H., Hwang, S.K., Lee, J.N.-K. & Andrus, R.D. (1995). Effects of various parameters on the stiffness and damping of soils at small to medium strains. *Pre-failure deformation of geomaterials*, Shibuya, Mitachi & Mura (eds). Rotterdam, Balkema, **2**, 785-816.

[6] Jardine, R.J. and Potts, D.M. (1988). Hutton tension leg platform foundations: an approach to prediction of pile behavior. *Geotechnique*, **38**(2), 231-252.

[7] Jardine, R.J. (1992). Some observations of the kinematic nature of soil stiffness. *Soils and Foundations*, **32**(2), 111-124.

[8] Vinale, F., d'Onofrio, A., Mancuso, C., Santucci de Magistris, F. and Tasuoka, F. (1999). The pre-failure behavior of soils as construction materials. *Proceedings of IS Torino 99*, Vol. 2, pp. 955-1010.

[9] Iwan, W.D. (1967). On a class of models for the yielding behaviour of continuous and composite systems. *J. Applied Mechanics*, **34**, 612-617.

[10] Mroz, Z. (1967). On description of anisotropic hardening. *J. Mech. Phys. Solids*, **15**, 163-165.

[11] Kreig, R.D. (1975). A practical two-surface plasticity theory. *J. Appl. Mech., Trans. ASME*, **E42**, 641-646.

[12] Dafalias, Y.F. & Popov, E.P. (1975). A model of non-linearly hardening materials for complex loadings. *Acta Mechanica*, **21**, 173-192.

[13] Kolymbas, D. (1977). A rate dependent constitutive equation for soils. *Mech. Res. Comm.*, **4**, 367-372.

[14] Puzrin & Houlsby (2001). A thermomechanical framework for rate-independent dissipative materials with internal functions. *International Journal of Plasticity*, **17**, 1147-1165.

[15] Santamarina, J.C. (2001). *Soils and waves.* John Wiley & Sons, Chichester, 488 p.

[16] Collins I. F. and Houlsby G. T. Application of thermomechanical principles to the modeling of geomaterials. *Proc. Royal Society of London, Series A* 1997; **453**: 1975-2001.

[17] Puzrin A. M. and Tatsuoka, F. (1998). Elastic strain energy for uncemented granular materials, *Soils and Foundations,* **34**(4), 267-275.

[18] Hashiguchi K. and Collins, I. F. (2001). Stress rate-elastic stretching relations in elastoplastic constitutive equations for soils. *Soil and Foundations,* **41**(2), 77-87.

[19] Einav I. and Puzrin A. M. (2002). Stress rate-elastic stretching relations in elastoplastic constitutive equations for soils. *Discussion. Soils and Foundations,* **42**(1), 159-160.

[20] Hashiguchi K. (2002). Stress rate-elastic stretching relations in elastoplastic constitutive equations for soils. *Closure. Soil and Foundations,* **42**(1), 160-163.

[21] Borja R. I. and Reguerio R. A. (2001). Strain localization in frictional materials exhibiting displacement jumps. *Comput. Methods Appl. Mech. Engng.,* **190**(20-21), 2555-2580.

[22] Borja R. I., Lin, C. H. and Montans, F. J. (2001). Cam-Clay plasticity, part IV: implicit integration of anisotropic bounding surface model with nonlinear hyperelasticity and ellipsoidal loading function. *Comput. Methods Appl. Mech. Engng.*, **190**(26-27), 3293-3323.

[23] Collins I. F. (2002). Associated and non-associated aspects of the constitutive laws for coupled elastic/plastic materials. *Int. J. Geomech.*, **2**(2), 259-267.

[24] Puzrin, A.M. and Einav, I. (2002). Associated and non-associated aspects of the constitutive laws for coupled elastic/plastic materials. *Discussion. Int. J. Geomech.*.

Experimental characterization of localized deformation in geomaterials

Jacques Desrues

CNRS - Université de Grenoble, Laboratoire 3S - BP 5X, Grenoble
Jacques.Desrues@hmg.inpg.fr

Summary. In geomaterials, among a large number of other solids, rupture is most frequently associated with localized deformation : shear bands, cracks and fissures. The transition from diffuse to localized deformation is known as "Strain Localization Phenomenon". Theoretical studies, initiated in the years 1970's by J. Rudnicki, J. Rice, I. Vardoulakis and others, have showed that the emergence of strain localization can be predicted on the basis of the constitutive law of the material, if the law incorporates enough of the complexity of real materials. This theoretical framework has motivated extensive experimental studies in Laboratoire 3S, CNRS - Université de Grenoble, France, on strain localization in granular soils and rocks. New methods for strain field measurement in a strained specimen have been developed, including false relief stereophotogrammetry (FRS) and Computed Tomography (CT). The present paper describes studies performed on different materials: sand, clay, soft rock, stiff marl, concrete. Experimental observations obtained on incipient and developed localization in "homogeneous" tests are presented. The discussion of the results addresses the issues of localization and peak strength, critical stress and strain, shear band orientation and thickness, and complex localization patterns. It is shown that previously assumed uniform deformation during triaxial compression of sand is actually quite complex patterns of strain localization. In the case of cohesive materials—clays, rocks, concrete—not only strain localization but also crack development are commonly observed. Special techniques developed to measure displacement discontinuities by means of FRS are presented and the results obtained for different cohesive geomaterials are discussed. Complementary information, including full-length thesis PDF copies, original data and videos can be downloaded from the web site of the team "Géomatériaux, Déformation et Rupture" in Laboratoire 3S, URL http://l3sphnum.hmg.inpg.fr/hps1/etagere.htm

1 Introduction

Geomaterials—soils, rock, concrete—are among the most illustrative materials as far as strain localization phenomenon is concerned. Indeed, Nature provides a lot of examples of localized deformation, for example in small or bigger trenches along roads, or in large cliffs in mountain landscapes. Seismic events may produce evident displacement discontinuities at the surface of the earth, and active faults are examples of localized deformation processes at the geographic scale. Localized deformation under tectonic stresses is also playing a great role in the stucturation of sedimentary basins, which is important for geologists working for oil companies. In the civil engineer's domain, practical situations in which geomaterials are loaded up to failure are rather common: natural or human-made slopes may become instable and collapse under certain circumstances, excavation works need to desagregate and remove a part of in-place material, collapsed foundations and retaining walls usually show clearly localized kinematical mechanisms. In fact, the early work by Coulomb in 1773 [12] on the stability of retaining walls was already taking into account, as an essential aspect of the problem, the localized nature of the collapse mechanism.

However, strain localization in geomaterials can take different aspects: shear bands are usually observed in granular masses and granular specimens—ductile materials, while cracks and fractures are more common in hard rocks and concrete—brittle materials. Nevertheless, a large transition zone lies in between the ductile and brittle fields. Occurrence of shear bands rather than cracks in a specimen or in a field situation depends not only on the material considered, but on loading process too (e.g. stress level, loading rate). Moreover, observation at different scales can lead to different diagnostics on the nature of localized deformation structures involved in a global deformation process.

Strain localization by shear banding has motivated a large amount of scientific work in Geomechanics since the middle of 1970's . Experimental, theoretical and numerical tools available in this framework have been continuously progressing. In the sequel, the main features of strain localization in geomaterials are presented, illustrated by a selection of experimental observations performed in laboratory 3S-IMG in Grenoble in both granular and cohesive materials.

Hostun "RF" sand is the reference sand in 3S-IMG laboratory. This sand is a granulometric class (0,5- 1,25mm) of a natural sand; it is obtained by sieving the material extracted from a natural sand deposit in Hostun, Drôme, France. The resulting material is a fine angular siliceous sand, described in details in [30]. This sand has been used in studies on localization in Grenoble by the author, and later by W. Hammad, M. Mokni and V. Roger in their thesis' work.

Beaucaire Marl is the cohesive material used by D. Tillard-Ngan, 3S-IMG for her tests devoted to localized failure in soft rocks. This material is extracted from a quarry near Beaucaire, Gard, France, at a depth approximately 70 meters below the natural surface; the geological stage is the Plaisancien stage of the tertiary era. It contains 36% quartz, 31% calcareous carbonate and 32% clay. A detailed description of the material is given in [53, 54].

False Relief Stereophotogrammetric (FRS) method is based on the analysis of a series of photographs taken during a deformation process imposed to a specimen. It takes advantage of the very accurate human sense of 3D perception. This sense is the result of an analysis of the slight geometrical differences between the left and right eye images. In natural perception, this process allows to perceive the three dimensional nature of the world, but in false relief method the relief is artificial because the two images proposed to the left and right eyes are not taken simultaneously from two different space point of view, but taken from a unique point of view at two successive times. This leads to a relief perception, directly linked to the incremental deformation undergone by the specimen. Quantitative measurements can be obtained using stereophotogrammetric apparatuses developed for topographic applications, together with a specific data processing developed for the false relief application (see [24, 16]).

Tomodensitometry is a method using X-ray attenuation through solid or bulk materials to quantify the compactness. For a granular or cohesive geomaterial, as far as the mineralogical composition is homogeneous and not changing during the process, the radiographic density can be calibrated versus the compactness. The X- ray scanner is an apparatus which allows to rebuild the internal distribution of radiographic density in 3D specimens, on the basis of a set of radiographic profiles realised in different directions (computerised tomography). The tests reported here have been performed in collaboration with LMA laboratory, Marseille, France. For an introduction to Tomodensitometry, see [23].

2 Shear bands in field cases and laboratory tests

Strain localization is a very common phenomenon in granular materials. It can be observed in natural soil masses, as well as in human built embankments, retaining walls and other soil engineers' works. When rupture occurs, the global mechanism usually involves a number of quasi-rigid blocks, sliding each on others along so called **rupture surfaces**.

This kind of phenomenon can be observed in other solids, like rocks, metals and alloys. The name of **shear bands** is given to the zones of localized deformation, in both mechanical engineering and Geomechanics field. Case histories described in soil mechanics' literature give a large illustration of that type of kinematics (among other references, "Failure" [50]).

In a number of cases, failure surfaces take place along pre-existing discontinuities, like interfaces between mechanically different layers or zones. In other cases, failure surfaces can be initiated by strong singularities in the geometry, in the loading, or in the mechanical properties.

However, localization can appear even when singularities and heterogeneities have been carefully avoided : this is the case when performing rheological tests. Such tests are performed on "homogeneous" specimens —i.e.

as homogeneous as possible, in order to characterise their mechanical properties. In a large range of test conditions (depending on material tested and on test techniques), the final state of rupture is reached with a completely non uniform kinematical field, involving so called rupture planes, one or a few, sometimes more.

The **axisymmetric triaxial test** is the basic test in soil and rock mechanics. Much work has been done to improve the homogeneity of this test. Both localized and diffuse modes of heterogeneity can be observed in the triaxial specimens; the loss of homogeneity is extremely sensitive to the quality of the test preparation and execution. For a discussion of these questions, see [11].

Localization is observed also in other rheological tests, like hollow cylinder test, or the Directional Shear Cell developed by Arthur ([2]), or the **true triaxial test**. In the latter, three independent stresses or strains can be imposed to a cubical specimen in three orthogonal directions normal to the faces (see [37] for more details). When this type of apparatus is displacement- controlled, loading is applied by six rigid platens; despite this kinematical restraint, it was shown by specific experiments in our laboratory that localization is active also in true triaxial tests ([26]).

Drainage conditions can affect significantly the response of granular specimens; in undrained tests, dilatancy (positive or negative) is constrained by the fluid but this constraint generates changes in pore fluid pressure, inducing effective stress paths completely different from the drained tests. Liquefaction can be observed in loose specimens. As far as localization is concerned, at first most experimental studies were addressing the case of drained test but after 1990, results in undrained condition have become available ([32, 39, 60, 61, 40, 47, 46]). The main conclusions of these studies are illustrated in the sequel.

Shear band analysis is a theoretical approach, related to **bifurcation theory**. Starting with the pioneering work of [48] and [45], it was developed during the last 25 years, in Geomechanics for soil-like materials (see [64] and [8]) and in mechanical engineering for metals. In this approach, strain localization is considered as a loss of uniqueness of a local boundary value problem. One important thing is that the constitutive equations play the central role in the analysis. In that sense, shear banding is considered as a material instability. For soil modellers, shear band analysis has become a test for constitutive equations; in near future, it can become a guide for identifying, or even for elaborating advanced models. A number of contibutions have been published by the author and colleagues in that spirit ([6, 21, 22, 9, 7, 5]).

3 Localization in granular bodies

Although strain localization in geomaterials has been recognised since the last 70s as an important research field, experimental studies have not been very

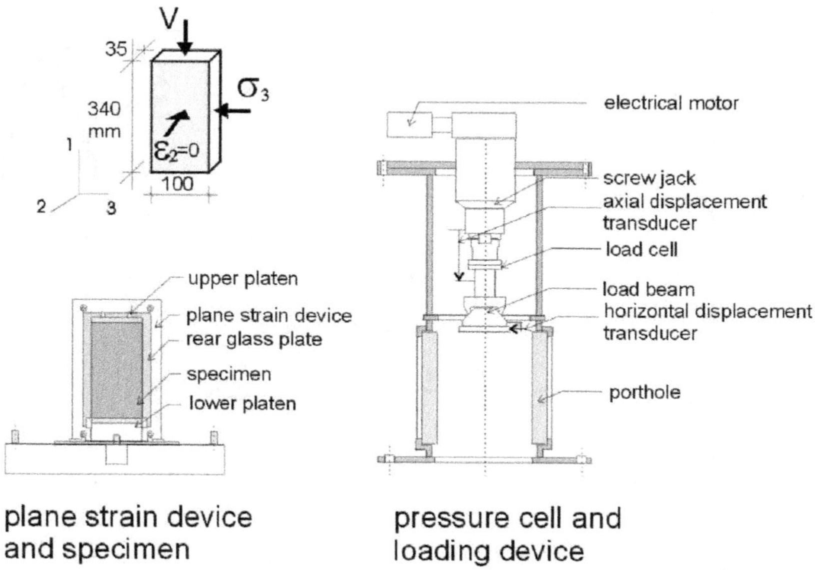

Fig. 1. *Biaxial test: geometry and loading conditions ([13])* .

numerous on that question; most of the results published have been obtained with specifically designed biaxial tests—Vardoulakis in Karlsruhe, Germany ([62, 63]), and later in Minneapolis, USA ([28, 32]), Desrues in Grenoble, France ([13, 25, 14, 15, 17, 18]), Tatsuoka in Tokyo, Japan ([52, 51, 68]), Arthur in London, UK ([2, 1]).

Plane strain biaxial test is very convenient to study strain localization, first because it is a true rheological test, and secondly because it allows a complete determination of the real strain field at any stage of the test. Figure 1 shows the geometry of the specimen, and the boundary conditions imposed to it. Zero strain is imposed in the direction 2, normal to the largest face of the specimen, by a plane strain device using two thick glass plates. Constant axial strain rate is applied by a mechanical jack, through a platen allowing free lateral translation. The specimen is wrapped into a rubber membrane, 0.4 mm thick. The contacts between the specimen and the plane strain device are lubricated; the same disposition is used for the contacts with the platens.

3.1 A typical test

As a first illustration of strain localization in granular bodies, the results of a typical biaxial test are presented. Figure 2 shows the axial load vs. displacement curve recorded during the test SHF06, performed on dense fine Hostun sand (relative density ≈ 90 %). The numbers on the curve indicate the serial

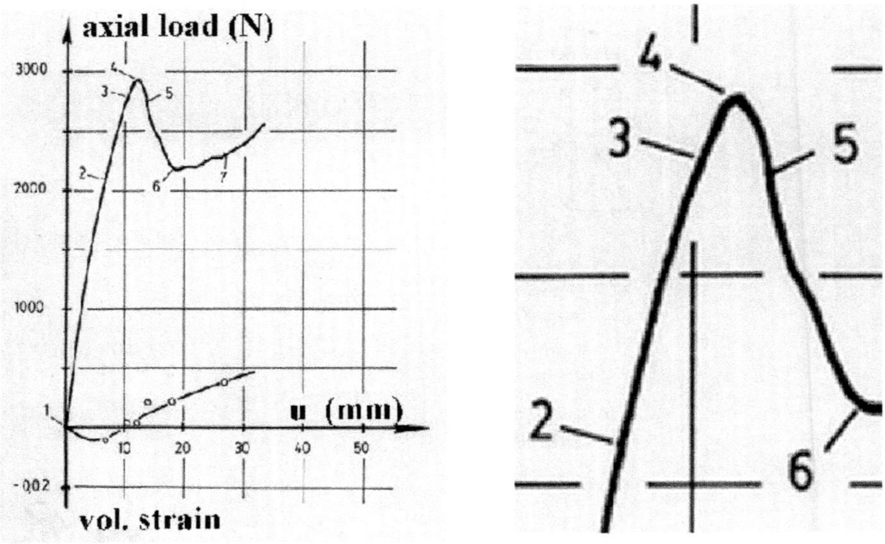

Fig. 2. *Test SHF06: axial load versus axial displacement.*

number of the photos. Figure 3 shows the results obtained by stereophotogrammetric method, for the set of increments 1,2 2,3 3,4 4,5 and 5,6. In this figure, upper pictures present the displacement vectors recorded. It can be seen that non homogeneous deformations are evident at increment 3,4 , and that a clear sliding block mechanism is operating at increment 5,6 (displacement scale is 5 times the geometric scale). In the same figure, the middle pictures present the incremental shear intensity, defined as $d\gamma = 1/2(d\varepsilon_I - d\varepsilon_{II})$. The symbols are sized proportionally to the local value of this scalar quantity. This set of pictures give a more clear visualisation of the initialisation of the shear band at increment 3,4 and its further evolution. It can be seen that in increment 3,4 the heterogeneous deformation has already the structure of a shear band; on the contrary, the increment 2,3 shows a non fully homogeneous deformation field, non clearly organised in a localized mode (the same, even less organised, for 1,2). A possible interpretation of these pictures is that localization is initiated by prior smooth heterogeneity; in such a case, one could see the process as a kind of "condensation" of the heterogeneity. In other cases, localization can be initiated by local strong heterogeneities, like inclusions; in those cases, the mode of development of the shear band is more like a propagation process. The bottom pictures in Figure 3 show the maps of incremental volume changes during the test; strong dilatancy is associated with the localization process.

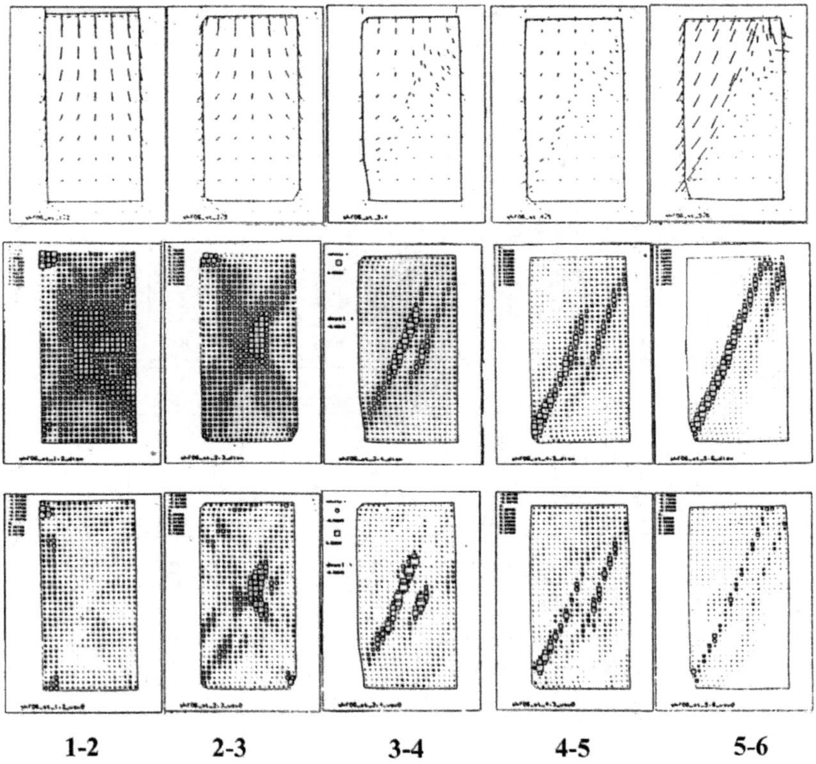

Fig. 3. *Stereophotogrammetric measures of incremental deformation during a plane strain test: (up) Incremental displacement; (middle) shear intensity; (bottom) volumetric strain.*

3.2 Inception and growth of localization

After this first example, the influence of different parameters of the test on the occurrence of strain localization is discussed. As a starting point, let us consider **drained, low pressure, biaxial tests** ($\sigma_3 \leq 100$kPa): a major result of biaxial experiments in that range is that localization always occurs. Incipient shear bands are observed after rather small quasi homogeneous deformation—about 3% axial strain for dense sand under 100 kPa confining pressure. No significant bulging is observed before.

Tests performed on specimens with different slenderness ratios (H/L ranging from 3.3 to 0.5) by W. Hammad and M. Mokni in Grenoble ([31, 39]) have shown strain localization in every case, even in very short specimens. Figure 4 shows the localization mechanism in the test shf89, $H/L = 1$: The shear band is "reflected" several times on the upper and lower platens, lead-

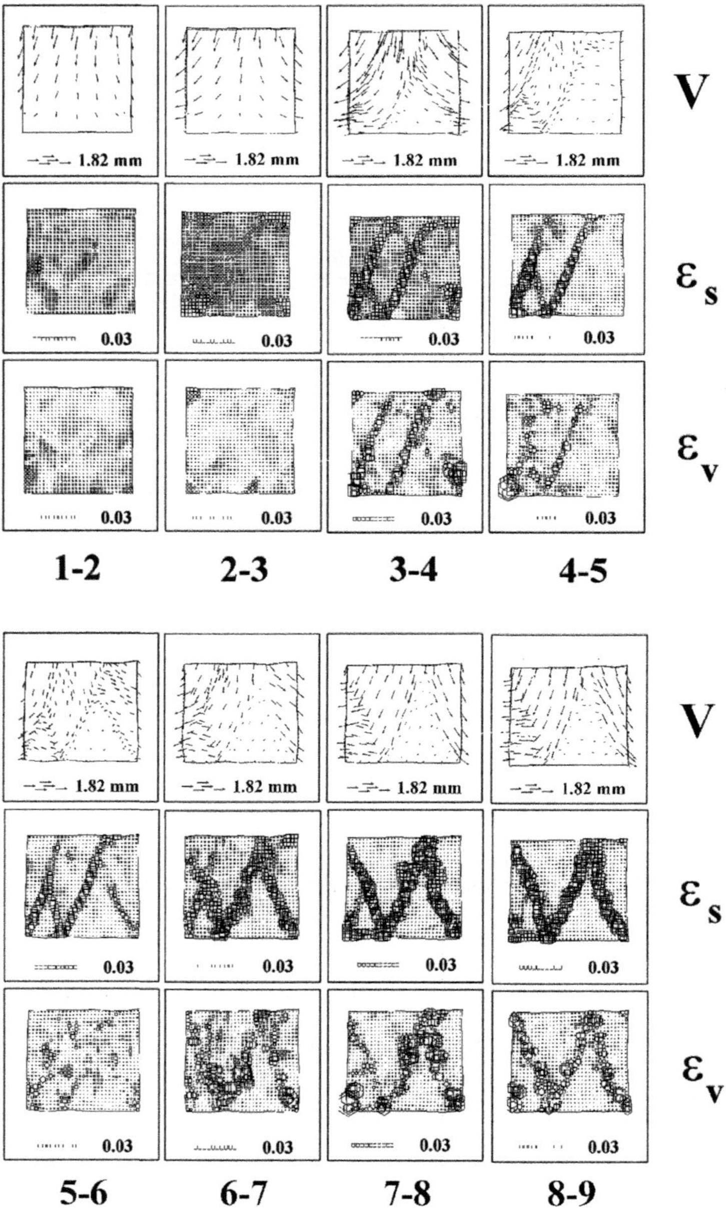

Fig. 4. *Stereophotogrammetric measures of incremental deformation during a plane strain test on a short specimen: (up) Incremental displacement; (middle) shear intensity; (bottom) volumetric strain (test shf89, Mokni 92).*

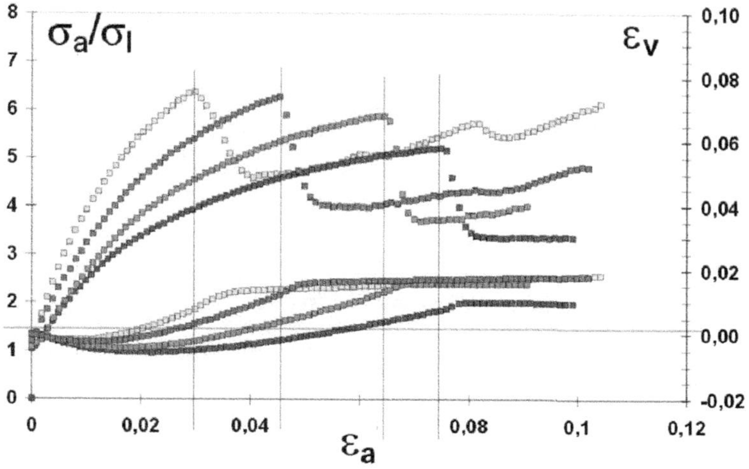

Fig. 5. *Influence of lateral pressure on the onset of localization in dense sand specimens: higher the pressure is, later the localized mode appears but it is not suppressed (after Hammad, 1991).*

ing at the end to a complete mechanism with a few triangular blocks sliding on well-developped shear bands. However, this test illustrates the fact that the primary patterns which take place are not always the final ones : in the present case, two parallel shear bands start together at the increment 3-4, then grow up in successive increments up to 5-6. But at 6-7 the patterns start to change and the final mechanism emerges at 7-8. The interpretation that we propose for this observation is that the onset of shear banding is essentially determined by the intrinsic behaviour of the material (constitutive law), but the final mechanism has to accommodate the boundary conditions of the test, which impose some changes to take place in the localized deformation patterns.

Loose specimens show localization too, but it comes later in term of axial strain.

When a **higher lateral pressure** is applied to the specimen, localization still occurs, but it comes much later; symmetric or antisymmetric diffuse modes of heterogeneity (bulging) can appear before, but localization remains the ultimate mode. Figure 5 illustrates the progressive delay of the onset of localization with lateral pressure. The same is true for loose specimens prepared with the same *Hostun RF* sand, as shown on figure 6: no peak is observed in the stress-strain curves, only a change in the slope and a sudden increase of the fluctuations of the curve indicates the onset of shear banding, clearly detected using stereophotogrammetry (not shown here). Other results obtained in 3S-

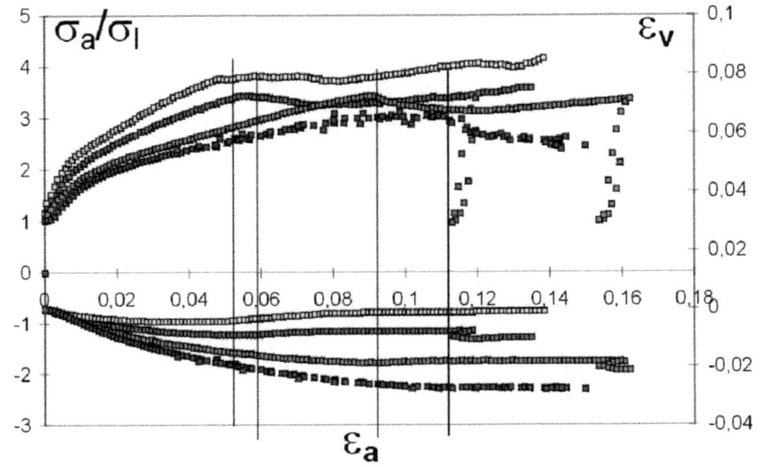

Fig. 6. *Influence of lateral pressure on the onset of localization in loose specimens: although no marked peak can be seen on the curves, strain localization is observed experimentally, and confirmed by stereophotogrammetry. The vertical lines indicate the onset of localization, which coincides with the beginning of the irregularities in the curves. The general trend is the same as in the dense specimens: the higher the pressure, the later the localized mode (after Hammad, 1991).*

IMG laboratory indicate that localization can occur in materials undergoing moderate and even strong contractancy, like calcareous sands ([31, 25]).

The situation described above for low pressures in plane strain is contrasting with classic observations in **axisymmetric tests** for the same materials and pressures: in the latter tests, bulging and other diffuse modes of bifurcation are almost always observed before localization; moreover, careful refinements of the test procedures can delay significantly the occurrence of localized deformations. M. Mokni has used Computed Tomography (CT) in Grenoble ([39]) to show that the occurrence of localized strain patterns in dense axisymmetric triaxial specimen can be delayed significantly beyond the stress peak strain, namely up to 20%, if very refined specimen preparation and test procedure are used. This establishes a specific stability of the axisymmetric test with respect to any other test condition. Theoretical predictions in the framework of bifurcation theory confirm this experimental observation ([48, 59]).

In **cubical triaxial tests**, even if it has been considered for a while in the soil testing community that strain localization would be precluded due to the rigid platens and the full kinematic control of the apparatus, it has been shown by J. Lanier and J. Desrues that shear bands do develop in cubical specimens subjected to various kind of stress paths in the triaxial apparatus in Grenoble [26]. The specimen showed in figure 7 reveals two conjugated mechanisms, each of them involving two shear planes intersecting the faces of the cube along figures resembling an inverted V on the photo.

Fig. 7. *strain localization in a cubical specimens tested in a rigid-platen true triaxial apparatus (after [26]).*

Undrained tests were performed in the biaxial apparatus in 3S-IMG Laboratory by M. Mokni in 1991 ([39, 40]) and later by V. Roger in 1998-2000 ([46, 47]). Similar tests were undertaken in other places ([32, 29]), some of them before, some between the two sets of tests performed in Grenoble. Eventually it was shown that strain localization has to meet drainage, at least locally i.e. in the band, the drained behaviour of the material in order to occur; otherwise the stress path followed does not reach the critical mobilized friction angle observed in drained tests at the onset of localization. In loose specimens sheared undrained, this condition is easily met and strain localizes normally; however, in dense, dilatant specimens, the drop in pore pressure may delay permanently the onset of localization, until eventually cavitation occurs in the pore fluid, triggering shear banding in the specimen (see figure 8). These findings are shown and discussed in details in [40, 47].

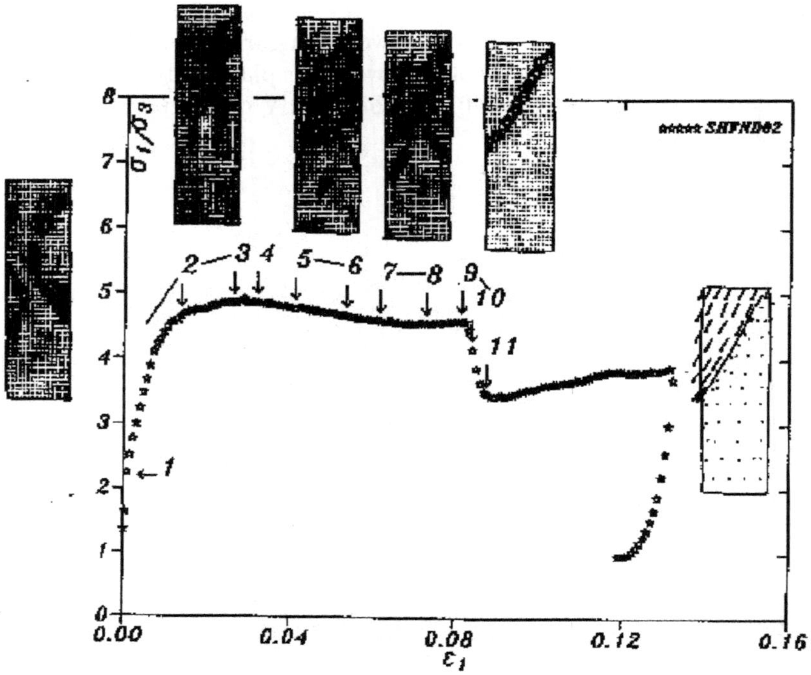

Fig. 8. *Strain localisation in an undrained test on dense Hostun RF sand. The onset of localisation is shown not to take place before the increment 9-10, when a sudden drop in the effective stress ratio σ_1'/σ_3' occurs (after [40]).*

3.3 Localization patterning

Conjugate shear bands: If coaxiality of the ends of the specimen is imposed by a biaxial apparatus, generally two shear bands will be observed, crossing each other. These two bands will develop simultaneously, or alternatively, in looser (then weaker) specimens ([13, 20]) in denser ones (then stiffer), the first shear band will be the only mechanism for a long while, occurring with a sharp peak as shown in figure 2; but finally a second band will appear, resulting in a large lateral load due to the apparatus' reaction. This lateral load mobilization was effectively recorded ([13]). When relative lateral displacement of the ends is free, single band mechanisms are likely to occur.

Parallel shear bands: More complex patterns, involving a number of parallel shear bands, are sometimes observed on sand specimens. The increment 3,4 in figure 3 may be considered as a simple example of patterning, but in that case the final figure of deformation was reduced to a single band. In shorter specimens (slenderness ratio 1.0 illustrated in Figure 4) tested in a plane strain device, localization patterns involves multiple reflections of the shear bands, but in addition the incipient localization in increment 3-4 shows

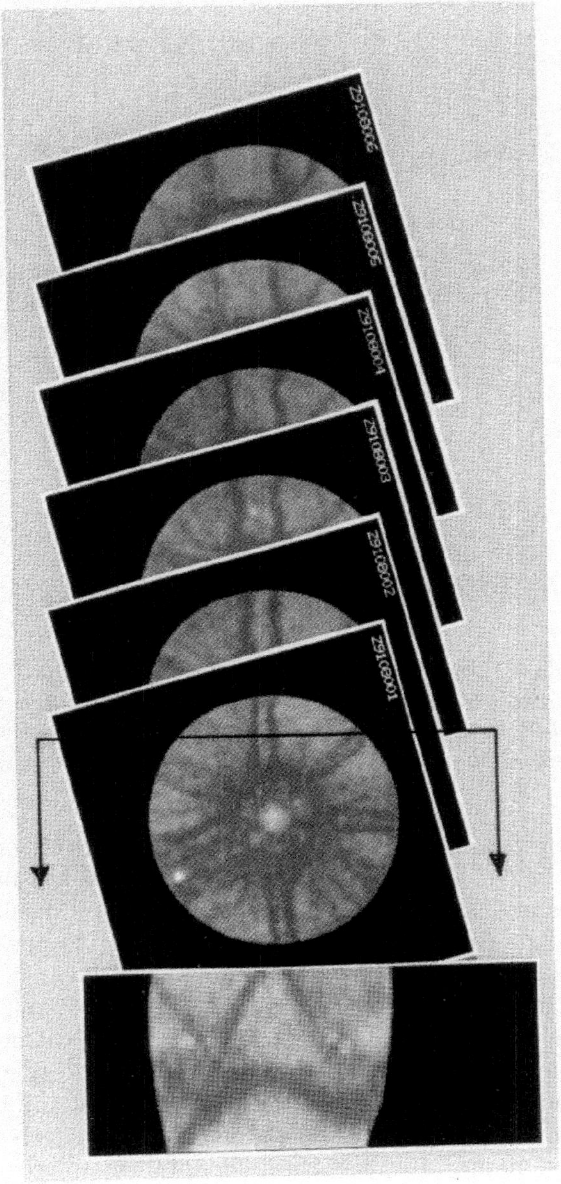

Fig. 9. *Illustration of the localization patterning in the specimen rfdt8: reconstruction of the density of the specimen on a plane parallel to the axis, and perpendicular to a pair of localization planes. The bottom picture shows the inverted V produced by the two associated planes intersecting on the top platen; the trace of the cone appears as a parabola oriented toward the lower platen [23]).*

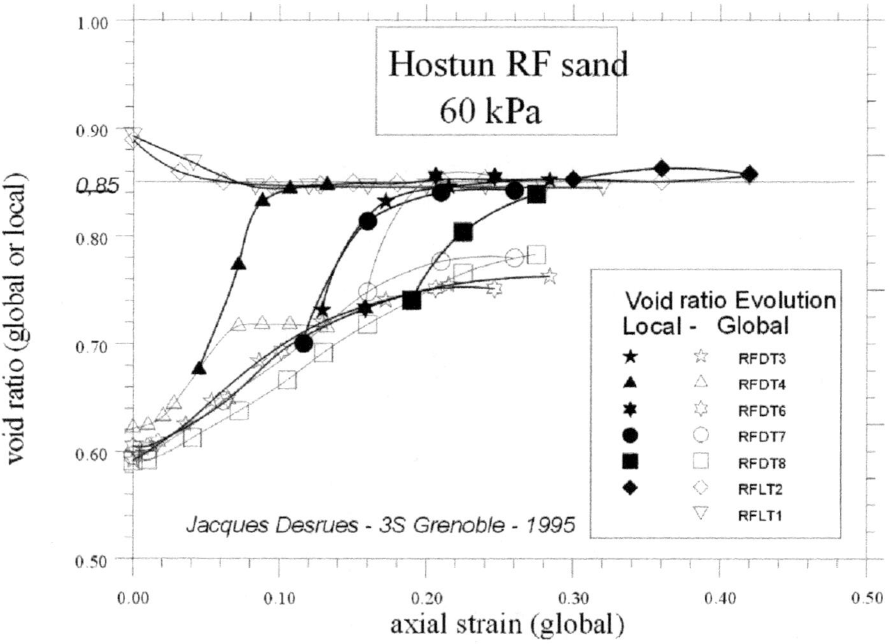

Fig. 10. *Compared global and local evolution of the void ratio in loose and dense specimens submitted to axisymmetric triaxial test under 60 kPa effective confining pressure. 'Global' means averaged over a cross section of the specimen, while 'local' stands for averaged over a zone identified as a localized shear zone [39, 23].*

two concurrent parallel shear bands, starting respectively one from the left lower corner of the specimen, and the other more or less in the middle; as in most cases in dense sand, one mechanism becomes dominant in post localization regime (increment 6-7), while the other disappears (in terms of incremental strain). Another localization figure which can be observed, more frequently in clays, is made of parallel and partially overlapping shear bands (so-called "en échelon" by geologists): the complete shear mechanism finally looks like a piecewise shear band, with diffuse strain bridges in-between ([51]).

In **triaxial axisymmetric tests**, the end restraint and imposed coaxiality can induce very complex patterns. A detailed image analysis of the density maps obtained during a test performed under tomographic control on a dense sand specimen revealed a pattern which seems to be somewhat generic. A complete set of tomograms was realised, covering the whole specimen. Six of them are shown, partially superposed, in Figure 9. In the first tomogram, entirely visible in the figure, one can notice, besides an approximate central black circle, a set of pairs of approximately straight and parallel black lines. The direction of each pair passes through the specimen axis and is approximately radial. The spacing between the twin lines in each pair increases from one tomogram to the lower one, as does the diameter of the circle (not visible here). On the sixth tomogram, the spacing is quite large but the lines remain approximately parallel by pairs. The individual lines are not passing through the specimen axis. From the set of slices placed next to each other, a perpendicular section can be reconstructed. Such a section, placed perpendicular to one of the pairs of lines, shows the trace of an inverted V. This can be seen on the bottom picture of Figure 9. The same reconstruction was performed on three different parallel planes perpendicular to the same pair (only one is shown on the figure). It was found that the V shapes were superposable, while the trace of the specimen contour was not, since it was cut at different distances from the central axis. The geometrical interpretation of these observations is the following : The localization pattern in this specimen involves two kinds of structures; first a cone centered on the axis of the specimen, with its tip outside the specimen, and its contour matching exactly the bottom section of the specimen; secondly, a set of plane strain mechanisms associated in pairs, each pair of plane crossing on a line which is a diameter of the top section. This spatial organisation can be seen as a generalization to the cylindrical specimens of the double mechanism shown in figure 7 for cubical specimens tested on axisymmetric stress paths. More details on this study can be found in [23].

3.4 Volumetric strains

Volumetric strains inside the shear band itself can be measured experimentally. Strongly dilating volume changes are reported by several authors, using various techniques, namely X-ray photography ([2, 63, 49]), gamma-ray local measurements ([26]), stereophotogrammetry, tomodensitometry or X-ray scanner ([10, 27, 23]). Comparison of the incremental shear intensity maps and incremental volume change in figure 3 (middle and bottom rows) shows clearly that large shear and isotropic deformations are affecting the shear band.

These observations have been confirmed by M. Mokni (3S-IMG Grenoble) using tomodensitometry on axisymmetric triaxial tests; it has been possible to measure the evolution of local void ratio in the shear bands, and it was proven by quantitative measurements that the void ratio was reaching a critical value in large deformation, the same for loose and dense specimens tested under the same confining pressure ([23]). Figure 10 illustrates these results.

So, although localization first questioned the significance of limit densities when observed as overall density in elemental tests, as a consequence of the recognised heterogeneity of the final state, finally the concept of critical density may be validated back at a more local level by the observations reported here.

3.5 Conclusion for granular bodies

Strain localization is likely to arise in most cases in granular bodies as far as limit loads are approached. The localized zone can be simply a single shear band, but it can also be much more complex than that, specially when kinematical constraints restrain its development. Shear surfaces are material structures, with a non vanishing thickness which depends on the grain size. A critical density is reached inside the shear bands in very large strain. Among the different influence factors which have been explored with respect to shear band occurrence in biaxial tests, the influence of initial compactness, grain size, specimen size and shape, lateral pressure can be summarised as follow: lower density, coarser granulometry, smaller specimens, less slender shape, and higher pressure will induce a delay in the onset of shear banding together with a steeper orientation.

4 Localization in cohesive geomaterials

4.1 Ductile versus brittle behaviour

There is quite a large range of cohesive geomaterials. From soft clays to hard rock, an extreme variety of rheological behaviour can be encountered. A distinction is usually made between a "ductile" and a "brittle" field. Following Paterson [43] , the definition of brittle fracture is: *"a discrete event in which the failure of the rock occurs, without significant prior deformation and without warning, at a particular stress"*. Conversely, rupture is said to be ductile when it comes after more or less large plastic deformation of the material. What is observed at rupture in axisymmetric triaxial tests ?

- In **clay specimens** strain localization is observed in most cases but it can take different forms: either a **single failure plane**, or **multiple parallel failure planes** are reported by the experimentalists, depending on the uniformity of initial density of the specimen, tendency to dilate or to compact, and boundary conditions ([36]). Dilatancy and contractancy are directly linked to the overconsolidation ratio of the clay specimen, that is to say to the ratio of the present stress level versus a previously applied one, memorised in the fabric of the specimen. So the overconsolidation ratio is a major indicator for the kind of failure to be expected.
- In **hard rock** specimens, failure is usually described as "brittle fracture". One may distinguish two principal types of brittle fracture:
  - shear fracture, in which a single fracture surface inclined to the principal stress direction is observed;
  - extension fracture, characterized by separation normal to the failure surface, which is generally oriented normal to the minor principal stress [43]. Axial splitting is a special case of extension fracture, which occurs in axisymmetric triaxial compression tests at low confining pressure.
- In **porous rock** specimens, failure has a larger spectrum with respect to the pressure level. At low confining pressure, extension bands occur. At higher confining pressure, shear bands are observed, associated with

dilative volumetric strain, or with compaction strain at high pressure [4]. At higher confining pressure again, compaction band seem to be the failure mode, i.e. localization bands with a pure compaction strain [41, 34]. Failure by cataclastic flow (diffuse grain crushing) is also reported at very high pressure [67].

From soft clays to hard rocks, differences are considerable: grain nature and shape, fabric, nature of the bonds between the grains, porosity, permeability... However, all these materials show a clear dependency of their failure modes on the confining pressure. Rocks cannot be called "brittle" or "ductile" whatever the circumstances are: pressure, temperature, strain rate. The same is true for consolidated clays, an essential difference with respect to rocks being the much lower critical confining pressure. In engineering problems, the relevance of ductile or brittle behavior depends on the problem considered: surface rocks will be brittle in most cases, but in deep mines, or in boreholes, the same rocks can behave in the ductile way. In earth science, large space and time scales are considered, together with large pressure and temperature; then, ductile behavior will be dominant.

Shear bands are well defined deformation structures for which a theoretical framework has been developed in the 20 past years; it is interesting to notice that among the first works in that development, some have been concerning rocks [45, 48]. There is a strong interest from both engineering and earth science field for bifurcation approach of localization in rocks.

Experimental observations show that rather different failure modes do exist in rocks, depending on pressure and other factors. For an application of the theory to an engineering problem, the relevance of the shear band approach must be checked with respect to the real failure mode in the actual situation.

4.2 From shear band to cracks

Strain concentration: from diffuse to localized cracking and damage

Localization results from a concentration of micro events. In clay, rotation or decohesion of particles are the dominant micro mechanisms inside bands [33, 44]. In rock, coalescence of microcracks, and other local micro mechanical events as those reported by Wong finally leads to macro cracking, with one or a few well-defined cracks [66, 38]. *"At higher pressure, the microcracks (now mainly inclined ones) tend to form inclined zones which develop into the macroscopic shear fracture, often by linking up* en echelon *inclined fractures"* [43]. At this stage, the localization process is clearly a displacement (velocity) discontinuity rather than a strain (rate) discontinuity. However, with Paterson one can estimate that *"at least in the shear failure case, the linking-up of the microcracks is probably generally a subsequent development to the local concentration of microcracking in a band of inelastic deformation."*

Shear bands observation in laboratory tests

There is a very wide range of scales at which deformation processes can be considered, from large continental structures to microscopic cracks. At some scales, shear band-like structures are described while observations at other scales seems to indicate a dominant cracking process. In laboratory studies, the specimen size is restricted to the centimetric to decimetric range, but the grain size versus specimen size ratio still can vary in a wide range, depending on the material considered: then the scale problem remains. Shear band observations in laboratory tests are reported by several authors; for instance, [42] have performed biaxial tests on Gosford Sandstone, which shows that in the vicinity of the final failure surface, a region exists in which the material of the blocks is somewhat disturbed (it shows a *"white, chalky appearance"*). The discrete fracture *"does not form instantaneously, but is the result of a gradual alteration of the microstructure"*. Microstructural observations show that, close to peak stress, *"only minor perturbation of the gross structure of the material"* can be detected. After complete failure, the global shear zone is made of *"well-defined sections* (having) *the same orientation, and separated by more diffuse and shorter shear zones(...)"*. Similar observations are reported by [35] on Grès de Fontainebleau .

Wawersik and Fairhurst working on Tennessee Marble report that at confining pressures higher than 2000 psi (about 14 MPa), the principal mechanisms are local shear failures, intergranular sliding and intercristalline glide on well-defined, regularly-spaced inclined planes [65].

When cracks coalesce, it is common to observe residual free particles which form a gouge layer inside the failure plane (e.g. [58] in Westerly granite).

Tillard-Ngan and Desrues [54] in 3S-IMG, Grenoble, have performed a series of plane strain tests on Beaucaire Marl, with a stereophotogrammetric survey of the strain field evolution in order to detect and characterise the onset of localization. The results of a typical test in this experimental set are presented hereafter. It was found that, under low confining pressure, macroscopic cracks were propagating through the specimen, with detectable but small shear strain concentration ahead the tip of the crack; the cracks were mode II cracks, as revealed by stereophotogrammetric measurements. No evidence of diffuse microcracking within the specimen before localization was found.

Torrenti and Benaija [56, 57] in LCPC, Paris, have performed plane stress simple compression tests on concrete specimens, with stereophotogrammetric survey too, using the same method in collaboration with J. Desrues. It was found that mode I axial cracks, observable at centimetric scale were opening throughout the specimen near the peak load, but these cracks were located in a band-like zone. Similar work has been performed in 3S-IMG on a siltite from Gard, France, [19]. Failure occur by the propagation of a shear crack and elastic unloading of a part of the specimen was measured around the crack.

4.3 Localization in Beaucaire marl

Marls are geomaterials which can be considered as transition materials between soils and rocks. It is interesting to study them, because they are concerned in a lot of difficult civil, environmental and petroleum engineering problems; moreover, they can be seen as a model material for deep rocks.

The unconfined biaxial test

The biaxial apparatus described in section 3 has been used to bring up to failure a parallelepipedic specimen in plane strain condition under zero lateral pressure. As in the case of sand specimens discussed before, the zero strain condition is imposed in the direction of the specimen thickness by a set of two rigid glass platens bolted together. A speed variator enables a vertical displacement controlled loading with a rate ranging between 2 μm/mn and 7900 μm/mn to be applied to the specimen through a screw jack. The axial displacement is measured by a potentiometer with a resolution of ± 2 μm. The force, which is measured by a 20 kN cell, is transmitted to the rock specimen through a suspended knee joint loading head.

The dimensions of all the specimens are 35 mm thick, 120 mm high and approximately 60 mm wide. In order to ensure that the history of deformation of the specimen could be retraced from the photos taken during the test by the stereophotogrammetric process, a thin layer of metallic paint is finely sprayed over the vertical face being photographed. The presence of the metallic granules in the paint enables reference points to be assigned on the surface during the stereophotogrammetric process. Furthermore the vertical faces of the specimen in contact with the glasses platens are lubricated with a silicone grease and anti-friction devices like greased teflon or latex bands are used to avoid the generation of shear traction at the specimen ends.

Crack propagation analysis using the FRS method

As discussed before, the application of the FRS method on photographs taken from a specimen undergoing a plane strain process enables the incremental displacement field all over the specimen to be quantified precisely. Then, the incremental strain field between different states in the loading process can be computed from the displacement data obtained. This allows to characterise the onset of strain localization in granular materials, as shown in previous sections. However, in cohesive materials, "strong" discontinuities i.e. displacement discontinuities can be the essential phenomenon to characterise at least for certain (late) steps of the test. Using an adequate strategy when gathering the data along the cracks, and a special data processing as far as the measured points are identified as belonging to a crack side, the relative displacement jump across the observed cracks can be computed [16]. The tangential component of the displacement jump across a crack can also be distinguished

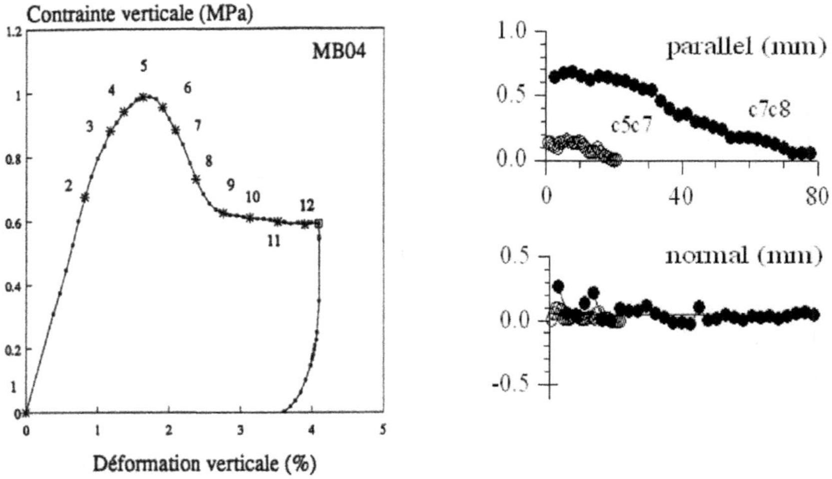

Fig. 11. *Axial stress-strain curve for the test MB04 (a) and displacement jump profiles along the left crack in test MB04 (b) (after Tillard-Ngan, 1992)*

from the normal component, which gives interesting information on the crack propagation.

Results

A series of biaxial tests have been carried out on Beaucaire Marl and stereophotogrammetric measurements on the photographs taken during these biaxial tests showed that macro-cracks with origins at the corners of the specimen propagate through the specimen at an angle of approximately 25° to the vertical. Figure 11 a presents the stress-strain curve recorded during the test MB04 which was carried out with lubricated teflon bands as anti-friction device. Figures 12 and 13 show the corresponding results from the stereophotogrammetric process applied on this test. It can be seen that first signs of strain localization appear between photographs 4 and 5 which were both taken before peak stress. An elongated shear zone with no significant signs of dilation or contraction is shown well before any crack initiation appears. This shear zone can also be found later in the vicinity ahead the advancing crack tip (figure 13). The displacement jump across the crack is aligned along the crack direction, this means that no normal opening of the crack has occurred in the basic process of propagation. Figure 11 b displays the profile of the components of the displacement jump along the left crack for the increments C5C7 (open symbols) and C7C8 (full symbols), respectively parallel to the crak (top line), and normal to the crack (bottom line). The profiles show clearly that in both increments, the normal jump is negligible compared to the tangential,

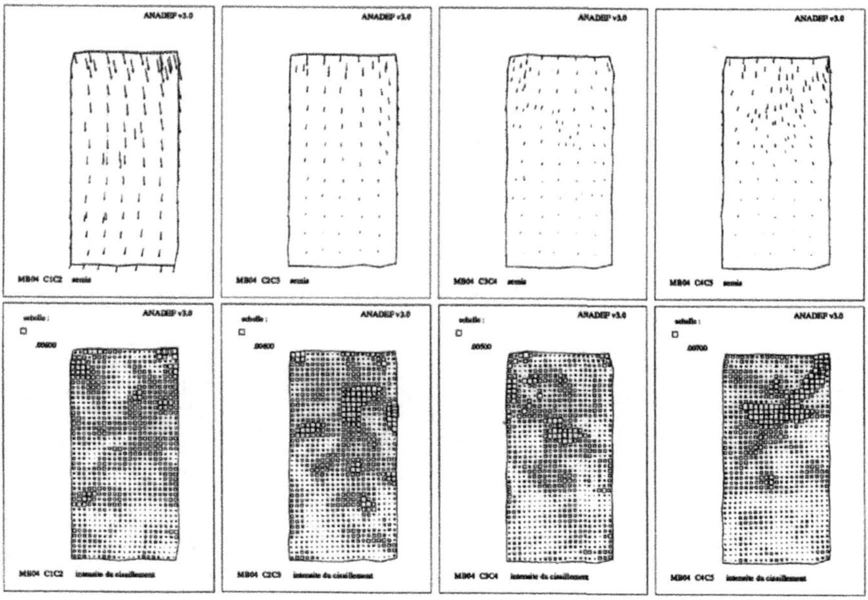

Fig. 12. *Test MB04 : incremental displacement fields (top line) and incremental shear strain fields (bottom line) between the photographs C1 and C2, C2 and C3, C3 and C4, C4 and C5 (after Tillard-Ngan, 1992)*

and the latter decreases rather progressively from the extremity, where the crack was initiated, to the active tip (vanishing at the tip as expected).

4.4 Localization in a concrete and a siltite

A contrasting behaviour with respect to the crack developpement mode has been observed by Benaija and Torrenti [3, 55] using stereophotogrammetry method in collaboration with J. Desrues [56, 57]. Figure 14 shows that, during an unconfined compression test performed on a parallelepidic specimen of concrete, localization is detected at the peak load using stereophotogrammetry. Moreover, the crack jump analysis shows that the detailled localization mechanism is a set of more or less axial cracks, organised in a band through the specimen. On each of the small crack, the displacement jump is essentially horizontal, i.e. normal to the crack direction. This observation is confirmed in the other tests performed by [3] in his thesis work.

The failure of siltite was observed with the same method on unconfined parallelepidic specimens by [19]. The siltite has a uniaxial compression strength of about 70 MPa, and is composed of carbonate (37%), quartz (37%) and clay. In order to polarize the deformation process in the plane perpendicular to the large face, which was surveyed by photographs, a cylindrical

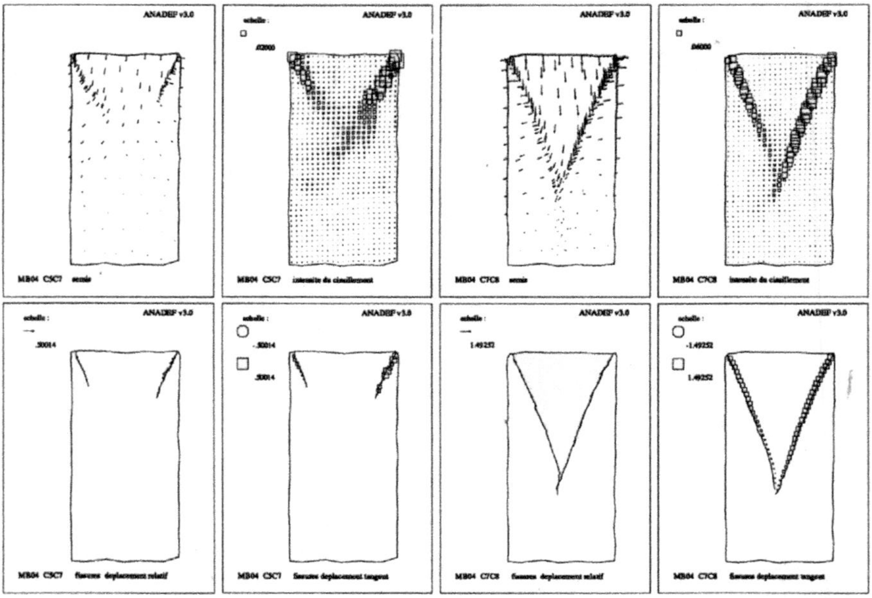

Fig. 13. *Test MB04 - top line: incremental displacement and incremental shear strain fields between photographs C5 and C7 (left)and C7- C8 (right) ; bottom line: relative displacement along cracks and its tangential component for the same increments C5- C7(left) and C7- C8 (right) (after Tillard-Ngan, 1992)*

hole was drilled in the center of the face (see Figure 15 right). Comparison with uniaxial tests performed on the same specimens without a hole showed no effect neither on the peak load nor on the peak strain. Macroscopic crack were initiated near the hole before the stress peak of the specimen response. Figure 15 left shows the stress-strain response of the specimen, with numbers indicating the photographs and time when they were taken. The curves show a continuous increase of the stress up to photograph 7, while the lateral strain responses in directions 2 and 3 (1 being the axial) remain very close. From photograph 7 to 11, the axial stress remains more or less constant while the strains in direction 2 and 3 diverge abruptly. Finally, after photograph 11, the specimen looses its strength and the axial stress drops suddenly. Figure 16 shows the loading steps 1-7, 7-11, and 11-12 corresponding to the three phases described above. The upper line of picture shows the increment displacement field, the second the incremental shear strain field, and the third the displacement jump across the cracks observed. In the increment 7-11, the displacement jump (line 3) shows a dominant shear component and also a normal component. In front of the crack (line 2), there is a concentration of the shear strain. Increment 11-12 is the step of complete failure of the specimen (the cracks crossed entirely the specimen). The right part of the specimen is

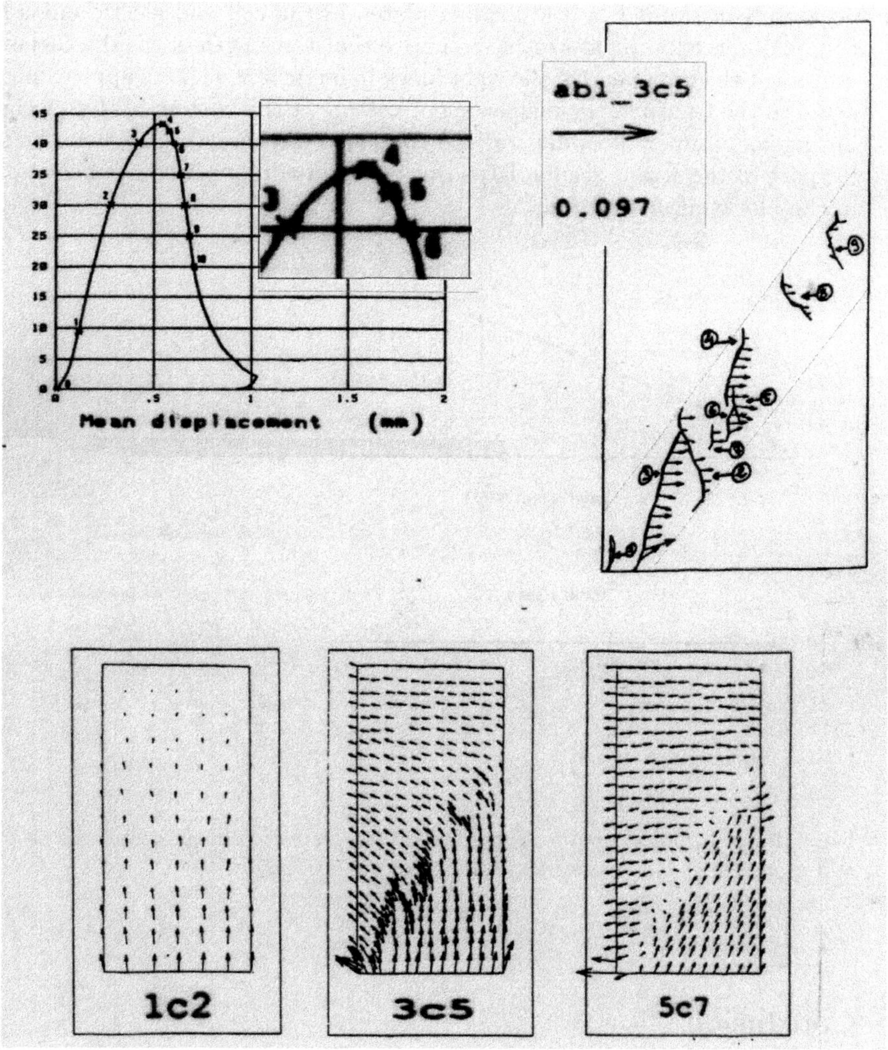

Fig. 14. *Unconfined compression of a concrete specimen. Load-displacement curve (top right), displacement jump during the load increment including the peak load (top left), incremental displacement fields before the peak, at peak, and after the peak (bottom line)* (courtesy of LCPC, Paris.)

no more constrained by the loading platen of the cell and elastic unloading can be observed: it is interesting to notice that the length of the displacement vectors at the upper end of the right block in increment 11-12 is approximately equal to the length of the displacement vectors at the bottom of the specimen during the pre-peak loading stage, increment 1-7; this indicates that the major part of the strain accumulated during the loading is recovered elastically in the blocks after rupture.

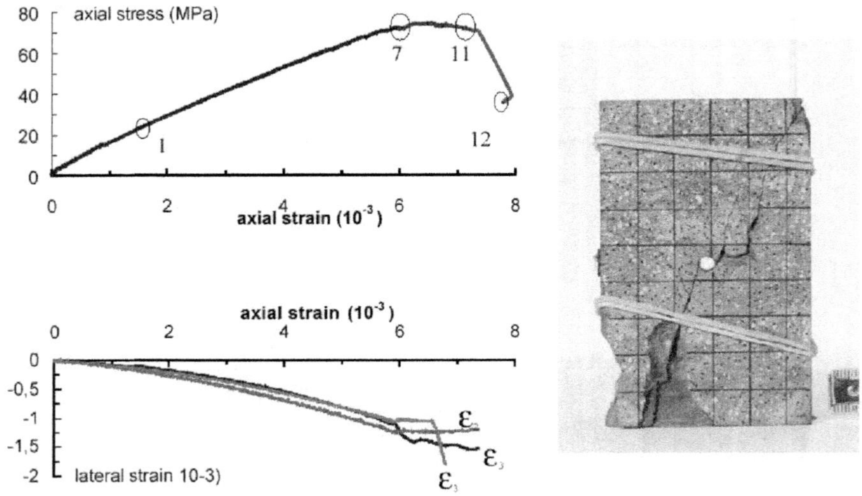

Fig. 15. *Unconfined compression of a siltite specimen. Stress-strain curves (left) and geometry of the specimen with a central hole (right).*

Conclusion

Geomaterials may vary in a large range with respect to their physical and mechanical properties: granular or cohesive, ductile or brittle, fined grained or coarse. Granular materials like sands have been shown experimentally to meet quite well the conditions for an application of shear band analysis to predict the onset of strain localization. However in a large number of cases, the failure mode reported for laboratory tests on rocks and other brittle cohesive materials consists of one or several, more or less plane, failure surfaces. The difference with granular bodies is that in the latter, the thickness of the shear bands does not vanish even at large strains.

On the other hand, one remark linked to scale problems must be considered: the ratio of the relevant problem dimension versus the shear band

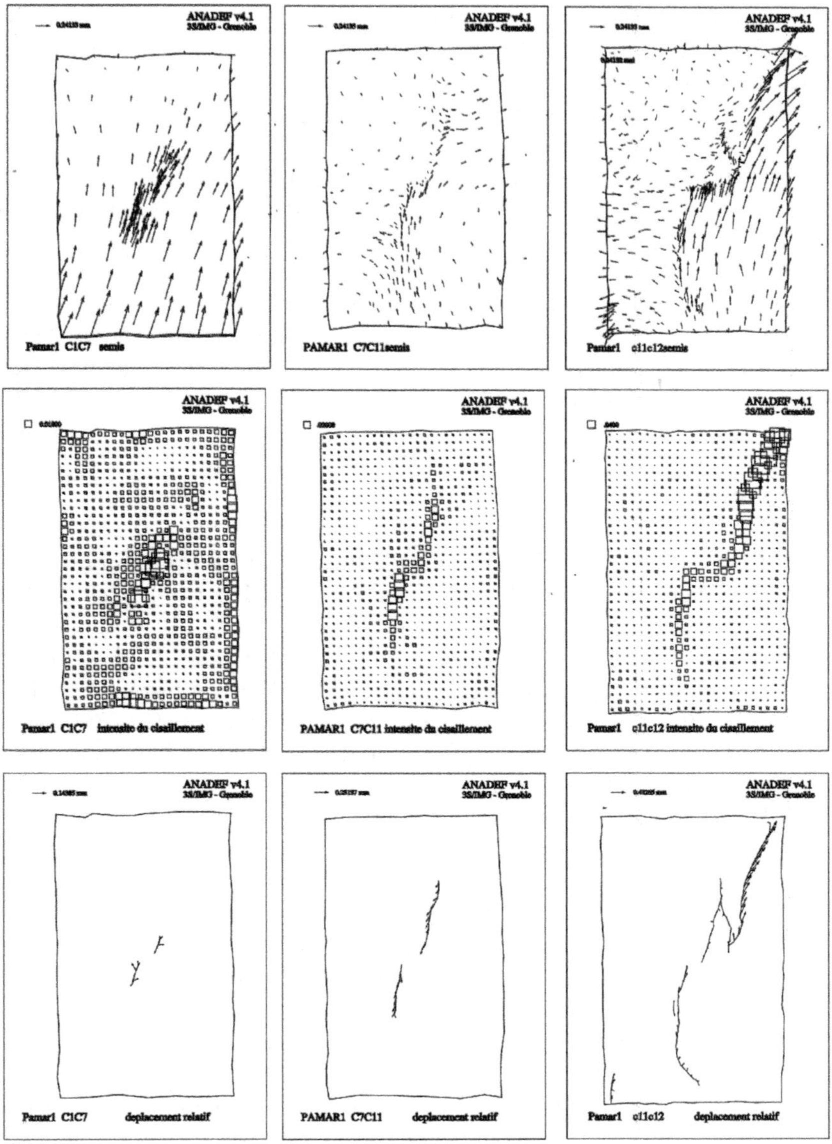

Fig. 16. *Unconfined compression of a siltite specimen, pre-peak (left row), peak (middle row) and post-peak behaviour (right row). Incremental displacement field (top line), incremenalt shear strain field (middle line) and incremental relative displacement jump on cracks (bottom line). Two macro cracks are propagating from the central hole. The displacement jump is in average parallel to the crack direction. Elastic unloading of the broken right piece of the specimen is observed in the last increment.*

thickness can easily tend to infinite in field problems even in granular bodies; large displacements can take place between the sides of these shear bands, in such a way that at the field scale, the situation looks no longer so different from those described above in cohesive materials.

Moreover, micro scale observations in some cases, and macro scale observations in other, suggest that, at some stage of its development, localization in rock takes the form of a band of concentration of local deformation. The final failure surfaces seem to result of a concentration of micro damages inside a non zero thickness zone, which evolve to a macro crack. Note also that in some cases, a gouge layer of granular material (residual particles of crushed grains) is formed inside the failure surfaces. This encourages in developing theoretical and numerical tools to model the localization phenomena in the shear band analysis framework.

References

[1] J.R.F. Arthur and T. Dunstan. Rupture layers in granular media. In P.A. Vermeer and H.J. Luger, editors, *IUTAM Conf. Def. Fail. Gran. Media* ,, pages 453–459. Balkema, 1982.

[2] J.R.F Arthur, T. Dunstan, Q.A.J.L. Al-Ani, and A. Assadi. Plastic deformation and failure in granular media. *Géotechnique*, 27:53–74, 1977.

[3] E.H. Benaija. *Application de la Stéréophotogrammétrie au Béton: cas de la compression simple.* Thèse de doctorat, Ecole Nationale des Ponts et Chaussées Paris, 1992.

[4] P. Bésuelle. Compacting and dilating shear bands in porous rock: Theoretical and experimental conditions. *J. Geophys. Res.*, 106(B7):13435–13442, 2001.

[5] R. Chambon, S. Crochepeyre, and J. Desrues. Localization criteria for non linear constitutive equations of geomaterials. *Mechanics of Cohesive-Frictional Materials*, 5:561–582, 2000.

[6] R. Chambon and J. Desrues. *Plastic instability*, chapter Bifurcation par localisation et non linéarité incrémentale: un exemple heuristique d'analyse complète, pages 101–19. Presses ENPC Paris, 1985.

[7] R. Chambon, J. Desrues, and D. Tillard. Shear modulus identification using experimental localisation data. In Vardoulakis I. Chambon R., Desrues J., editor, *Localisation and bifurcation theory for soils and rocks*, pages 101–112. Balkema, 1994.

[8] R. Chambon, J. Desrues, and I. Vardoulakis. *Localisation and Bifurcation Theory for Soils and Rocks.* BALKEMA Rotterdam, 1994.

[9] R. Charlier, R. Chambon, J. Desrues, and W. Hammad. Shear band bifurcation in soil modelling: a rate type constitutive law for explicit localisation analysis. In C.S. DESAI and E. KREMPL, editors, *Third int. conf. on Constitutive laws for Engng materials, Tucson* ,, pages 399–402. ASME Press, 1991.

[10] J.L. Colliat-Dangus, J. Desrues, and E. Flavigny. Avantages et inconvénients de l'utilisation d'un système d'antifrettage dans l'essai triaxial de compression. *Revue Française de Géotechnique*, (34):34–55, 1986.

[11] J.L. Colliat-Dangus, J. Desrues, and P. Foray. *Advanced Triaxial Testing for Soil and Rocks - ASTM STP 977*, chapter Triaxial testing of granular soil under elevated cell pressure, pages 290–310. ASTM, 1988.

[12] C.A. Coulomb. Sur une application des règles de maximis et minimis à quelques problèmes de statique, relatifs à l'architecture. *Académie Royale des Sciences*, pages 1–40, 1773.

[13] J. Desrues. *La localisation de la déformation dans les matériaux granulaires.* thèse de doctorat es sciences, USMG - INPG Grenoble, 1984.

[14] J. Desrues. *Geomaterials constitutive equations and modelling*, chapter Shear Band Initiation in Granular Materials: Experimentation and Theory, pages 283–310. Elsevier, 1990.

[15] J. Desrues. *Bilan et Perspectives du Greco Géomatériaux - Paris 25-26 octobre 89*, chapter Rupture localisée dans les géomatériaux : nouvelles approches, pages 151–178. Imprimerie de Grenoble, 1992.

[16] J. Desrues. Analyse stéréophotogrammétrique de la fissuration progressive. In Taroni M. Eds. Berthaud Y., Paraskevas D., editor, *Etudes du comportement des matériaux et des structures ,Photomécanique 95, Cachan ,14-16 mars 1995*, pages 149–162. Eyrolles, 1995.

[17] J. Desrues. Experimental strain localisation in plane strain tests on sands and weak rocks. In *Dynamic Plasticity and Structural Behaviors ,Plasticity'95, Osaka ,July 1995*, pages 415–418, 1995.

[18] J. Desrues. *Mécanique des Géomatériaux*, volume 2, chapter Détection expérimentale de la localisation, pages 130–138. Hermes, 1995.

[19] J. Desrues, P. Bésuelle, M. Kntz, and G. Viggiani. Etude de la cinématique de la fissuration et de la rupture d'échantillons de siltite. In *Recueil des communications, Journées Scientifiques 1997, Bagnols-sur-Cèze*, pages 69–70, 1997.

[20] J. Desrues and R. Chambon. *Large deformation of solids: physical basis and mathematical modelling*, chapter Bifurcation par localisation de la déformation: étude expérimentale et théorique à l'essai biaxial sur sable, pages 433–459. Elsevier Applied Sc., 1986.

[21] J. Desrues and R. Chambon. Shear band analysis for granular materials: the question of incremental non linearity. *Ingenieur Archiv*, 59:187–196, 1989.

[22] J. Desrues, R. Chambon, W. Hammad, and R. Charlier. Soil modelling with regard to consistency: Cloe, a new rate type constitutive model. In DESAI C.S. and KREMPL E., editors, *Third int. conf. on Constitutive laws for Engng materials, Tucson ,*, pages 395–398. ASME Press, 1991.

[23] J. Desrues, R. Chambon, M. Mokni, and F. Mazerolle. Void ratio evolution inside shear bands in triaxial sand specimens studied by computed tomography. *Géotechnique*, 46(3):529–546, 1996.

[24] J. Desrues and B. Duthilleul. Mesure du champ de déformation d'un objet plan par la méthode stéréophotogrammétrique de faux relief. *Journal de Mécanique Théorique et Appliquée*, 3(1):79–103, 1984.

[25] J. Desrues and W. Hammad. Shear banding dependency on mean stress level in sand. In Z.Sikora E.Dembicki, G.Gudehus, editor, *Proceedings , 2nd Int.Workshop on Localisation and Bifurcation ,Gdansk ,1989*, pages 57–68. Techn. Univ. Gdansk, 1989.

[26] J. Desrues, J. Lanier, and P. Stutz. Localization of the deformation in tests on sand sample. *Engineering fracture mechanics*, 21:909–921, 1985.

[27] J. Desrues, M. Mokni, and F. Mazerolle. Tomodensitométrie et localisation dans les sables. In Associazione Geotechnica Italiana, editor, *Proceedings ,ECSMFE X ,Florence (It.) ,26-30 May 1991*, pages 61–64. Balkema, 1991.

[28] A. Drescher and I. Vardoulakis. Geometric softening in triaxial tests on granular material. *Géotechnique*, 32(4):291–303, 1982.

[29] R.J. Finno, W.W. Harris, M.A. Mooney, and G. Viggiani. Strain localization and undrained steady state of sands. *Journal of Geotechnical Engineering, ASCE*, 122(6):462–473, 1996.

[30] E. Flavigny, J. Desrues, and B. Palayer. Le sable d'hostun "rf" - note technique. *Revue Française de Géotechnique*, (53):67–69, 1990.

[31] W. Hammad. *Modélisation non linéaire et étude expérimentale de la localisation dans les sables.* Thèse de doctorat, UJF - INPG Grenoble, 1991.

[32] C. Han and I. Vardoulakis. Plane-strain compression experiments on water-saturated fine-grained sand. *Géotechnique*, 41(1):49–78, 1991.

[33] P.Y. Hicher and H. Wahyudi. Microstructural analysis of strain localisation in clay. *Computers and Geotechnics.*, 16:205–222, 1994.

[34] K.A. Issen and J.W. Rudnicki. Conditions for compaction bands in porous rock. *J. Geophys. Res.*, 105:21529–21536, 2000.

[35] D. Kondo and J.P. Henry. Etude expérimentale de la localisation de la déformation. In GRECO Géomatériaux, editor, *Rapport scientifique du GRECO Géomatériaux ,*, pages 201–204, 1990.

[36] P.V. Lade and J. Tsay. Effects of localization in triaxial tests on clay. In *Proc. XI ICSMFE San Francisco ,*, volume 1, pages 549–552, 1985.

[37] J. Lanier. *Constitutive Equations and Modelling*, chapter Recent trends in laboratory testing. Elsevier, 1989.

[38] B. Menéndez, W. Zhu, and T.-F. Wong. Micromechanics of brittle faulting and cataclastic flow in berea sandstone. *J. Struct. Geol.*, 18(1):1–16, 1996.

[39] M. Mokni. *Relations entre déformations en masse et déformations localisées dans les matériaux granulaires.* Thèse de doctorat, UJF - INPG Grenoble, 1992.

[40] M. Mokni and J. Desrues. Strain localisation measurements in undrained plane-strain biaxial tests on hostun rf sand. *Mechanics of cohesive-frictional materials*, 4:419–441, 1999.

[41] W.A. Olsson. Theoretical and experimental investigation of compaction bands in porous rock. *J. Geophys. Res.*, 104:7219–7228, 1999.

[42] A. Ord, I. Vardoulakis, and R. Kajewski. Shear band formation in gosford sandstone. *Int. J. Rock Mech. Min. Sci. Geomech Abstr.*, 28(5):397–409, 1991.

[43] M.S. Paterson. *Experimental Rock Deformation - The Brittle Field.* Springer-Verlag, 1978.

[44] S. Raynaud. Fracturation et dissolution sous contrainte des roches, 1995. Thèse de doctorat.

[45] J.R. Rice. *Theoretical and Applied Mechanics*, chapter The Localization of Plastic Deformation, pages 207–220. North-Holland Publishing Company, 1976.

[46] V. Roger. *Etude expérimentale et théorique de la localisation des déformations dans les matériaux granulaires en condition isochore.* thèse de doctorat, UJF, 2000. 20 Jan 2000.

[47] V. Roger, J. Desrues, and G. Viggiani. Experiments on strain localisation in dense sand under isochoric conditions. In Oka F., editor, *Localisation and Bifurcation Theory for Soils and Rocks*, pages 239–248. Balkema, 1998.

[48] J.W. Rudnicki and J.R. Rice. Conditions for the localisation of deformation in pressure sensitive dilatant materials. *JMPS*, 23:371–394, 1975.

[49] G. Scarpelli and D.M. Wood. Experimental observations of shear band patterns in direct shear tests. In P.A. Vermeer and H.J. Luger, editors, *Proc. IUTAM Conf. Def. Fail. Gran. Media* ,, pages 473–484. Balkema, 1982.

[50] R.F. Scott. Failure. *Géotechnique*, 37:423–466, 1987.

[51] F. Tatsuoka, T. Nakamura, C.C. Huang, and K. Tani. Strenght anisotropy and shear band direction in plane strain test of sand. *Soils and Foundations*, 30(1):35–54, 1990.

[52] F. Tatsuoka, M. Sakamoto, T. Kawamura, and S. Fukushima. Strenght and deformation characteristics of sand in plane strain compression at extremely low pressures. *Soils and Foundations*, 26(1):65–84, 1986.

[53] D. Tillard. *Etude de la rupture dans les géomatériaux cohésifs. Application à la marne de beaucaire.* Thèse de doctorat, Université Joseph Fourier - Grenoble I, 1992.

[54] D. Tillard-Ngan, J. Desrues, S. Raynaud, and F. Mazerolle. Strain localisation in beaucaire marl. In Anagnostopoulos A. et al., editor, *Geotechnical engineering of hard soils - soft rocks* ,, pages 1679–1686. Balkema, 1993.

[55] J.M. Torrenti. Comportement mécanique du béton bilan de six années de recherches. Technical report, Etudes et recherches des laboratoires des ponts et chaussées, 1996.

[56] J.M. Torrenti, J. Desrues, EH. Benaija, and C. Boulay. Stereophotogrammetry and localization in concrete under compression. *Journal of Engineering Mechanics*, 117(7):1455–1465, 1991.

[57] J.M. Torrenti, J. Desrues, E.H. Benaija, and C. Boulay. Application of stereophotogrammetry on concrete compression test. In Silva Gomes J.F. et al. (eds), editor, *Recent Advances in Experimental Mechanics ,Experimental Mechanics ,Lisbonne ,July 1994*, pages 63–68. Balkema, 1994.

[58] J. Tullis and R.A. Yund. Experimental deformation of dry westerly granite. *J. Geophys. Res.*, 82(36):5705–5718, 1977.

[59] I. Vardoulakis. Bifurcation analysis of the triaxial test on sand samples. *Acta Mechanica*, 32:35–54, 1979.

[60] I. Vardoulakis. Deformation of water-saturated sand: I. uniform undrained deformation and shear banding. *Géotechnique*, 46(3):441–456, 1996.

[61] I. Vardoulakis. Deformation of water-saturated sand: Ii. effect of pore water flow and shear banding. *Géotechnique*, 46(3):457–472, 1996.

[62] I. Vardoulakis, M. Goldscheider, and Q.G. Gudehus. Formation of shear bands in sand bodies as a bifurcation problem. *Int. J. Num. Anal. Meth. Geom.*, 2:99–128, 1978.

[63] I. Vardoulakis and B. Graf. Calibration of constitutive models for granular materials using data from biaxial experiments. *Géotechnique*, 35(3):299–317, 1985.

[64] I. Vardoulakis and J. Sulem. *Bifurcation analysis in Geomechanics*. Blackie, 1995.

[65] W.R. Wawersik and C. Fairhurst. A study of brittle rock fracture in laboratory compression experiments. *Int. J. Rock Mech. Min. Sci.*, 7:561–575, 1970.

[66] T-F Wong. Micromechanics of faulting in westerly granite. *Int. J. Mech. Min. Sci. Geomech. Abstr.*, 19:49–64, 1982.

[67] T.F. Wong, C. David, and W. Zhu. The transition from brittle faulting to cataclastic flow in porous sandstones: Mechanical deformation. *J. Geophys. Res.*, 102:3009–3025, 1997.

[68] T. Yoshida, F. Tatsuoka, M.S.A. Siddique, and Y. Kamegal. Shear banding in sand observed in plane strain compression. In Vardoulakis I. Chambon R., Desrues J., editor, *Localisation and bifurcation theory for soils and rocks ,*, pages 165–179. Balkema, 1994.

Discontinuous character of rock masses: some open questions related to rock fracture mechanics and scale effects

Marta Castelli, Valeria Saetta, Claudio Scavia

Dipartimento di Ingegneria Strutturale e Geotecnica, Politecnico di Torino, Italy

1 Introduction

Rock Mechanics deals with materials whose mechanical behavior is largely influenced by the presence of natural discontinuities. Due to this, it is necessary to refer to mechanical models which allow to introduce displacement discontinuities into the continuous medium.

Resort to numerical techniques for the analysis of cracked rock structures proves necessary because of the geometrical complexity of most application problems. Several numerical methods have been proposed to solve these problems, which can be subdivided in the following categories:

- *Continuum methods*: are not able to take into account the presence of discontinuities from a mechanical point of view
- *Discontinuum methods*: allow to incorporate discontinuities in the displacement field
- *Equivalent continuum methods:* due to the large number of natural discontinuities it is always necessary to substitute, at a certain scale, the discontinuous medium with a continuous one. The mechanical characteristics of the continuous medium must be such that its behavior is equivalent, from a mechanical point of view, to that of the discontinuous medium

The choice of one of the methods described above depends on the size of the structure relative to the discontinuity spacing, the imposed stress level, and the orientation and strength of the discontinuities. As an example, for the shallow tunnel in Fig. 1, if discontinuities are not present or largely spaced and don't influence the behavior of the structure, a continuum approach is justified (Fig. 1a); if the rock mass is highly fractured and the spacing is small with respect to the structure size, a continuum equivalent approach can be used (Fig. 1c), considering mechanical characteristics reduced with respect to the intact rock (i.e., the characteristics obtained in the laboratory have to be reduced); finally, if the discontinuity spacing are comparable to the size of the structure, the behavior of a single discontinuity or a single block is determinant and it is necessary to take into account separately the behavior of both the rock and the discontinuities, through a discontinuum method (Fig. 1b).

In rock engineering attention has mainly been dedicated to the solution of practical problems, frequently recurring to empirical or oversimplified approaches (Bieniawski 1974; Hoek and Brown 1980). However, many aspects of the mechanical behavior of the continuum equivalent and the discontinuities are often of interest, and require a long and complex research work.

In the following, a discontinuum method is presented and two examples of application to rock structures are discussed, where some open questions still exist: the propagation of natural discontinuities in rock masses and the scale effects on the stiffness modulus of a continuum equivalent rock mass.

2 Discontinuum methods: the Displacement Discontinuities Method

Several problems related to fractured rock masses can be solved advantageously through the BEM technique of the Displacement Discontinuity Method (DDM, Crouch and Starfield 1983), which allows to simulate natural discontinuities by the direct discretisation of a displacement discontinuity distribution along a plane crack.

The DDM is based on the assumption that the distribution of a displacement discontinuity û along a line (natural discontinuity, crack) in a body can be approximated by subdividing the line into N Displacement Discontinuity (DD) elements (Fig. 2) and assuming û to vary over the individual element, according to a predetermined (constant, linear, square root, etc.) mode (Fig. 3). An accurate evaluation of the stress and strain state at the crack tips can be achieved by assuming the tips of the crack to consist of two square-root elements (Scavia 1992) and by subdividing the central part into constant elements.

If we consider one DD element to occupy the portion $|x| \leq a$, $y = 0$ in the x, y plane of an infinite elastic body, we can distinguish its two surfaces by saying that one surface is on the positive side of $y = 0$ ($y = 0_+$) and the other is on the negative side ($y = 0_-$), as depicted in Fig. 4.

If one assumes, for example, that the $\hat{u}_x$ and $\hat{u}_y$, components of $\hat{u}$ are constant and equal to D_x, and D_y in the interval [-a, a], passing from one side of the DD element to the other, the displacements undergo a constant change in value $D_i = (D_x, D_y)$ which is given by:

$$\begin{aligned} D_x &= u_x\,(x, 0_-) - u_x\,(x, 0_+) \\ D_y &= u_y\,(x, 0_-) - u_y\,(x, 0_+) \end{aligned} \tag{1}$$

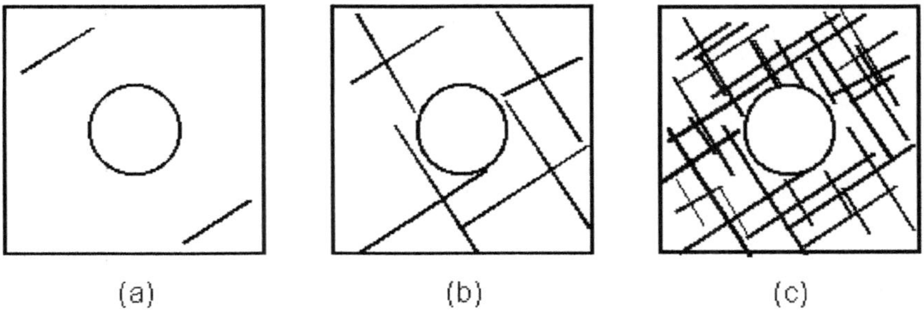

Fig. 1. tunnel in fractured rock mass: **(a)** continuum methods; **(b)** discontinuum methods; **(c)** continuum equivalent methods

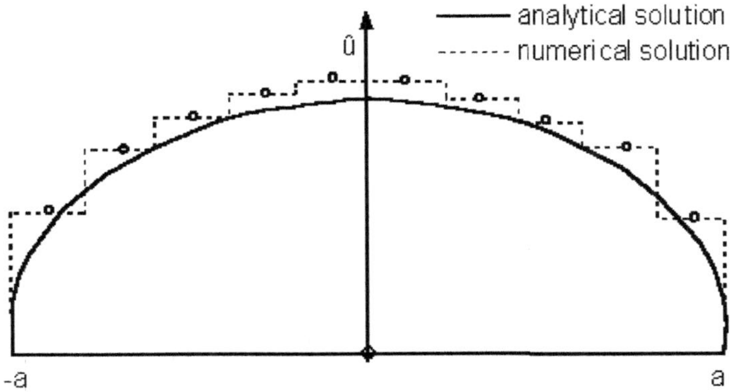

Fig. 2. Discretisation of a displacement discontinuity û through constant DD elements

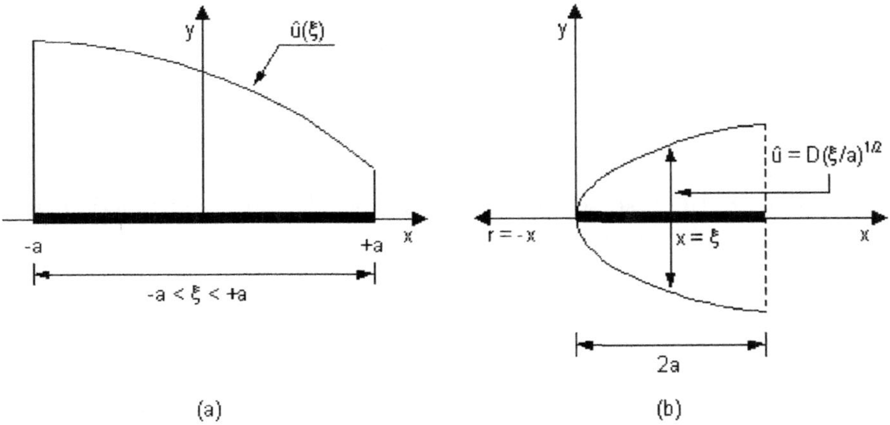

Fig. 3. **(a)** DD element; **(b)** square root DD (after Scavia 1992)

where D_x and D_y are positive according to the sign convention shown in Fig. 4.

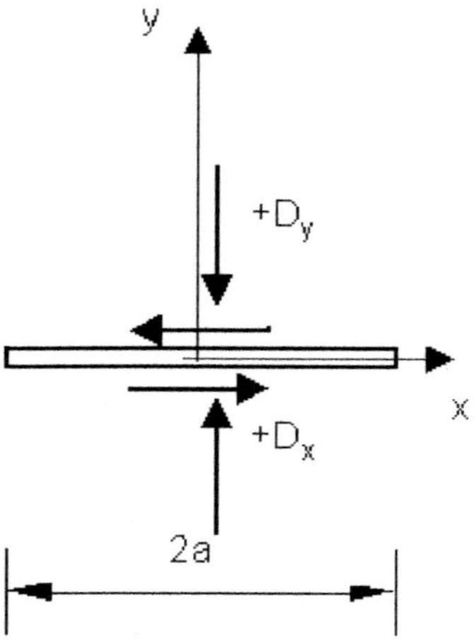

Fig. 4. Representation of a constant DD element (after Crouch and Starfield 1983)

For any type of DD element, the analytical solution for stresses and displacements, due to the presence of a $\hat{u}$ distribution, at any point in the x, y plane is given by:

$$\begin{aligned} \sigma_x &= 2G\left(\mathrm{f}_{,yy} + y\mathrm{f}_{,yyy}\right) + 2G\left(2\mathrm{g}_{,xy} + y\mathrm{g}_{,xyy}\right) \\ \sigma_y &= 2G\left(\mathrm{f}_{,yy} + y\mathrm{f}_{,yyy}\right) - 2Gy\mathrm{g}_{,xyy} \\ \tau_{xy} &= -2Gy\mathrm{f}_{,xyy} + 2G\left(\mathrm{g}_{,yyy}\right) \end{aligned} \tag{2}$$

$$\begin{aligned} u_x &= -\left(1-2\nu\right)\mathrm{f}_{,x} - y\mathrm{f}_{,xy} + 2\left(1-\nu\right)\mathrm{g}_{,y} - y\mathrm{g}_{,yy} \\ u_y &= 2\left(1-\nu\right)\mathrm{f}_{,y} - y\mathrm{f}_{,yy} + \left(1-2\nu\right)\mathrm{g}_{,x} - y\mathrm{g}_{,xy} \end{aligned} \tag{3}$$

where G is the shear modulus, ν is Poisson's ratio, and $f_{,i}$, $f_{,ij}$, $f_{,ijk}$ and $g_{,i}$, $g_{,ij}$, $g_{,ijk}$ are the first, second and third order derivatives of functions $\mathrm{f}(x, y)$ and $\mathrm{g}(x, y)$ with respect to x and y.

Functions $f(x, y)$ and $g(x, y)$ and their derivatives can be evaluated analytically using Green's function approach (Crouch and Starfield 1983).

Starting from the analytical solution for a single DD element, it is possible to obtain the numerical solution that relates to stresses and displacements of the body containing the crack by summing the effects of all the N elements that make up the crack (Fig. 5). As the stresses applied to the crack surfaces are known (for instance, on the surfaces of an open crack they are zero), it is possible to determine, at the midpoint of each DD element, the unknown displacement discontinuities that are necessary to produce such stresses by solving the $2N$ equations:

$$\begin{aligned}\sigma_s(i) &= \sum_{j=1}^{N} A_{ss}(i,j)\cdot D_s(j) + \sum_{j=1}^{N} A_{sn}(i,j)\cdot D_n(j)\\ \sigma_n(i) &= \sum_{j=1}^{N} A_{ns}(i,j)\cdot D_s(j) + \sum_{j=1}^{N} A_{nn}(i,j)\cdot D_n(j)\end{aligned} \tag{4}$$

where: $D_s(j)$, $D_n(j)$ are the unknown displacement discontinuities in the tangential and normal directions, in the center of the j-th element of the crack;

$A_{ss}(i,j)$, $A_{sn}(i,j)$, etc. are the coefficients of influence of $D_s(j)$ and $D_n(j)$ on stresses in the i-th element, as derived from Eqs. (2) and (3) suitably modified to take into account the position and orientation of the i and j elements (Crouch and Starfield 1983);

$\sigma_s(i)$, $\sigma_n(i)$ are the known tangential and normal stresses acting on the i-th element.

$\sigma_s(i)$ and $\sigma_n(i)$ are referred to as induced stresses as they represent the variation of stresses in the body that contains the crack because of the presence of the DD elements. This variation from the original states of stress, $\sigma_{s0}(i)$, $\sigma_{n0}(i)$ (caused, for example, by the weight of the material) to the final states $\sigma_{sf}(i)$, $\sigma_{nf}(i)$ (i.e. nil stresses on the surfaces of the open crack) can be determined as follows:

$$\begin{aligned}\sigma_s(i) &= \sigma_{sf}(i) - \sigma_{s0}(i)\\ \sigma_n(i) &= \sigma_{nf}(i) - \sigma_{n0}(i)\end{aligned} \tag{5}$$

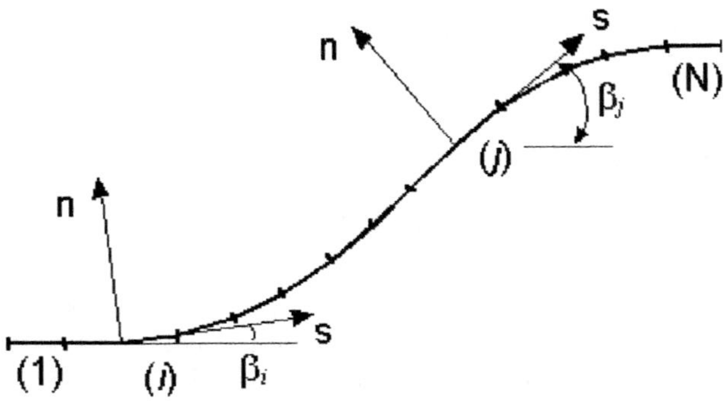

Fig. 5. Representation of a crack through a set of N DD elements, s and n being local co-ordinates (after Crouch and Starfield 1983)

2.1 Analysis of open cracks in tensile and compressive stress fields

Only negative (i.e. opening) normal displacement discontinuities, D_n, occur in tensile stress fields, and hence the normal and shear stresses on the crack surfaces during loading are nil.

In compressive stress fields, instead, the displacement discontinuities $D_n$are positive and the two surfaces of the DD element overlap, leading to a physically impossible situation. It is however possible to overcome this conceptual difficulty by considering that the element has a finite thickness in its undeformed state, which is small compared to its length, but larger than D_n. The normal and shear stresses are nil as in tensile stress fields (Crouch and Starfield 1983).

2.2 Analysis of closed cracks in a compressive stress field

The previously described method is based on the assumption that cracks remain open throughout the entire loading process. In compressive stress fields, however, the crack can be closed, as a function of its genesis and geometry. In these circumstances, crack behavior can be modeled as follows:

- the edges of a crack that were in contact in the absence of stresses, are deformed in the normal direction without giving rise to displacement discontinuities ($D_n = 0$);
- in the tangential direction, if a shear stress is applied too, frictional stresses develop on the crack faces. An elastic-ideally plastic behavior of crack surfaces is assumed (Fig. 6). In the elastic stage, the behavior is governed by the value of stiffness H_s, and the shear stresses are, as a consequence, proportional to the latter. The shift from the elastic to the plastic stage is assessed using Coulomb's failure criterion. In the plastic stage, the shear stresses are equal to the maximum available shear strength.

A detailed description of the procedure adopted for closed cracks is given in Scavia 1992).

2.3 Discontinuum approach: simulation of crack propagation

Rock Fracture Mechanics is a research field related to the study of fractured rock masses through a discontinuum approach. In this connection, the DDM is able to simulate both the propagation of open natural discontinuities located in tensile stress field and closed ones located in compressive stress fields.

Closed cracks in brittle materials, subjected to compression and shear, may spread according to two mechanisms:

- induced tensile propagation (Fig. 7a): developing outside the crack plane in the direction of the maximum tensile stress induced by slip between the crack faces;
- propagation by shear (Fig. 7b): developing in a plane that is critical from the standpoint of shear strength.

In the case of open cracks subjected to tensile stress fields (tensile propagation) and closed cracks subjected to induced tensile stresses (induce tensile

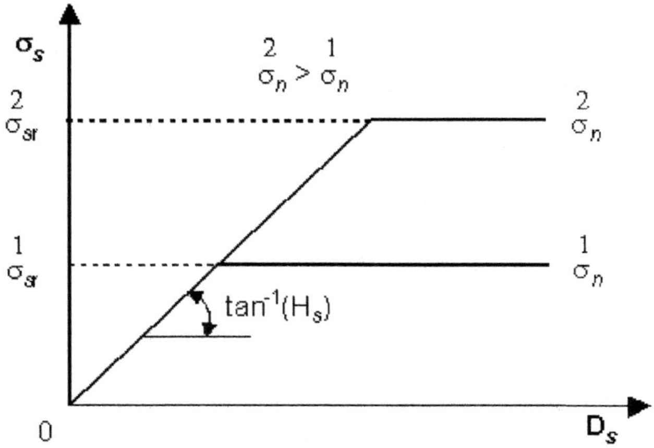

Fig. 6. Model of crack surface behavior under shear: shear resistance $\overset{1}{\sigma}_{sr}$ and $\overset{2}{\sigma}_{sr}$ for two levels of normal stress $\overset{1}{\sigma}_n$ and $\overset{2}{\sigma}_n$ (after Crouch and Starfield 1983)

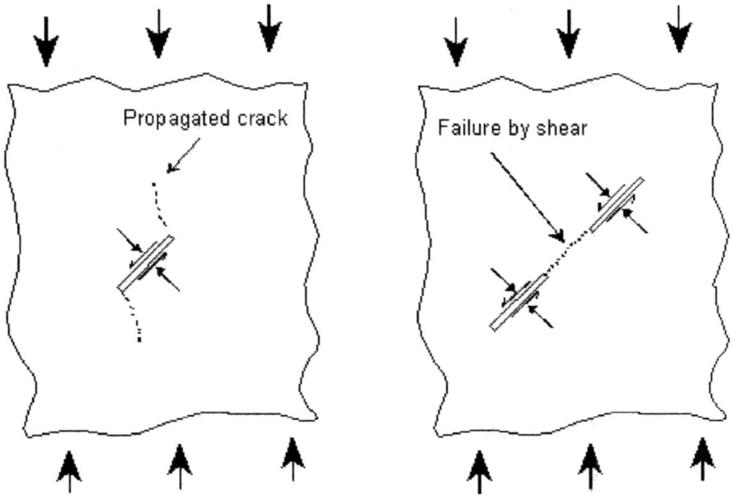

Fig. 7. Propagation of closed and compressed cracks: **(a)** by induced tensile stress; **(b)** in a plane that is critical in terms of shear strength

propagation), knowing the values of the DDs (D_n and D_s) for the square root elements placed at the crack tips, the values of the stress intensity factors in mode I and mode II, K_I and K_{II}, can be determined as:

$$K_{\mathrm{I}} = D_n \frac{G\,(1-\nu)\,\sqrt{2\pi}}{4\sqrt{\bar{x}}}$$
$$K_{\mathrm{II}} = D_s \frac{G\,(1-\nu)\,\sqrt{2\pi}}{4\sqrt{\bar{x}}} \tag{6}$$

where $\bar{x}$ is the distance between the crack tip and the mid point of the square root element.

The possibility of propagation is taken into account by means of the criterion proposed by Erdogan and Sih (1963), which has been validated by numerous experimental investigations and can be easily implemented in computation codes (Ingraffea 1987).

The criterion can be summarized as follows:

- cracks spread radially starting from their tips;
- the direction of propagation, defined by an angle θ_0, is perpendicular to the direction along which the maximum tensile stress, $\sigma(\theta_0)$, is found;
- cracks begin to spread when $\sigma(\theta_0)$ reaches a critical value $\sigma(\theta_0)_C$.

By expressing $\sigma(\theta_0)$ and $\sigma(\theta_0)_C$ as functions of the stress intensity factors, we can write the propagation criterion in this form:

$$K_{\mathrm{IC}} = K_{\mathrm{eq}} = \cos\frac{\theta_0}{2}\left[K_{\mathrm{I}}\cos^2\frac{\theta_0}{2} - \frac{3}{2}K_{\mathrm{II}}\sin\theta_0\right] \tag{7}$$

where θ_0 can be determined by solving the following equation:

$$K_{\mathrm{I}}\sin\theta_0 + K_{\mathrm{II}}\,(3\cos\theta_0 - 1) = 0 \tag{8}$$

Pure mode I conditions ($K_I \neq 0$, $K_{II} = 0$), give $\theta_0 = 0$, i.e., crack propagation takes place in the pre-existing crack plane.

Pure mode II conditions ($K_I = 0$, $K_{II} \neq 0$), give $\theta_0 = \pm 70.53\,^\circ$, i.e., propagation takes place outside the crack plane.

Mixed mode conditions ($K_I \neq 0$, $K_{II} \neq 0$) give θ_0 between 0 and $\pm 70.53\,^\circ$.

It should be noted that the assumption of propagation taking place when tensile stresses (whether direct in mode I or induced in mode II) exceed a predetermined critical value entails the need to determine the only experimental parameter obtainable in mode I, that is to say, material toughness, K_{IC}.

On the contrary, shear propagation develops in a plane that is critical from the standpoint of shear strength (Fig. 7b).

Several observations in laboratory compression tests indicate that somewhere near the peak load, rock deformation becomes localized. The zone of localized deformation is referred to as a fracture, fault, shear band, rupture zone or simply a failure plane.

Within the framework of Linear Elastic Fracture Mechanics, the possibility of shear induced propagation in the uncracked material can be assessed by determining the state of stress at the tips of the crack and comparing this value with the one obtained on the basis of the Mohr-Coulomb strength criterion, which is expressed as:

$$\sigma_{1R} = \sigma_3 \frac{(1+\sin\varphi)}{(1-\sin\varphi)} + 2c\frac{\cos\varphi}{(1-\sin\varphi)} \tag{9}$$

where σ_3 is the minimum principal stress, φ and c are the friction angle and the cohesion of the intact material.

The comparison can be made in terms of principal stresses and propagation takes place in the direction in which the maximum acting principal stress (σ_1) is the same as the maximum admissible principal stress (σ_{1R}).

Example of application: simulation of shear band evolution

The proposed method has been applied to the simulation of plane strain compression tests carried out by Tillard (1992) on intact samples of Beaucaire marl, a sedimentary rock coming from a natural deposit in Southern France (see Scavia et al. 1997, for a detailed description). The marl is fine grained and quite homogeneous. Microscopic observation did not reveal any evidence of pre-existing micro-cracks; from the technical point of view, Beaucaire marl can be described as a soft rock.

The tests considered have been performed in a bi-axial apparatus: a prismatic rock specimen of height 120 mm, width 60 mm, and length 35 mm is mounted between two rigid walls inducing plane strain conditions under displacement control, at zero confining pressure. Photographs of the deforming specimen have been taken throughout the test to record the development and progression of localized deformation, and analyzed using the method of stereophotogrammetry, based on the photogrammetric analysis of successive pairs of photographs. Details about this method can be found in Desrues (1984, 1995).

The results of only one test will be considered hereafter. Fig. 8a shows the curve of axial load as a function of axial strain. The moments of the test when photographs were taken are noted on this curve by numbers 1 through 12. No fissures were detected prior to photograph 5, near the peak load. When photographs 5 and 7 are viewed in stereo, two distinct fissures are visible which originate from the upper edges of the specimen. Afterward, fissures evolve in such a way that a quite developed cone can be observed in the upper portion of the specimen. The stereoscopic view guided the selection of points for measuring displacements.

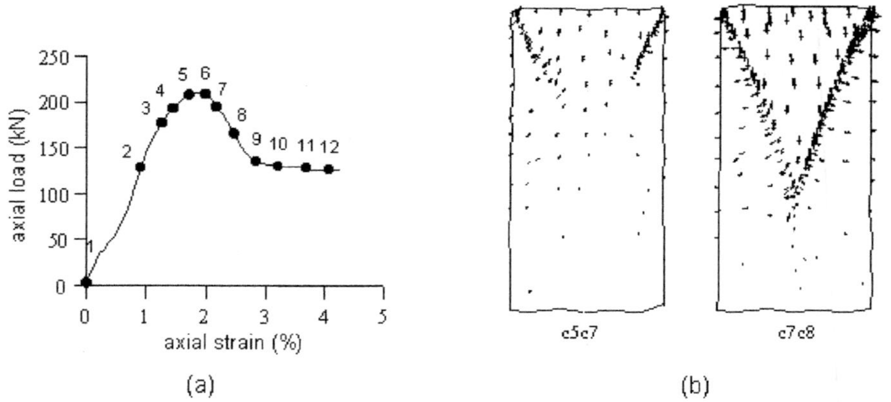

Fig. 8. Plane strain compression test on Beaucaire marl samples: **(a)** load response; **(b)** measured displacements

Several measurements were made on the two sides of the visible discontinuities. Only two photographic increments are shown in Fig. 8b, where the displacement vectors have been magnified by a factor of 20. The complete set of material parameters adopted in the numerical analysis, derived from Tillard (1992), is given in Table 1.

In the numerical simulation, two initial notches, 2 mm long and inclined 28 ° to the vertical, are inserted at the upper corners of the specimen. The orientation of the notches is approximately equal to the initial orientation of the experimentally observed cracks. Crack propagation is triggered at a vertical applied stress of 0.9 MPa. The slight difference between this value and the experimental one (0.99 MPa) can be ascribed to the presence of the notches and the resulting stress concentration at the tips. Shear propagation of the two cracks takes place in an unstable mode and, accordingly, the applied load is progressively reduced.

Fig. 9 compares the crack trajectories for the load increments corresponding to points 5 to 7 and 7 to 8 in Fig. 8. In the experiment, an asymmetry is observed in the propagation trajectories of the left and right cracks in the increment c5c7. This cannot be detected numerically, because the problem has been assumed to be symmetric. Accordingly, measured and computed displacement discontinuities (D_s) along the two cracks are compared for the two load steps at which the computed propagation length equals the measured one (21 and 28 mm, for the left and right crack, respectively).

Results of such a comparison are shown in Fig. 10, where s is a curvilinear co-ordinate measured along the crack.

In the increment c7c8, both the experiment and the numerical simulation show coalescence of the two cracks. However, measured crack lengths obtained experimentally are greater than the computed ones. In order to compare the results in terms of displacement discontinuities for the maximum observed

Young modulus	$E = 81.0$ MPa
Poisson ratio	$\nu = 0.35$
Cohesion	$c = 0.33$ MPa
Friction angle of intact material	$\varphi_p = 28\,^\circ$
Friction angle of crack surface	$\varphi_r = 20\,^\circ$
Uniaxial compressive strength	$C_0 = 1.10$ MPa

Table 1. Mechanical parameters assumed in the numerical simulation

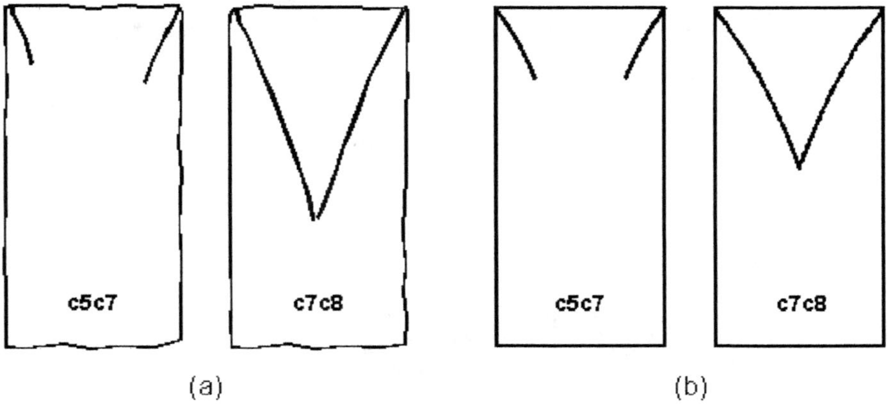

Fig. 9. Propagation trajectories: **(a)** experimental; **(b)** numerical

crack length (82 mm), an additional (non symmetrical) analysis has been performed by inhibiting the propagation of the left crack, while the right crack propagated up to the required length. Results are shown in Fig. 11.

On the basis of the above results, it can be concluded that a good agreement between numerical and experimental results was found as for the propagation trajectory of the shear cracks and the distribution of tangential displacement discontinuities along the crack. However, the numerical model is unable to simulate the global response of the specimens under load, since the direct comparison of computed and experimental load-displacement curves is not significant. In fact, while in the experiment the load reduces after peak as the specimen shortens steadily (Fig. 8a), in the numerical test a reduction of the vertical strain results from the imposed load reduction.

A non-linear approach: the Slip Weakening Model

A possible way to try to simulate the overall specimen deformation behavior is to resort to a non-linear approach. From the experimental point of view, in fact, shear propagation is accompanied by considerable energy dissipation due to friction, and compressed cracks evolve along shear fracture planes only after a long process involving the formation of micro cracks, their propagation and coalescence and eventually large-scale shear failure, which is frequently accompanied by cataclasis of the material between the micro cracks (see, for example, Reches and Lockner 1994).

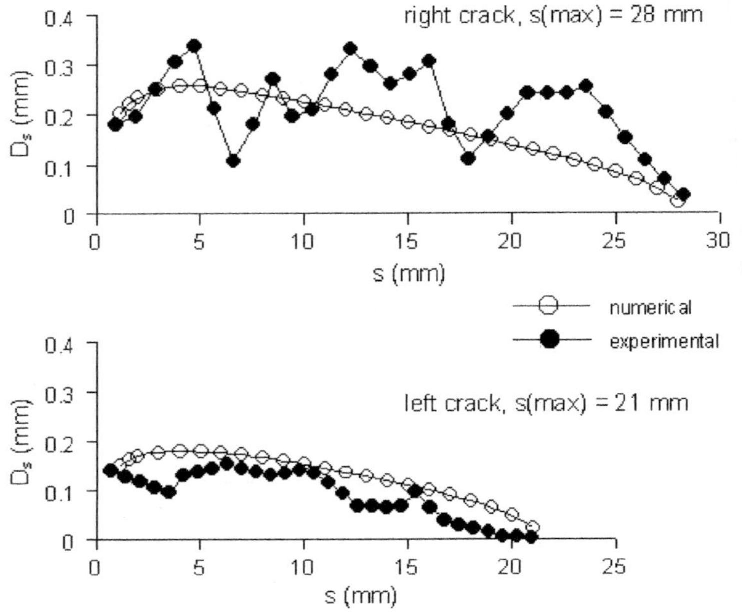

Fig. 10. Displacement discontinuities along left and right cracks (symmetric case): increment c5c7

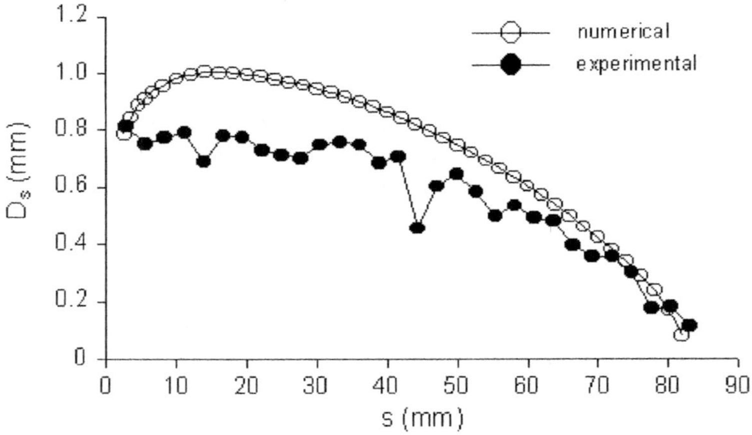

Fig. 11. Displacement discontinuities along the right crack (asymmetric case): increment c7c8

The micro structural process of breakdown near the crack tip can be interpreted by assuming that it gives rise to cohesive stresses, which oppose the action of applied loads. Thus, a Non Linear Fracture Mechanics approach, based on the Slip-Weakening Model (Palmer and Rice 1973) has been proposed by Allodi et al. (2002) and implemented in a computation method based on the Displacement Discontinuity Method.

According to the Slip Weakening Model, the crack can be subdivided in a real part, where residual conditions are reached due to the slip between crack surfaces δ, and a fictitious part (corresponding to a process zone), where cohesive stresses decrease gradually from the peak to the residual value with increasing relative displacements. This decrease and the extension of the process zone ω are greatly affected by the critical value of the slip (δ^*). Due to this, the crack and the non-linear process zone at the advancing tip are represented, in the DDM-based method, through a set of Displacement Discontinuity Elements (DDEs), where the Displacement Discontinuities in the tangential direction are determined on the basis of a friction law (Fig. 12). The friction angle decays from the peak (φ_p) to the residual (φ_f) value and the cohesion from the value c_p (representative of the intact material) to zero, for a certain value of the relative displacement along the crack (i.e. the critical tangential displacement discontinuity D_s^*). As a result, shear strength decreases from the peak (σ_s^p) to the residual (σ_s^f) value as a function of D_s^* on the DDEs making up the process zone (Fig. 13).

With reference to Fig. 13, the first element of the process zone is a tip element with the intact material strength (peak value). The propagation occurs when the shear stress at the tip element exceeds the shear strength. When such condition is met, the step is repeated and the following tangential stress is prescribed for the tip element:

$$\sigma_s(\text{tip}) = \sigma_{sr}(\text{tip}) = c^{\text{P}} + \sigma_n(\text{tip}) \tan \varphi^p \tag{10}$$

A new element can then be added at the process zone tip, in the critical direction with regards to shear failure, following the Mohr–Coulomb criterion. The peak values of material characteristics are associated to this element. For the previous tip, shear reduced parameters are calculated as a function of the computed value of D_s.

Propagation stops when the advancing tip intersects the structure edge elements, while the analysis stops when all the elements making up the shear band reach their residual strength and the structure is considered as failed.

Sensitivity analyses. Sensitivity analyses of triggering and propagation of a shear band are presented in this section in order to verify the influence of the critical slip value δ^* on the overall stress-strain behavior and the extension ω of the process zone. A uniaxial compressive test in plane strain conditions is simulated on a specimen having the mechanical characteristics of a soft rock (Table 2). The geometrical configuration of the specimen and the boundary conditions are shown in Fig. 14.

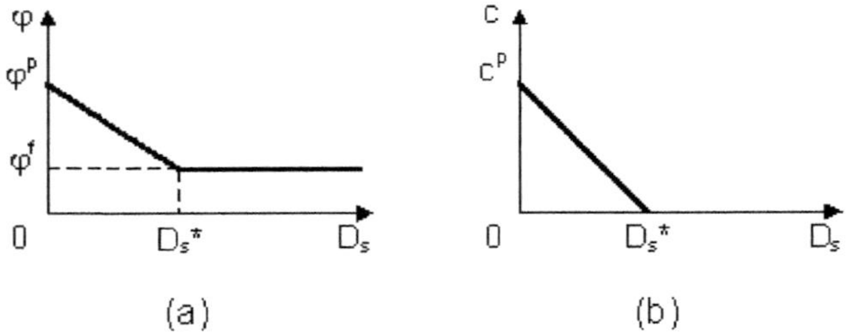

Fig. 12. Adopted linear slip-weakening law: **(a)** friction angle φ; **(b)** cohesion c

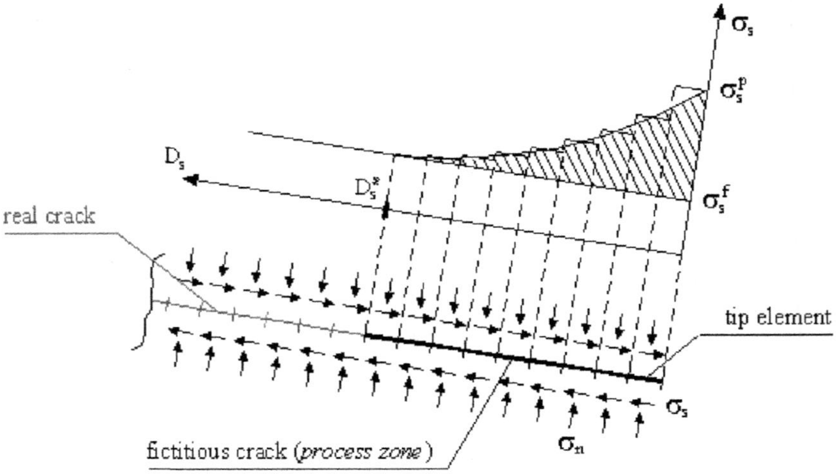

Fig. 13. DDEs making up a process zone, with indication of the decreasing shear strength from the tip (σ_{sp}) to the real crack (σ_{sf}). D_s^* is the critical displacement discontinuity.

Specimen size	60 × 120 mm
Length of the initial defect	3 mm
Inclination of the initial defect	59 °
Young modulus E	81 MPa
Poisson coefficient ν	0.35
Intact material friction angle (peak) φ^p	28 °
Discontinuity surface friction angle (residual) φ^f	20 °
Peak cohesion c^p (intact material)	0.33 MPa
Critical slip value δ^*	1div20 mm

Table 2. Sensitivity analyses: mechanical and geometrical characteristics of the specimen

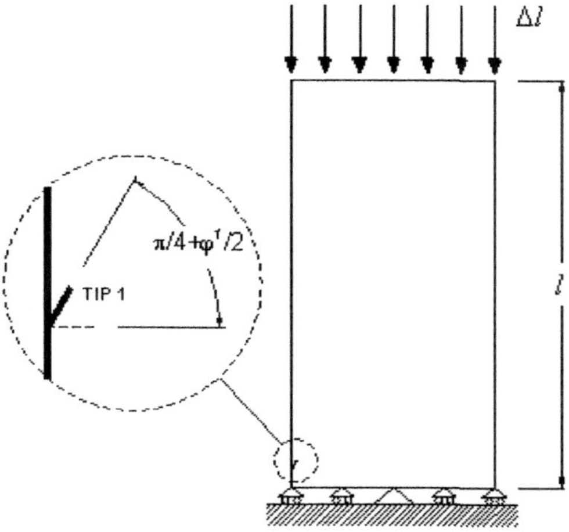

Fig. 14. Sensitivity analyses: geometrical scheme of the uniaxial compression test

The simulations are carried out by prescribing, at the upper surface of the specimen, constant displacement steps Δl in the longitudinal direction. The lower surface is constrained in order to avoid rigid motions caused by numerical asymmetries. At each step, the corresponding mean axial stress σ_m acting on the upper surface is computed. Shear propagation is triggered at the bottom left of the specimen, where some DDEs with the mechanical characteristics of the intact material and an inclination to the horizontal $\alpha = \pi/4{+}\varphi^f/2$ are located in order to simulate a defect in the material.

The results are shown in Fig. 15. Whatever the critical slip value, the triggering occurs (point A) for an applied displacement equal to Δl = 1.3 mm ($\varepsilon^A = (\Delta l/l)\%$= 1.08%; σ_m^A = 0.99 MPa). For $\sigma_m = \sigma_m^P$ = 1.15 MPa (ε^P= 1.25%), the propagation of the process zone becomes unstable and the shear band evolves at constant applied displacement (with inclination $\alpha = \pi/4{+}\varphi^f/2$ to the horizontal) until the intersection with the specimen edge occurs. Once the shear band reaches the edge, the propagation has completely occurred and the applied displacement increases to let the real crack evolve.

From Fig. 15 it can be seen that the stress-strain behavior after the peak becomes more and more ductile with the increment of the critical slip value. As an example, for a critical slip value equal to δ^* =1 mm the curve is vertical (Fig. 16a), while, for δ^* =5 mm, after a short vertical branch, it shows a softening behavior (Fig. 16b). The vertical branch of the curve indicates that probably a snap-back occurs. At the moment such behavior cannot be followed by the numerical model.

Fig. 17 illustrates the deformed specimen at points A (triggering of the shear band), B (triggering of the real crack) and C (failure of the specimen) with regard to the case depicted in Fig. 16a (δ^* =1 mm). The process zone extension ω at point B is equal to 67 mm.

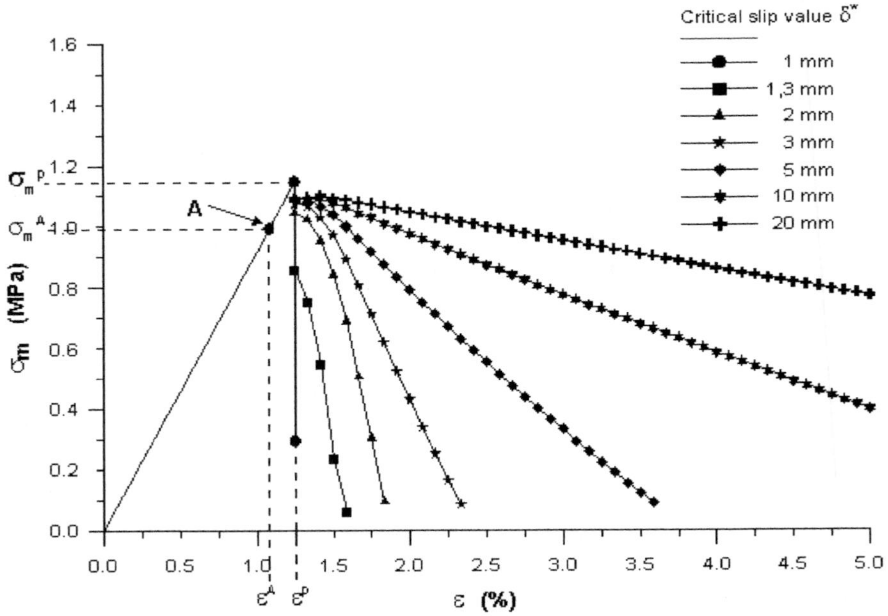

Fig. 15. Sensitivity analyses: stress-strain behavior resulting from the numerical tests

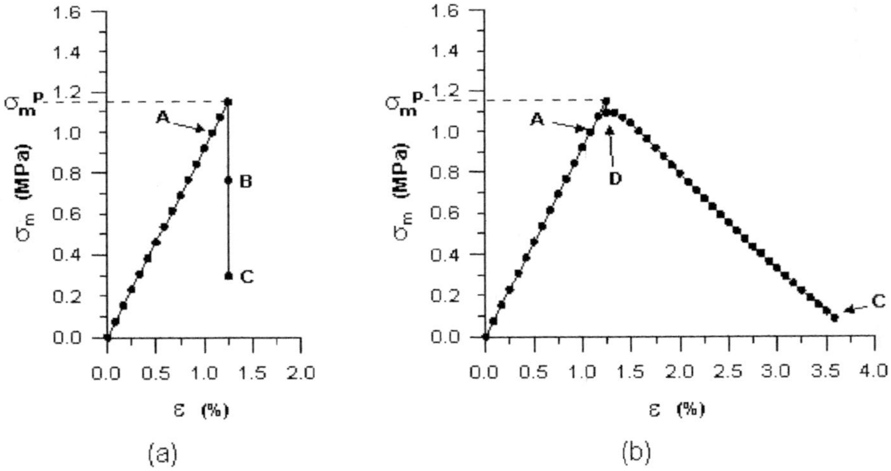

Fig. 16. Sensitivity analyses: comparison between the stress-strain behaviors: **(a)** δ^{*}=1 mm; **(b)** δ^{*}=5 mm

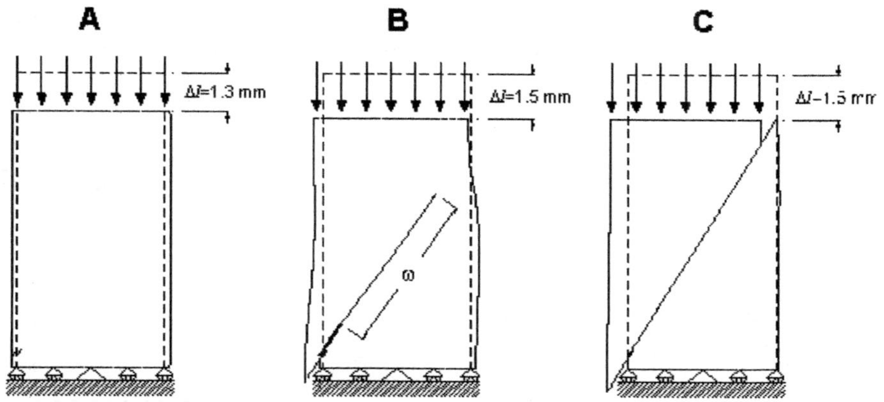

Fig. 17. Sensitivity analyses: deformed specimen at points A, B and C in Fig. 16a (deformations are magnified by a factor of 12)

In the case depicted in Fig. 16b ($\delta^* = 5$ mm), when the process zone intersects the specimen edges, displacements along the discontinuity are still too small to trigger the real crack, which therefore develops only after intersection (point D).

Although the work is at its initial stage, the first results seem promising in the simulation of shear band formation in a rock-like material in compression. However, the problem of the simulation of a snap-back behavior after the peak is still open. Due to this kind of extremely brittle behavior (Wawersik and Fairhurst 1970) in fact, energy actually must be removed after the peak, and the displacement decreases, with the softening branch taking on a positive slope. This cannot be simulated by the proposed method.

2.4 Continuum equivalent approach: stiffness modulus of cracked rock volumes

When using a continuum equivalent approach, problems arise for the choice of the mechanical parameters (strength, deformability) which are affected by strong scale effects (Pinto da Cunha 1993). In fact, the results of laboratory tests conducted on small sized specimens, containing only micro cracks, are different from those of large scale tests because of the presence of natural discontinuities of different orders of magnitude in large rock volumes. For this reason, the parameters determined in the laboratory cannot be used directly in the analyses of real rock structures. In this case, in order to determine the continuum equivalent mechanical parameters, the Representative Elementary Volume (R.E.V.), that is the minimum size of volume which can be taken as representative of the mass at a certain scale, have to be evaluated.

The problem is analyzed here with reference to the stiffness modulus of the rock mass and through the Displacement Discontinuity technique described before.

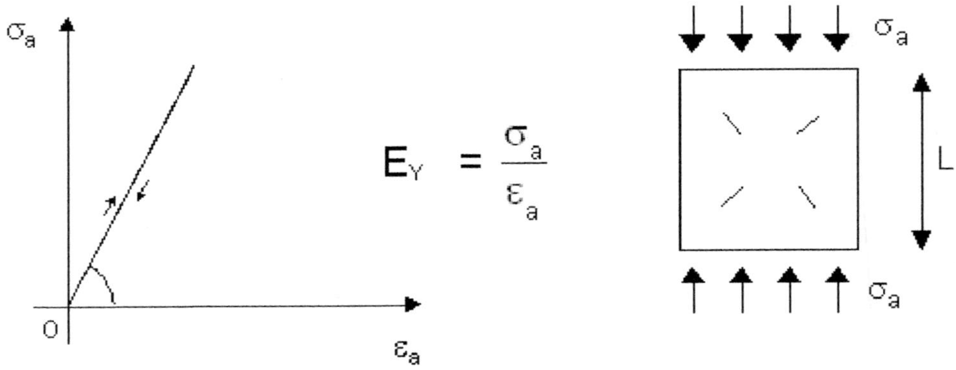

Fig. 18. Example of a specimen subjected to uniaxial compressive load and elastic axial stress-strain behavior of the specimen before crack propagation

The stiffness modulus is determined through a numerical simulation by applying a uni-axial compressive load on square shaped specimens that contain a finite number of discontinuities (cracks), by assuming a plane strain state of deformation (Fig. 18). The behavior of the material is assumed to be linear-elastic. The cracks have all the same length and their centers are located at the nodes of a square mesh; they are set at a random inclination for each generated specimen, and do not intersect each other.

The stiffness modulus is usually determined in situ or by laboratory tests at low load levels so that the applied load-global displacement relationship tends to be linear. This means that the surfaces of the discontinuities contained in the rock mass can be in plastic conditions, but the discontinuities themselves have not yet propagated (Jaeger and Cook 1976). For this reason no simulation of propagation is carried out during the numerical test.

In these conditions, the global behavior of the cracked solid can be considered isotropic. The authors are fully aware that the hypotheses of non intersecting and randomly oriented discontinuities are simplifications of the real geometry of the discontinuities in rock masses. However, these simplifications are considered essential to compare the results obtained at different scales among them and with the analytical solution. Furthermore, open or closed cracks can be considered. At the end of the axial load application, the mean value of the displacements measured in a number of elements located along the upper (u_t) and lower (u_b) bases of a specimen with side L is determined. The strain, σ_a, in the direction of application of the load is calculated as:

$$\varepsilon_a = \frac{|u_t| - |u_b|}{L} \tag{11}$$

from which the global stiffness modulus, $\overline{E}$ of the cracked specimen is obtained as:

$$\overline{E} = \frac{\sigma_a}{\varepsilon_a} \tag{12}$$

where σ_a is the level of applied axial stress for which $\overline{E}$ has been determined (Fig. 18).

For the sake of simplicity reference will be made, in the following, to the stiffness modulus in plane strain condition.

Procedure for the determination of the stiffness modulus of the Representative Elementary Volume (REV)

With regard to rock volumes containing open cracks, stiffness moduli are first computed through an analytical solution (Nemat-Nasser and Hori 1993). The analytical procedure is based on the Dilute Distribution method, that is extended to a system of cracks with density D which is defined as (Yin and Ehrlacher 1996):

$$D = \frac{N \cdot l^2}{4 \cdot L^2} \tag{13}$$

where l is the length of the cracks, N is the total number of cracks contained in the specimen with side L assumed to be square shaped. The two-dimensional cracks have the same length l, do not intersect, and have random dips.

The expression of the stiffness modulus of a cracked volume, $\overline{E}$, normalized with respect to the value of the stiffness modulus of the rock matrix (i.e. of the material without cracks), E, is given by:

$$\frac{\overline{E}}{E} = \left[1 + D\left(1 - \nu^2\right)\pi\right]^{-1} \tag{14}$$

This equation, where ν is Poisson's ratio, applies to randomly arranged cracks, in the assumption of a plane strain state of deformation.

The analytical solution, which assumes that only normal opening DDs can develop in tensile stress fields, can also be used in compressive stress fields and for normal closing DDs. In other words, by analogy with the DDM, it is assumed that the flat cracks are sufficiently open, in their undeformed conditions, to prevent the crack surfaces from overlapping.

The problems associated to the determination of the stiffness modulus that has to be attributed to the REV through the DDM in the case of open cracks can be shown by resorting to the following examples. Assuming a fracture density value $D = 0.11$, different sets of cracked volumes with the geometric characteristics summarized in Table 3, are produced where, with increasing rock volume size, the crack length l is constant and the number of cracks increases. For the sake of simplicity, the value of l has been kept constant to avoid the introduction of the parameters that describe the length variations into the analyses. However, no conceptual difficulty exists for the REV evaluation if one assumes, for instance, different density functions of the crack length.

An example of the random generation of cracks in volumes of increasing size is illustrated in Fig. 19. The stiffness modulus of the rock matrix E is assumed to be 10^5 MPa. Each specimen is obtained with a random distribution of crack dips.

The axial load application ($\sigma_a = -10$MPa) is simulated under the assumption of the cracks remaining open and $\overline{E}$ is evaluated. The results are given in Fig. 20 and Table 3.

From the examination of Fig. 20, it can be seen that the scatter in the results decreases with increasing specimen size. In particular, scatter is high for volumes that contain a few cracks and decreases with increasing the number of cracks. This depends on the fact that, in a volume containing a small number of cracks, the dip of each individual crack has decisive effects on the value of the stiffness modulus. As the number of cracks increases, the dip of individual discontinuities becomes less influential. It can be observed that the scatter in the results decreases but does not vanish. This means that the REV size has not been reached for the two considered fracture density values.

Fig. 20 also give the arithmetic mean of the stiffness modulus for each set of analyses and the value of $\overline{E}$ determined through the analytical solution (Eq. 14) for a random distribution of cracks. It can be seen that the mean value of $\overline{E}$ for volumes that are not representative of the REV comes close to the theoretical value. This fact suggests the possibility of determining the stiffness modulus of a cracked rock volume by resorting to the mean value of the stiffness moduli of a set of specimens containing a limited number of cracks. This consideration allows to skip the direct computation of the REV stiffness modulus which is very time-consuming because of the high number of cracks, represented by DD elements.

The same numerical analyses have been carried out assuming $D = 0.22$, maintaining the number of cracks and their length equal to those of the case of $D = 0.11$. The results show that the distribution of $\bar{E}/E$ is similar to the one shown in Fig. 20, obtained for $D = 0.11$, even though the values of the stiffness modulus are lower on account of the higher value of fracture density.

Numerical analyses have also been performed, using the same procedure, for closed cracks by selecting a fracture density value $D = 0.083$ and speci-

Rock volume size L (mm)	Crack length l (mm)	No. cracks (-)	No. specimens (-)	Mean $\bar{E}$ (MPa)	s. d. $\bar{E}$ (MPa)	Analytical value (MPa)
9	3	4	80	75933	6404	
13.5	3	9	30	79760	4223	
31.5	3	49	15	80893	2578	75089
49.5	3	121	15	79445	1218	
76.5	3	289	6	76315	1621	

Table 3. Geometric characteristics of the rock volumes ($D = 0.11$, open cracks)

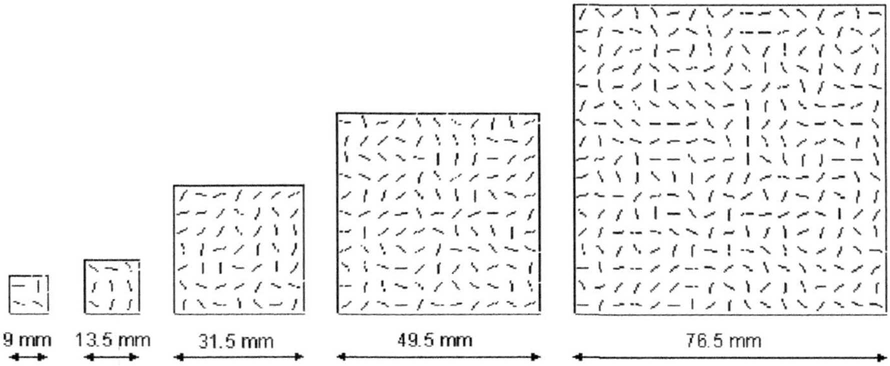

Fig. 19. Generation of specimens for $D = 0.11$ (open cracks)

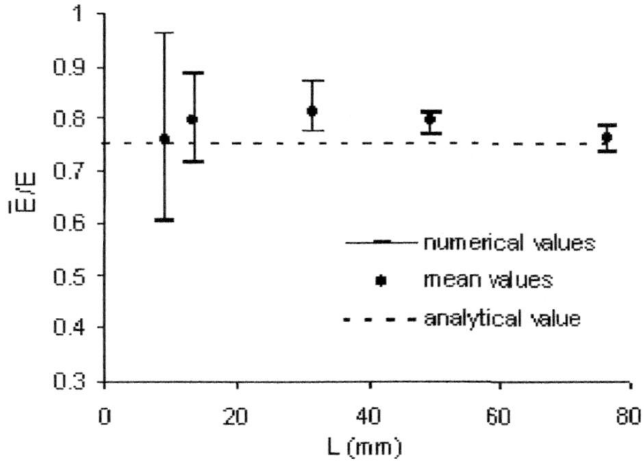

Fig. 20. Variation in the ratio between the cracked ($\overline{E}$) and the uncracked (E) moduli as a function of rock volume size, $L(D = 0.11$, open cracks)

mens whose characteristics are listed in Table 4. An example of the random generation of cracks inside volumes of increasing size is shown in Fig. 21.

The results are given in Fig. 22 and Table 4.

The distribution of $\bar{E}/E$ in Fig. 22 is similar to the curve that are depicted in Fig. 20. In this case, it proves impossible to perform an analytical comparison, as analytical solutions for closed cracks are not available. However, by analogy with open cracks, the mean value of $\overline{E}$ can be attributed to the REV with a reasonable degree of approximation.

Rock volume size L (mm)	Crack length l (mm)	No. cracks (-)	No. specimens (-)	Mean E (MPa)	s. d. E (MPa)	Analytical value (MPa)
6.4	3	4	80	52993	10918	
9.6	3	9	30	60282	8359	
22.4	3	49	15	59079	4637	60144
35.2	3	121	15	59192	2665	
54.4	3	289	6	56297	1538	

Table 4. Geometric characteristics of the rock volumes ($D = 0.083$, closed cracks)

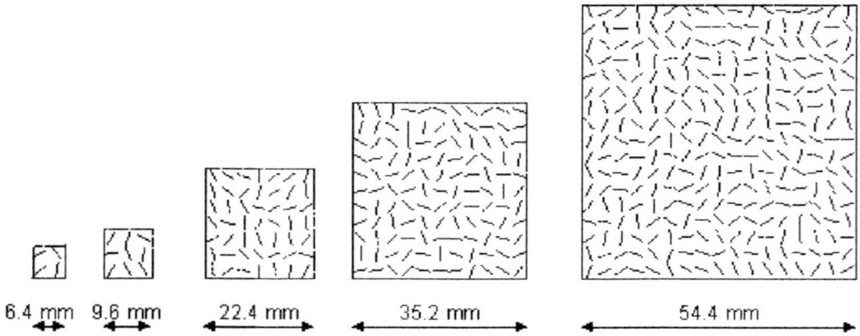

Fig. 21. Specimen generation for $D = 0.083$ (closed cracks)

Evaluation of scale effects on the stiffness modulus

The study is performed by assuming that scale phenomena are due to the presence of large sized discontinuities in the rock mass, in addition to the small cracks that are included in the laboratory specimens. Accordingly, with increasing volume size, the fracture density also increases. No other possible causes of scale effects are considered here. For the sake of simplicity, the

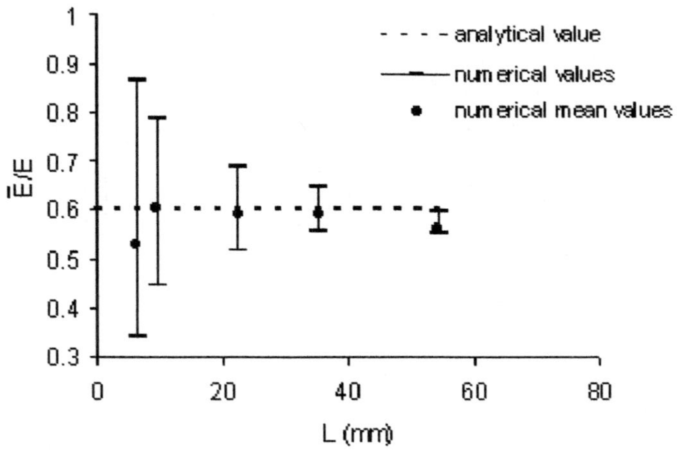

Fig. 22. Variation in the ratio between the cracked ($\overline{E}$) and the uncracked (E) moduli as a function of rock volume size, $L(D = 0.083$, closed cracks)

mechanical properties of the discontinuities are assumed not to vary for any size, that is, from micro cracks to large natural discontinuities. However, a variation in friction angle with the size of the discontinuities could easily be included in the numerical analyses.

Several statistical distributions of the length of the discontinuities in rock masses can be assumed (negative exponential, log-normal, power function...). Here a truncated negative exponential statistical distribution is assumed, according to the results of several observations on real length distributions in rock masses (Robertson 1970; Call et al. 1976; Priest and Hudson 1981). In this case the probability density function is expressed as follows:

$$\mathrm{d}\,(l) = \frac{\mathrm{a}}{\mathrm{e}^{-\mathrm{a}\cdot l_{\mathrm{min}}} - \mathrm{e}^{-\mathrm{a}\cdot l_{\mathrm{max}}}} \cdot \mathrm{e}^{-\mathrm{a}\cdot l} \tag{15}$$

with $l_{\mathrm{min}} \leq l \leq l_{\mathrm{max}}$, where l_{min} and l_{max} are the minimum and maximum length of the cracks observed in the rock mass and a is the inverse of the mean of the distribution $\overline{l}$.

Thus, the study of scale effects is conducted by determining the mean value $\overline{E}$ on rock volumes of increasing size, i.e., from the size of the laboratory specimen to that of the rock mass (Fig. 23).

The laboratory specimen is defined as follows: crack lengths smaller than the minimum value (l_{min}) of the truncated negative exponential distribution, crack density $D_0 = 0$, stiffness modulus $= E$. The rock mass is represented by the entire distribution of the lengths of the cracks. The intermediate rock volumes are characterized by intervals of the same negative exponential distribution. These intervals are defined by the minimum length value, l_{min}, and

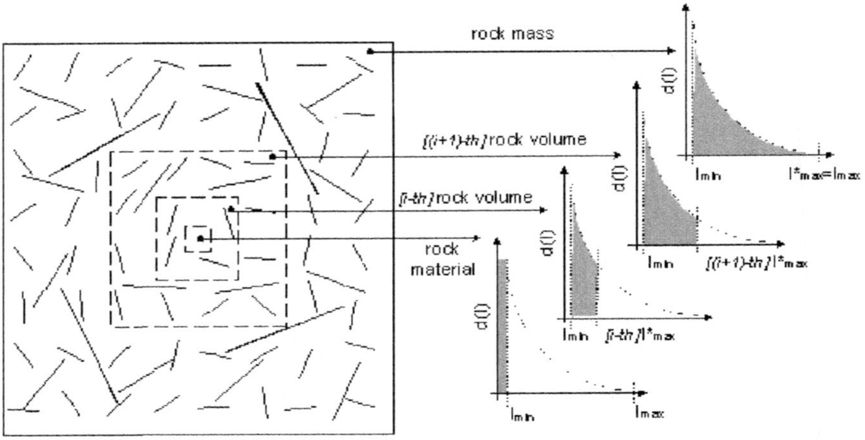

Fig. 23. Rock volumes of increasing size and the respective lengths of the discontinuities

a maximum value, l^*_{max}, which increases with the size of the considered rock volume up to the maximum length of the discontinuities observed in the rock mass, l_{max}.

The fracture density of a rock mass that contains a number N_{TOT} of cracks of n different lengths is obtained starting from Eq. 13:

$$D = \frac{\sum\limits_{\alpha=1}^{n} N_\alpha \cdot l_\alpha^2}{4 \cdot L^2} \tag{16}$$

where:

L is the size of the rock mass (assumed to be square shaped)
l_α is the length of the cracks
N_α is the number of cracks of length l_α:

$$\sum_{\alpha=1}^{\text{n}} N_\alpha = N_{\text{TOT}} \tag{17}$$

In the case of a continuous distribution of crack lengths, it proves necessary to refer to the frequency density distribution d(l) of the lengths of the discontinuities (Eq. 15). If N_α is set equal to N_{TOT}·d(l), the fracture density of the rock mass will be:

$$D = \frac{N_{TOT}}{4 \cdot L^2} \cdot \frac{a}{e^{-a \cdot l_{\min}} - e^{-a \cdot l_{\max}}} \cdot \int_{l_{\min}}^{l_{\max}} \left(e^{-a \cdot l} \cdot l^2\right) \cdot dl \tag{18}$$

By solving the integral, Eq. 18 becomes:

$$D = -\frac{N_{TOT}}{4 \cdot L^2} \cdot \frac{l}{e^{-a \cdot l_{\min}} - e^{-a \cdot l_{\max}}} \cdot \left[e^{-a \cdot l_{\max}} \cdot \left(l_{\max}^2 + \frac{2}{a} \cdot l_{\max} + \frac{2}{a^2} \right) - e^{-a \cdot l_{\min}} \cdot \left(l_{\min}^2 + \frac{2}{a} \cdot l_{\min} + \frac{2}{a^2} \right) \right] \tag{19}$$

hence, for the i-th interval $(l_{min}, l^*_{max,i})$, that is, for the i-th intermediate rock volume, crack density, D_i, is given by:

$$D_i = -\frac{N_{TOT}}{4 \cdot L^2} \cdot \frac{l}{e^{-a \cdot l_{\min}} - e^{-a \cdot l_{\max}}} \cdot \left[e^{-a \cdot l^*_{\max,i}} \cdot \left({l^*_{\max,i}}^2 + \frac{2}{a} \cdot l^*_{\max,i} + \frac{2}{a^2} \right) - e^{-a \cdot l_{\min}} \cdot \left(l_{\min}^2 + \frac{2}{a} \cdot l_{\min} + \frac{2}{a^2} \right) \right] \tag{20}$$

In the assumption of closed cracks, the stiffness modulus, $\overline{E}$, of the i-th rock volume, with fracture density D_i, is determined through the DDM as follows. For the i-th interval, the value of the fracture density D_i, is calculated from (20); a value of crack length, l_i^*, comprised in the interval, is selected. Then, an i-th fictitious rock volume containing 9 random oriented cracks is produced whose side L_{Di} is calculated from Eq. 13:

$$L_{\mathrm{Di}} = 1.5 \cdot l_{\mathrm{i}}^* \cdot D_{\mathrm{i}}^{-1/2} \tag{21}$$

For each i-th fictitious rock volume, a series of simulations on specimens with side L_{Di}, 9 random cracks of length l_i^* are carried out, to determine the stiffness moduli and their mean. The authors are conscious that the low number of cracks is somewhat arbitrary, however this simplification makes it possible to perform a considerable number of numerical analyses whilst ensuring that the arithmetic mean value comes sufficiently close to the REV value. Thus, the obtained mean value represents the stiffness modulus that has to be attributed to the REV, following the assumptions made in the previous sections.

To define the scale effect, it is necessary to link the crack length (and the fracture density) to the correspondent real rock volume size (Fig. 23). This size is different from the fictitious rock volume because it is a function of the maximum length of the i-th interval l^*_{max}. The connection is made by assuming that the size of the real rock volumes corresponds to the size of the REV with regards a given fracture density and crack length. On the basis of the numerical simulation carried out by Yin and Ehrlacher (1996), a REV side (L_i) is selected for each i-th interval equal to ten times the crack length l^*_{max}. The variation in the mean values of the stiffness modulus as a function of L_i is taken to be the scale effect.

The stiffness modulus of each rock volume is calculated analytically for volumes containing open cracks, through Eq. 14, the fracture density being calculated through Eq. 20.

Explanatory example. For a clearer explanation of the foregoing procedure, an application to a rock mass having the geometrical and mechanical characteristics listed in Table 5 is given.

The truncated negative exponential distribution of the discontinuities in the rock mass is shown in Fig. 24. 5 rock volumes characterized by the crack length intervals and the relative fracture density values listed in Table 6 have been considered. The stiffness modulus of each volume is determined by performing a series of 30 analyses on 30 specimens including 9 cracks of equal length, l^*_i, with random dips and side L_{Di} as determined from Eq. 21. Table 7 gives the geometric characteristics of the specimens and the results of the numerical analyses.

The last series of 30 analyses, included in Table 7 (series 6), is performed on specimens with the same crack length interval as series 5 by assuming a different value of l^*_i (i.e., corresponding to the mean of the negative exponential statistical distribution).

Total number of cracks contained in the rock mass, N_{TOT}	1000
Size of the rock mass containing N_{TOT} cracks, L	2 m
Minimum length of the cracks in the rock mass, $l_{\min}$	10 mm
Maximum length of the cracks in the rock mass, $l_{\max}$	200 mm
Mean value of the statistical distribution of crack length, $\bar{l}$	40 mm
Stiffness modulus of the rock material, E	100000 MPa
Poisson's coefficient of the rock material, ν	0.2
Friction angle between the surfaces of the crack, φ	45 °
Stiffness of the crack surfaces, K_s	1 MPa /mm
Applied load, σ	-10 MPa

Table 5. Explanatory example: geometrical and mechanical characteristics

Figs. 25 and 26 show the intervals of the negative exponential distribution for the series of analyses 1 and 5, and one of the 30 specimens included in each considered series. For each specimen, the cracks are generated through a random process.

The diagrams in Figs. 27 and 28 show the values of the $\overline{E}/E$ ratios and the correspondent mean values (listed in Table 7), as determined through the numerical method for each set of specimens as a function of fictitious rock volume size and fracture density.

Volumes	l_{min}	l^*_{max}	D
(-)	(mm)	(mm)	(-)
1	10	50	0.033
2	10	80	0.083
3	10	110	0.133
4	10	150	0.186
5	10	200	0.223

Table 6. Explanatory example: geometric parameters of the natural discontinuities in the considered volumes

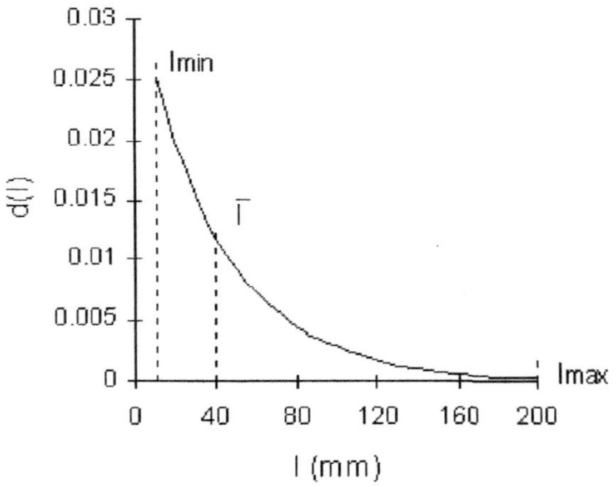

Fig. 24. Explanatory example: truncated negative exponential curve for the lengths of the discontinuities contained in the rock mass

Each mean value is taken to be representative of the value that would be obtained for the RVE. Mean values are seen to decrease with increasing fracture density (which entails an increase in the size of the fictitious rock volume size analyzed). Furthermore, for low density values, the scatter in the results is small and the values are close to the modulus for the rock material; this is mainly due to the fact that the cracks are far apart and have limited influence on the state of stress and strain. Conversely, the influence of the cracks becomes very significant with increasing fracture density, and crack dip and length play a decisive role in the evaluation of the stiffness modulus, resulting in higher scatter.

Diagrams 27 and 28 also show the values of the $\overline{E}/E$ ratios obtained for the two sets of specimens with the same fracture density (series of analyses 5 and 6). The comparison confirms that the variation in the stiffness modulus does not depend solely on crack length, but it is affected by fracture density, also when the cracks are closed.

The same example is repeated in the assumption of open cracks. Table 8 gives the intervals in which the negative exponential distribution curve is subdivided, the fracture density values and the stiffness modulus of rock volumes, as determined through Eq. 18.

In Fig. 29 the results obtained analytically for open cracks are compared with the numerical results concerning closed cracks (in the assumption of L_i=10 $l^*_{\max}$,). Again the variation in the mean values of the stiffness modulus as a function of L_i is taken to be the scale effect.

In this application example, the rock mass stiffness value obtained is seen to be reduced by about 40% in the assumption of open cracks and by about 15% in the assumption of closed cracks, with respect to that of the rock material.

Series of analyses (-)	$l_{\min}$ (mm)	$l^*_{\max}$ (mm)	D (-)	l^*_i (mm)	L_{Di} (mm)	Mean $\bar{E}$ (MPa)
1	10	50	0.033	25	204	97713
2	10	80	0.083	45	234	94194
3	10	110	0.133	60	246	91090
4	10	150	0.186	85	294	88110
5	10	200	0.223	105	333	85755
6	10	200	0.223	40	126.6	85405

Table 7. Explanatory example: geometric parameters of the specimens subjected to numerical analyses and the mean of the $\bar{E}$ values

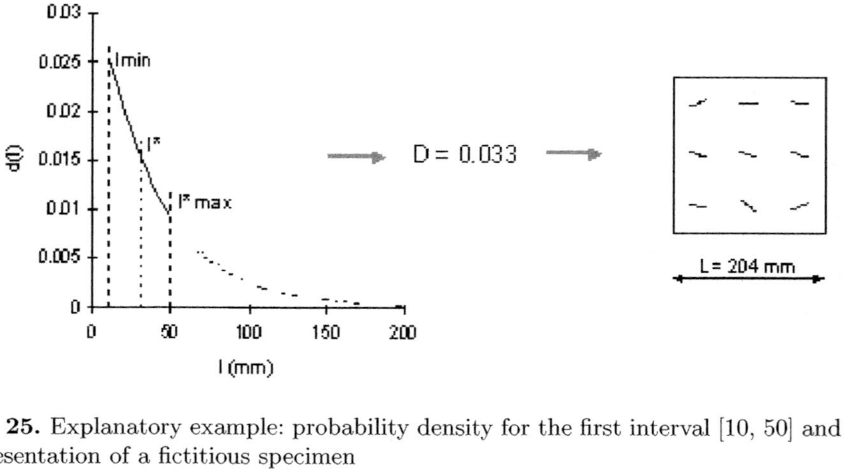

Fig. 25. Explanatory example: probability density for the first interval [10, 50] and representation of a fictitious specimen

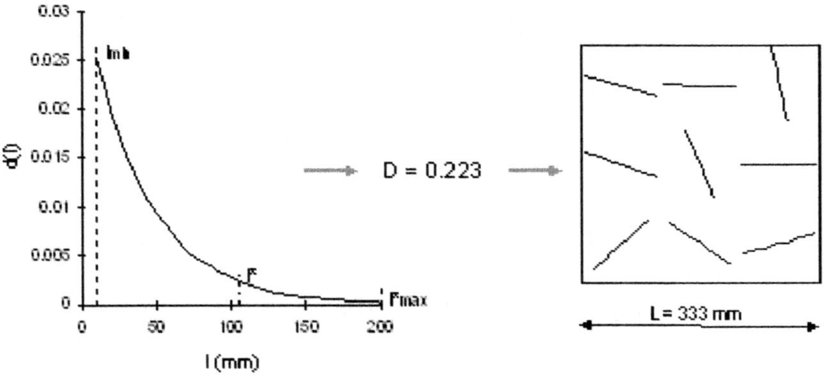

Fig. 26. Explanatory example: probability density for the sixth series of analyses, [10, 200] interval with l^*=105 mm, and representation of a fictitious specimen

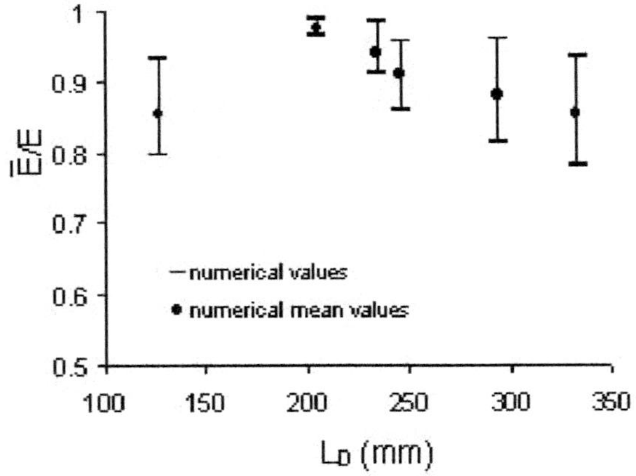

Fig. 27. Explanatory example: variation of $\overline{E}/E$ as a function of the fictitious rock volume size, L_D

Volume series	$l_{\min}$ (mm)	$l^*_{\max}$ (mm)	D (-)	$\overline{E}$ (MPa)
1	10	50	0.033	90992
2	10	80	0.083	79987
3	10	110	0.133	71480
4	10	150	0.186	64407
5	10	200	0.223	59702

Table 8. Negative exponential distribution intervals and values of the stiffness modulus analytically determined (open cracks)

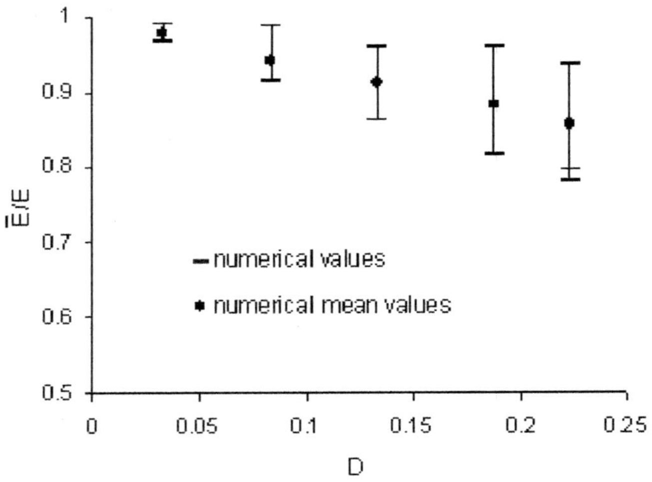

Fig. 28. Explanatory example: variation of $\overline{E}/E$ as a function of fracture density D

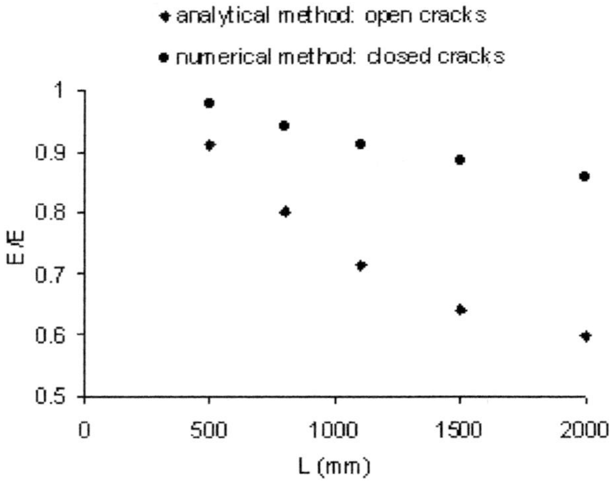

Fig. 29. Variation in the values of the stiffness modulus, $\overline{E}/E$, as a function of the REV size L: comparison between the results obtained for open and closed cracks

Some unsolved problems

The investigation on scale effects on the stiffness modulus of rock masses as a function of the increase in the discontinuity lengths and the fracture density with an increase in the size, shows that stiffness decreases for both open and closed cracks with increasing rock volume size.

Some unsolved problems still exist. In particular, it proved impossible to perform numerical analyses on volumes of the same size as the REV, on account of the great number of cracks that must be taken into account. A methodology to determine the REV with respect to the problem significant dimensions would therefore be of great interest.

However, a comparison with the analytical solution for open cracks has shown that the value of the stiffness modulus of the REV can be approximated, with sufficient accuracy, by the mean values obtained from specimens with the same fracture density but containing a smaller number of discontinuities.

A similar methodology is needed for the evaluation of rock mass strength characteristics. Experimental evidences show in fact that failure of rock structures is due to the unstable propagation and coalescence of discontinuities. The difficulty to simulate crack intersections due to discontinuity propagations makes the proposed method inappropriate for such purposes.

3 Conclusions and further developments

A discontinuum method for the analysis of cracked rock structures has been presented, and two different approaches have been discussed as examples of application to rock structures: the propagation of natural discontinuities in rock masses (discontinuum approach) and the determination of the stiffness parameters to be assigned to a continuum equivalent rock mass (continuum equivalent approach).

With regard to the discontinuum approach, the numerical implementation of the Slip-Weakening Model has been proposed for the simulation of the propagation of shear bands, and the influence of the critical slip value on the stress-strain behavior of the specimen and the process zone extension has been discussed, highlighting the open questions that still exists: the difficulty in the simulation of a very brittle snap-back behavior.

With regard to the continuum equivalent approach, an investigation on scale effects on the stiffness modulus of rock masses has been presented. Again, the results shows that some unsolved problems still exists. In particular, a methodology is needed for the determination of the REV with respect to the problem significant dimension. Finally, the application of the same approach to the evaluation of the strength characteristics of rock masses should take into account the possibility of crack propagation and coalescence in the proximity of failure.

References

[1] Allodi A, Castelli M, Scavia C (2002) Implementation of the Slip-Weakening Model in a Displacement Discontinuity Method based numerical technique. In: Proc Int Conf BEM 24, 17-19 June 2002, Sintra, Portugal

[2] Bieniawski ZT (1974) Estimating the strength of rock materials. J South Africa Inst Min Metall, vol 74, pp 312 320

[3] Call RD, Savely J, Nicholas DE (1976). Estimation of joint set characteristics from surface mapping data. In: Proc19th US Symposium on Rock Mechanics

[4] Crouch SL, Starfield AM (1983). Boundary Element Methods in Solid Mechanics. George Allen and Unwin, London

[5] Desrues J (1984). La localisation de la déformation dans les matériaux granulaires (in French). Thèse de Doctorat es Sciences, USMG & IMG, Grenoble

[6] Desrues J (1995). Analyse stéreophotogrammétrique de la fissuration progressive (in French). Photomécanique 95, Berthaud Y, Paraskevas D, Taroni M (eds), Eyrolles, Paris, pp 149 162

[7] Erdogan F, Sih GH (1963). On the Crack Extension in Plates under Plane Loading and Transverse Shear. J of Basic Eng, ASME, vol 85, pp 519 527

[8] Hoek E, Brown ET (1980). Underground excavation in rock. J South Africa Inst Min and Metall, George Allen and Unwin, London

[9] Ingraffea AR (1987). Theory of Crack Initiation and Propagation in Rock. Fracture Mechanics of Rock, Academic Press, London, pp 71 108

[10] Jaeger JC, Cook NGW (1976). Fundamentals of Rock Mechanics, Chapman and Hall, London

[11] Nemat-Nasser S, Hori M (1993). Micromechanics: overall properties of heterogeneous materials. North-Holland, Amsterdam, pp 3 23, 27 31, 85 94, 113 143

[12] Palmer AC, Rice JR (1973). The growth of slip surfaces in the progressive failure of over-consolidated clay. In: Proc Roy Soc Londond A. 332, pp 527 548

[13] Pinto da Cunha A (1993). Scale effects in rock engineering. Pinto da Cunha (ed) Balkema, Rotterdam, pp 3 14

[14] Priest SD, Hudson J (1981). Estimation of discontinuity spacing and trace length using scanline surveys. Int J Rock Mech Min Sci, vol 13, pp 135 148

[15] Reches Z, Lockner DA (1994). Nucleation and growth of faults in brittle rocks. J Geophys Res, vol 99(B9), pp 18.159 18.173

[16] Robertson AmacG (1970). The interpretation of geological factors for use in slope theory. In: Proc Symp on Plannig Open Pit Mines, van Rensburg (ed), Balkema, Rotterdam

[17] Scavia C (1992). A numerical technique for the analysis of cracks subjected to normal compressive stresses. Int J Num Meth in Geomech, vol 33, pp 929 942
[18] Scavia C, Viggiani G, Castelli M, Desrues J (1997). An experimental and numerical study of shear fracture propagation in rock. In: Proc Int Symp on Deformation and Progressive Failure in Geomechanics, IS NAGOYA '97, pp 175 180
[19] Tillard D (1992). Etude de la rupture dans les géomatériaux cohésifs. Application à la marne de Beaucaire (in French). Thèse de Doctorat es Sciences, UJF, Grenoble
[20] Wawersik WR, Fairhurst C (1970). A study of brittle rock fracture in laboratory compression experiments. Int J Rock Mech Min Sci vol 7, pp 561 575
[21] Yin HP, Ehrlacher A (1996). Size and density influence on overall moduli of finite media with crack. Mechanics of materials, vol 23, pp 287-294

Similarity in soil and rock mechanics

Dimitrios Kolymbas

Institut für Geotechnik und Tunnelbau, Universität Innsbruck,
Dimitrios.Kolymbas@uibk.ac.at

1 Introduction

Why is similarity an interesting issue in mechanics and, in particular, in soil and rock mechanics? The reason is that similarity introduces equivalence classes, i.e. the considered items are grouped into classes in such a way that if item A and item B belong to the same class, then they are in a certain sense equivalent, i.e. it is sufficient to consider only item A. This item is representative of all other items of the same class. This brings in an enormous simplification. Virtually, recognition (and, possibly, also cognition) is nothing but detecting similarities.

All items that belong to the same equivalence group have in common some invariant properties and are related to eachother by some transformations (e.g. rotations) that characterize this group. *Scaling* is such a transformation. Scalar and tensorial quantities can be stretched i.e. $x \rightarrow \lambda x$, scaling is the pertinent transformation and the related similarity is called scale invariance. Scale invariant (or simply *scaling*) functions have the property $y(\lambda x) = \lambda^\alpha y(x)$, i.e. they are homogeneous.

Similarity refers not only to boundary value problems but also to materials such as soil and rock (think of isotropy groups, which are not addressed here). Materials exhibiting invariance with respect to scaling, are called self-similar or scale invariant (or simply 'scaling'). Three scalings are of particular interest:

1. Scaling of size
2. Scaling of stress
3. Scaling of time

The properties of soil and rock with respect to these scalings will be considered with reference to the mathematical properties of the underlying constitutive laws.

2 Scaling of size

A basic notion in analyzing material behaviour is the concept of a *simple material.* The upshot is that all relevant properties of the material are condensed into a material point, i.e. a zero-dimensional object. Experiments with small samples can reveal the behaviour of a material point, provided that the stress and strain is uniformly distributed within the sample (so-called element test). The same behaviour (in terms of relations between stress and strain) is then expected from larger samples or whole structures (e.g. an earth dam). This idea is the rationale of all laboratory tests in soil mechanics: Samples are extracted from the field and meticulously tested in the laboratory. The obtained results serve to predict the behaviour of large structures. Not only the material testing in the laboratory but also numerical simulations are mostly based on the concept of simple materials, i.e. on the absence of any size effect. In fact, the boundary value problems we consider consist of boundary conditions and field equations that describe the behaviour of individual *material points.*

Being widely used, the concept of simple material is tacitly taken for granted. Deviations from this behaviour are wondered. These are mainly observed in rock mechanics (the scale effect is the main open question in rock mechanics, as it refers to the transition from small rock samples to large rock mass) but also with soil. An interesting issue of simple materials is that they do not have any internal length. Assuming soil as simple material means to disregard its granular structure. This has in fact been the case for decades until the attention of the scientific community in soil mechanics was focussed at strain localization into thin shear bands (see fig. 1 and 2). A zero thickness was rejected as being connected with zero dissipation and the thickness of the shearband was commonly accepted to be ca. $10\, d_{50}$. This view was underpinned e.g. with COSSERAT theories. It is, however, interesting to note that a material may lack any internal length though being composed of grains of finite dimensions. This is the case if the maximum grain size scales with the size of the sample in such a way that the grain size distribution is scale invariant.

Fig. 1. Shear bands in nature (Photograph: *International Society of Rock Mechanics*)

Fig. 2. Shear bands in rock (Courtesy Prof. I. Vardoulakis)

2.1 Scale invariant grain size distributions

Consider a cubical soil sample with mass M_0 and volume l^3. The grain size distribution $M(\delta)$ gives the mass of grains with diameter less than or equal to δ. The distribution curve $y(\delta)$ gives the ratio $y := M(\delta)/M_0$, where δ is logarithmically plotted. We denote with $\delta_y := \delta(y)$ the grain diameter which is not exceeded by $100 \times y$ percent of the sample mass. A grain size distribution is scale invariant if it does not have any internal length. This is the case if y depends only on the ratio δ/δ_{max}, i.e. $y = \eta(\delta/\delta_{max})$.

Consider a soil sample of the mass M_0 with self-similar grain size distribution. Let the grain diameters be within the range $0 < \delta < \delta_{max}$. If we truncate this distribution by removing all grains with $\delta > \delta_1$, we obtain the grain size distribution $\eta(\delta/\delta_1)$ of the remaining sample from the initial distribution $\eta(\delta/\delta_{max})$, see also Fig. 3:

$$\eta(\delta/\delta_1) = \frac{1}{\eta(\delta_1/\delta_{max})}\eta(\delta/\delta_{max}) \qquad 0 < \delta < \delta_1 \tag{1}$$

The factor $1/\eta(\delta_1/\delta_{max})$ is to re-normalize the truncated distribution. The functional equation (1) is fulfilled by any power law

$$\eta(\delta/\delta_{max}) = (\delta/\delta_{max})^\beta \quad ,$$

where β is a real number.

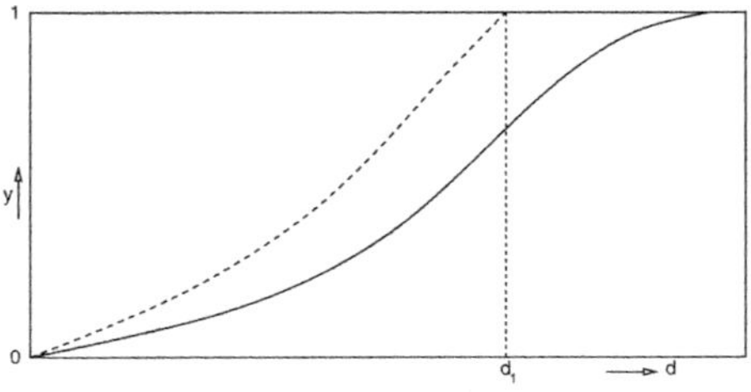

Fig. 3. Grain size distribution of a sub-sample $\delta \leq \delta_1$ (dashed)

We see thus that power laws $\left(y = \text{const} \cdot \delta^{\beta}\right)$ can represent scale invariant (or self-similar) grain size distributions. Power laws are typical for fractals, which are sets consisting of parts similar to the whole, and which can be described by a fractional dimension. In a log-log-plot this relation appears as a straight line ($\ln y = \text{const}_1 + \text{const}_2 \ln \delta$) whereas in the usual log-plot it appears as an exponential function $\left(y = \text{const}_3 \cdot 10^{\text{const}_4 \cdot \lg \delta}\right)$. In Fig. 4 are plotted in a log-log-diagramm the grain size distributions of a morainic soil and a deposited soil. It should be added, that morainic material is characterized by its geological genesis: It origins from rock broken by moving glaciers which pushed it ahead without any grain size separation. A separation can be accomplished e.g. by flowing water which exerts upon the grains drag forces that depend on their size. As a result, fluviatile sand deposits do not exhibit a power law distribution, in contrast to non-separated clastic (=fracture) sediments.

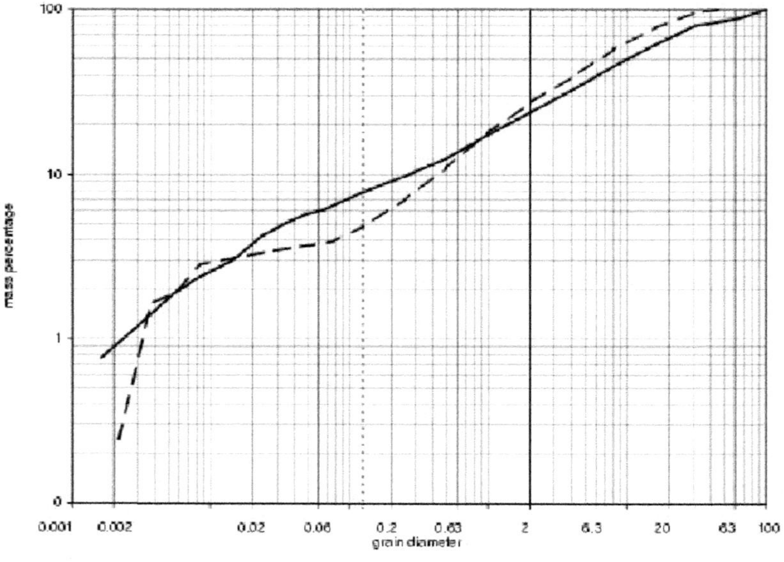

Fig. 4. Grain size distribution of morainic and deposit (dashed) soils.

TURCOTTE and HUANG have listed [1] a series of fragmented objects with fractal character, i.e. power law distributions with fractal dimensions varying between 1.44 and 3.54. See also [15]. For morainic debris has been reported $D = 2.88$ [2]. Despite some naive models there is no convincing explanation for the fact that fragmentation leads to power law distributions. It is a fact, however, that power laws are ubiquitous and govern many distributions in many different fields (e.g. economy). They are also called PARETO distributions according to the power law which describes the distribution of wealth. LÉVY distribution is yet another name for power laws.

There is also another aspect of similarity with respect to granular soils. It is accepted that the shape of the grains plains a role in the overall mechanical behaviour of a soil (see e.g. [20]) and it is known that there are several shapes of grains. Taking into account the irregularity of the grains, how can we define similarity of two grains? Of course, such a similarity can only exist in a statistical sense. Several quantities have been introduced to indicate the shape and roughnes of a grain (or its cross section). One possibility is to evaluate the FOURIER spectrum of the cross section given in polar coordinates $r(\phi)$. Another possibility is to evaluate the fractal dimension of this curve [3]. It has been found that the so obtained fractal dimension correlates with the friction angle [4].

The notion 'dimension' stems from geometrical fractals, which are geometrical objects with dimensions exceeding their topological dimension. A geometrical fractal is an object of irregular geometry whose scaling properties are described by the fractal dimension D. D can range between topologic and Euclidian dimension. E.g., a profile of a rough surface is topologically a line ($D = 1$), but is defined in Euclidian two-space, and the fractal dimension is between 1 and 2 [5]. A fractal object may be covered by small objects (boxes). The fractal dimension D describes how the total size of the set (i.e. the number of boxes) depends on the box size. To cover a fractal curve or a fractal surface with boxes of the edge δ we need N boxes (see Fig. 5). Obviously,

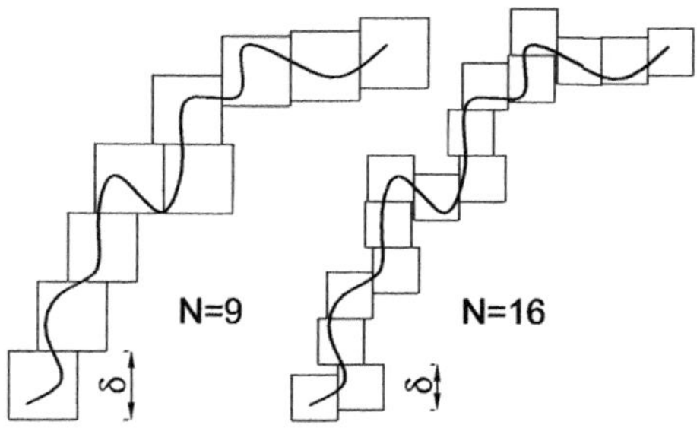

Fig. 5. The number N of boxes needed to cover a curve depends on their edge length δ.

N depends on δ: $N = N(\delta)$. The smaller δ, the larger N is. For usual (non-fractal) curves we have $N \propto \frac{1}{\delta}$ and for non-fractal surfaces we have $N \propto \frac{1}{\delta^2}$. In general $N \propto \frac{1}{\delta^D}$, where D (the so-called fractal dimension) is a fraction for fractals. E.g. the shoreline of Great Britain is reported to have the fractal dimension $D = 1.3$.

The length of a fractal curve is obtained as $L \approx N\delta = \text{const}\cdot\delta^{1-D}$. Plotting L (or N) over δ in a log-log plot gives a straight line the slope of which gives D.

The fractal nature of geological materials is revealed if we consider natural or artificial rupture surfaces. Such surfaces are self-similar in the sense that a part of them is similar to the whole: If such a surface is subdivided in smaller parts, each part looks like the original surface. This is why photographs of rock surfaces and of granular soils do not reveal the size of the objects shown, unless another object of known size (e.g. a hammer or a coin) is added. Fig. 6 gives the impression of a high mountain although it shows a sand heap of ca 20 cm height.

Fig. 6. Heap of sand (Courtesy: Prof. G. Gudehus)

The fact that rock surfaces are very often fractal is also revealed by the following fact: Rocks belong to the few objects that can be often painted better by computers than by professional painters (cf. fig. 7 - fig. 9).

2.2 Power laws

Here are some examples of power laws:

1. The relation between the cumulative number of earthquakes and their magnitude, known as GUTENBERG-RICHTER law
2. Relation between river length and area of the drainage basin
3. Relation between number of oil fields and oil volume
4. Distribution of faults. According to SORNETTE [10], however, the pure fractal description is too naive and more sophisticated quantifiers must be introduced that reconcile the existence of hierarchical structures in many scales (see fig. 10).

Fig. 7. Rock surfaces in arts (painter: F. Kontoglou)

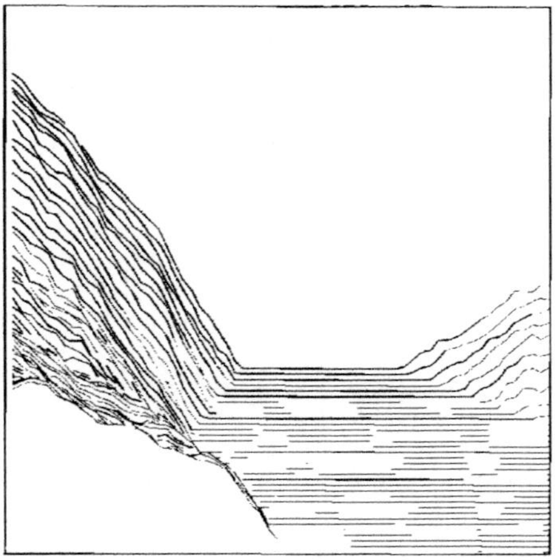

Fig. 8. Fractal rock landscape [11]

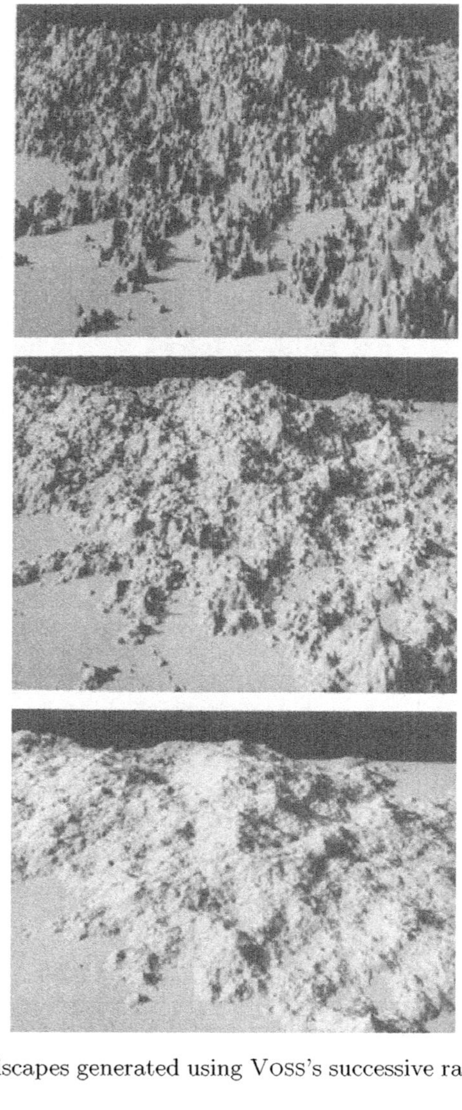

Fig. 9. Fractal landscapes generated using VOSS's successive random algorithm [11]

2.3 Size effect on mechanical behaviour

Regarding the size effect on the mechanical behaviour of soil and rock, the following aspects should be taken into account:

1. If there is a size effect upon the mechanical behaviour, our approach to problems with laboratory and computer should be thoroughly checked.
2. The small range of sample sizes accessible to laboratory testing restricts our possibility to detect a size effect.

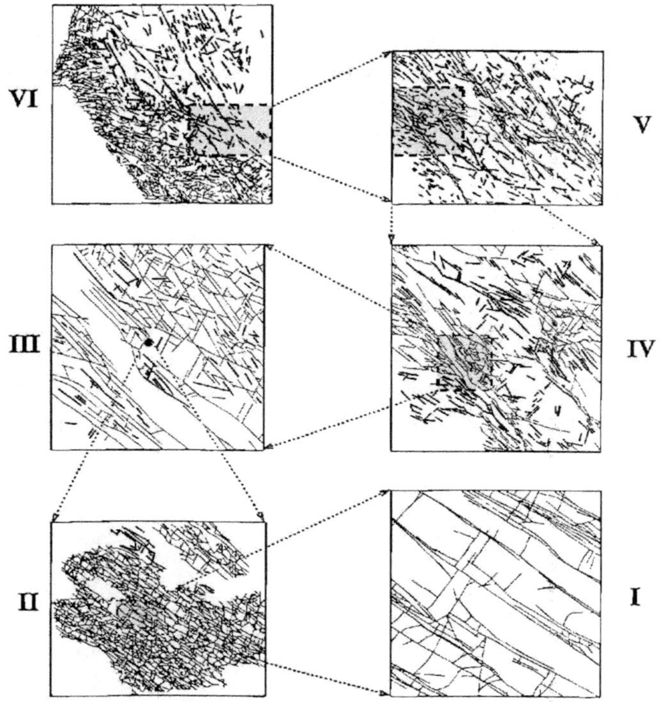

Fig. 10. Series of fracture networks from the field (satellite images of the western Arabian plate) [10]

3. Another factor that narrows our possibility to detect size effects is the large scatter of the experimental results. Especially results referring to shear strength (e.g. friction angle) should be considered with caution since they are usually influenced by the onset of bifurcations. The pattern of such bifurcations may also be size dependent.

Size effect in rock

Speaking of rock, one always has to distinguish between (intact) rock and (jointed) rock mass. In the laboratory, only intact rock is accessible to series of tests. It should be noted, however, that the long term objective is to find a scaling law that enables to predict the behaviour of rock mass from laboratory tests with a small sample of jointed rock.

Considering intact rock, it was observed that the fracture energy increases with sample size. This effect was attributed to the fractality of the fracture surface. Assuming that the fracture energy G is proportional to the fracture surface and assuming the latter proportional to d^D, where d is the diameter of the sample, we obtain that the ratio G/d^2 is not constant but increases with

d^{D-2}. Fractal rock rupture surfaces have the dimension $2 < D < 3$. The fractal dimension D can be obtained from the fractal dimension of its two-dimensional profile plus one. To obtain the latter, the roughness of the surface is scanned with a mechanical pin or with a laser beam [6], [7]. The fractal dimension of the traced curve has been obtained for marmor as: $1.11 < D < 1.76$ and for sandstone $1.02 < D < 1.41$.

Another aspect of scaling of jointed rock refers to the water inflow into tunnels. This can be a severe obstacle of tunnel heading (see fig. 11) but has also importance in assessing the environmental impact of a tunnel. The inflow is mainly governed by open joints, the location of which can hardly be predicted. Of course, only the largest joints determine the amount of inflow. If their spatial distribution obeys a power law, then it is conceivable that the water inflow per meter tunnel increases with length of tunnel. The underlying scaling law could be obtained—in principle—from a small piece of rock.

Fig. 11. Water inrush in tunnel

Size effect in soils

The cohesion of soil samples is scale dependent: smaller samples have a larger cohesion. This can be attributed to the fact that the voids of the sample cross section have a fractal dimension [8]. Plotting the cohesion of London clay over the sample diameter gives the fractal dimension $D = 1.64$. The sizes of the investigated samples ranged between 38 and 300 mm in diameter for fissured Oxford clay, and 13×25 and 300×600 mm for London clay.

As regards non-cohesive soils, it is not known yet whether the aforementioned scale invariant grain size distribution of debris implies a particular scaling of its mechanical behaviour (e.g. of the friction angle). Knowledge of this scaling would be beneficial in view of the fact that bulky debris cannot be investigated in the laboratory but is very important from a geotechnical point of view (see fig. 13-15).

Fig. 12. Water inrush in tunnel

Fig. 13. Rock debris in mountains

Fig. 14. Debris

Fig. 15. Clay particles

3 Scaling of stress

The related question is: What happens with a material if we increase all stress components by the same factor? Of course, it is reasonable to expect the density to increase with increasing hydrostatic stress. There are, however, also some further aspects. Regarding a linear-elastic material, e.g., we notice that the stiffnesses (i.e. E and/or G) are material constants and, consequently, do *not* depend on stress. A piece of rubber or steel has the same stiffness on atmospheric pressure and at the bottom of the ocean. This has far-reaching consequences: If our underground were linear-elastic (as many colleagues do assume) this would mean that a line-load (e.g. a rail road) would infinitely sink into the underground due to elastic deformations. Luckily, this is not the case. The stresses due to gravity increase linearly with depth: In a vertical depth of 1 km prevails a vertical stress of 200 at (it is remarkable that tunnelling and mining is possible in even larger depths!). However, also the stiffness increases with depth (think of how stiff is a vacuum packaged coffee powder).

Let us consider a constitutive equation of the rate type, i.e. $\overset{\circ}{\mathbf{T}} = \mathbf{h}(\mathbf{T}, \mathbf{D}, \ldots)$, and let us assume that this equation scales with respect to stress. This means that $\mathbf{h}(.)$ is homogeneous in $\mathbf{T}$, i.e.

$$\mathbf{h}(\lambda\mathbf{T}, \mathbf{D}) = \lambda^{\alpha}\mathbf{h}(\mathbf{T}, \mathbf{D})$$

Of particular interest is the case $\alpha = 1$ which implies that increasing all stress components by a factor would also increase the stiffness by the same factor. A consequence would be that normalized stress strain curves of triaxial tests run at different cell pressures would coincide. This implies that the friction and dilatancy angles are independent of the stress level. This is however not true, since both angles are considerably reduced at increased stress levels. This fact implies that the constitutive function $\mathbf{h}$ does *not* scale with $\mathbf{T}$ and constitutes the so-called barotropy of sand and explains why model tests with reduced size of sand bodies must be run within centrifuges (see fig. 16) at increased gravity (see next session). The barotropy of soils implies also that the ductility of soil is increased with increasing stress level, just as this is the case for rock (cf. the experimental results mentioned by PATERSON, see fig. 17 and 18).

Fig. 16. Centrifuge of the University of Bochum

It is interesting to observe that barotropy is inherently related to another property of granular media, the pyknotropy. In a nutshell, pyknotropy means that dense sand has a different behaviour than loose sand. This means that a quantity indicating the packing, e.g. the void ratio e, is an additional independent variable of the function $\mathbf{h}(\ldots)$. The relation between barotropy and

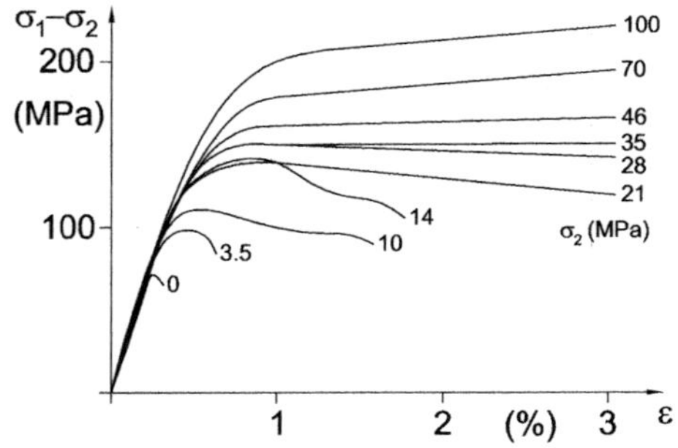

Fig. 17. Stress strain curves from triaxial tests with marble with various lateral pressures (according to PATERSON).

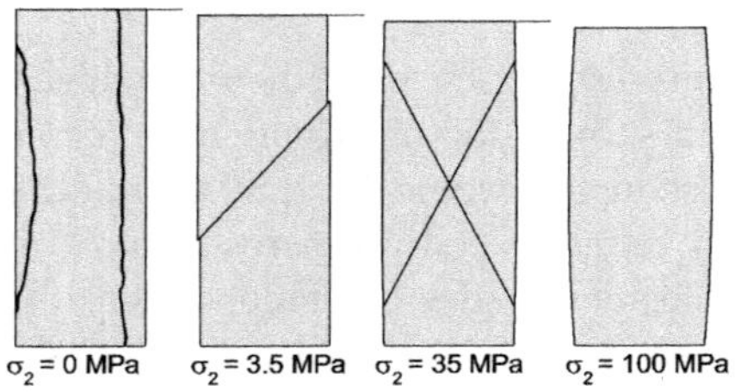

Fig. 18. Pattern of failure of marble samples with various lateral pressures (according to PATERSON).

pyknotropy is manifested by the fact that decreasing the density of sand has the same effect as increasing the stress level. The reason is that the attribute 'dense' or 'loose' is only meaningful in combination with the corresponding stress level (see fig. 19 and 20).

Another reason for the departure of granular media from scaling with stress is grain crushing. Considering oedometric compression of granular materials, it has been observed that the bending of the e vs. $\ln \sigma$ curve indicating a reduction of stiffness can be attributed to grain crushing. This occurs when σ obtains the value of the grain's *tensile* strength [9].

3.1 Centrifuges in geotechnical engineering

With centrifuges we can carry out model tests under increased gravity. In this way we can simulate the stress levels that prevail in the depth of, say, 100 m. This is, e.g., the depth of a pile foundation or the hight of an earth dam. Depending on the radius r and the angular velocity ω there can be obtained accelerations $\omega^2 r$ of several hundreds of g (e.g. $440g$ at the centrifuge of IGT/ETHZ; this centrifuge has a payload of 2 tonnes). Model tests in the centrifuge are meaningful e.g. when we analyse the stability of a slope in cohesive soil. From soil mechanics (or, alternatively from the Π-theorem) we know that the dimensionless variable $\frac{\gamma h}{c}$ plays an important role and should

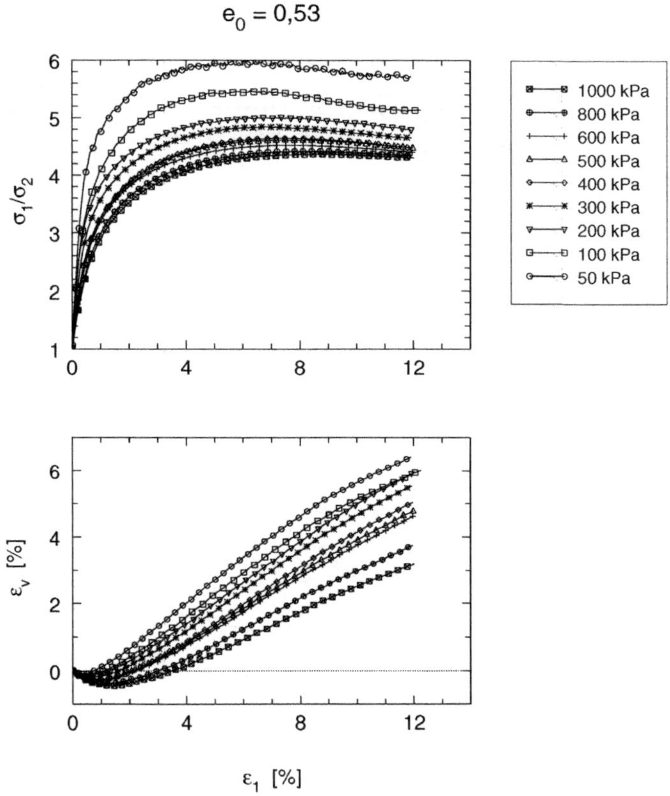

Fig. 19. Results of triaxial tests with various lateral pressures. e_0 is the initial void ratio.

therefore be identical for model and prototype. Herein c is the cohesion, γ the specific gravity and h the height of the slope. If we reduce the geometric scale by the factor $1/N$, i.e. $h_M = h_P/N$ (h_M = height of the slope in model, h_P = height of the slope in the prototype), then we have either to reduce c by the same factor $1/N$ or to increase γ by the factor N. To reduce the cohesion is difficult and it means that we have to run the test with a different material, which most probably will have completely different properties as the original one (with respect to stiffness, dilatancy etc.)[1]. Therefore it is advisable to use the same material and to increase the volume force within a centrifuge. Errors however, may result from CORIOLIS forces (if radial velocities occur in the test) and also from the inhomogeneity of volume force within the drum or basket (dependence of volume force on the radius) and the related the nonlinear

[1] A wealth of scaled model tests have been carried out in the past with artificial materials (produced using various recipes) to simulate rock or soil behaviour under reduced stress. The behaviour of materials is however so complex that it is illusory to believe that geomaterials with desired properties can ever be produced.

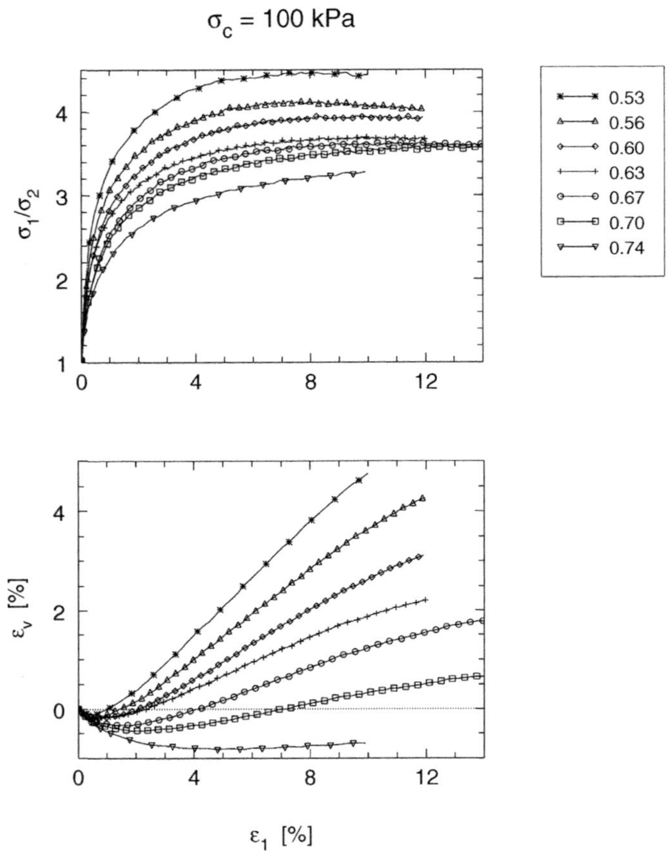

Fig. 20. Results of triaxial tests with various initial void ratios. $\sigma_c = \sigma_2$ is the lateral stress.

distribution of the vertical stress with depth (g depends on r). Bear also in mind how difficult it is to miniaturize construction machines and components within a moving centrifuge.

4 Time scale

The largest scalings in geology and geotechnical engineering refer to the time domain. A rock sample, e.g. can deformed in the laboratory with a rate of, say, 1 mm/h. The same deformation in nature may take some thousand or even millions of years, if imposed by tectonic forces, or some μs, if imposed by the impact of, say, a meteorite. It should not be wondered why the results of such slow deformations are much different than the results of fast deformation. The geology abounds of examples of folded rock strata (see fig. 21). With rates applicable in laboratories we can achieve such foldings only with viscous fluids, but not with rigid rocks.[2]

Fig. 21. Folded rock

Fig. 22. Ptygmatic vein in granite gneiss [14]

Invariance of the material behaviour (i.e. of the constitutive equation) with respect of changes of time scale is called rate independence. In terms of the constitutive equation of the type $\overset{\circ}{\mathbf{T}} = \mathbf{h}(\mathbf{T}, \mathbf{D}, \ldots)$, where $\mathbf{T}$ is the stress and $\mathbf{D}$ is the stretching[3], rate independence implies that the function $\mathbf{h}(.)$ is positively homogeneous of the first degree in $\mathbf{D}$, i.e.

$$\mathbf{h}(\mathbf{T}, \lambda\mathbf{D}) = \lambda\mathbf{h}(\mathbf{T}, \mathbf{D})$$

for $\lambda > 0$. A further implication is that this constitutive equation does not contain any material parameter with the dimension of time. The theories of elasticity and plasticity are devoted to rate independent materials. In particular, dry friction is assumed, to a first approximation, to be independent of

[2] Some remarkable simulations of folding have been obtained with dry sand.

[3] The stretching $\mathbf{D}$ is the symmetric part of the velocity gradient. For rotation-free deformations it can be set $\dot{\mathbf{E}} = \mathbf{D}$, where $\mathbf{E}$ is the logarithmic strain.

the velocity. In reality, all dissipative materials exhibit a more or less pronounced rate dependence (which is a generalized form of viscosity).[4] This fact has been since long known in soil mechanics [17] and has been recently rediscoverd by physicists [18]. PRANDTL [16] has found that shear stress due to friction depends logarithmically on the slip velocity v:

$$\sigma = \sigma_0 + c \cdot \ln(v/v_0)$$

Such a logarithmic viscosity can also be built in into constitutive equations that connect stress, stress rate and stretching. In terms of physics it can be explained by means of thermal excitation and activation energies as described by the ARHENIUS-equation. It is interesting to note that the resulting relation $\dot{\varepsilon} = a \exp(b\sigma)$ predicts a sort of yield stress, *the value of which depends on time scale*, see Fig. 23.

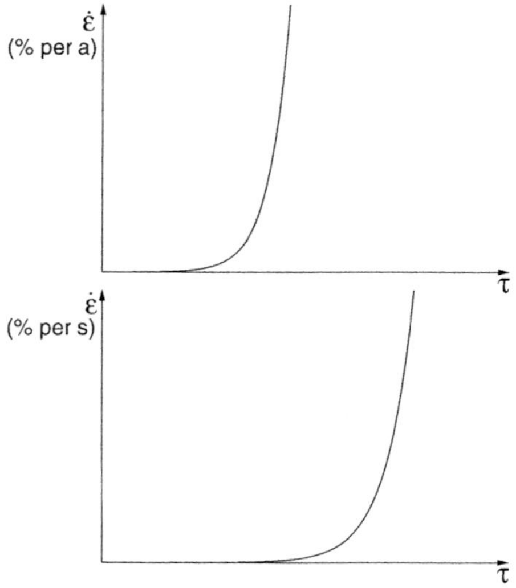

Fig. 23. 'Yield stress' in dependence of time scale ($\dot{\varepsilon} = \alpha \exp(bx)$)

The logarithmic viscosity is not only valid for dry friction between two rigid bodies. It prevails also in continuous deformation of clay. A jump of the strain rate from $\dot{\varepsilon}_a$ to $\dot{\varepsilon}_b$ implies a stress change $\Delta\sigma$ which is proportional to $\log(\dot{\varepsilon}_b/\dot{\varepsilon}_a)$. To describe the friction along faults, the law of DIETERICH-RUINA has been launched according to which *step increases in slip rates of steadily sliding surfaces lead to sudden increases in the frictional stress followed by exponential decay towards a substantially smaller value.* Its mathematical formulation reads:

[4] It should be noted that the relation between viscosity, creep and relaxation is straightforward only for the case of linear viscoelasticity.

$$\mu = \mu_0 + A \ln \frac{v}{v_0} + B \ln \frac{\theta}{\theta_0},$$

where μ is the friction coefficient and θ is a state variable. This behaviour deviates from what PRANDTL has observed and what has been observed with clay but—interestingly— describes exactly what has been observed with continuous deformation of dry sand when applying jumps of the strain rate (TATSUOKA [19], EICHHORN [13]), see fig. 24. There is no doubt that rate dependence and irreversible (or dissipative) behaviour are interrelated in some way which has not yet been discovered. For geotechnical engineering, rate dependence of soil is of particular interest because it is the key to understanding soil viscosity, relaxation and creep processes. Open questions referring to time scaling are e.g.

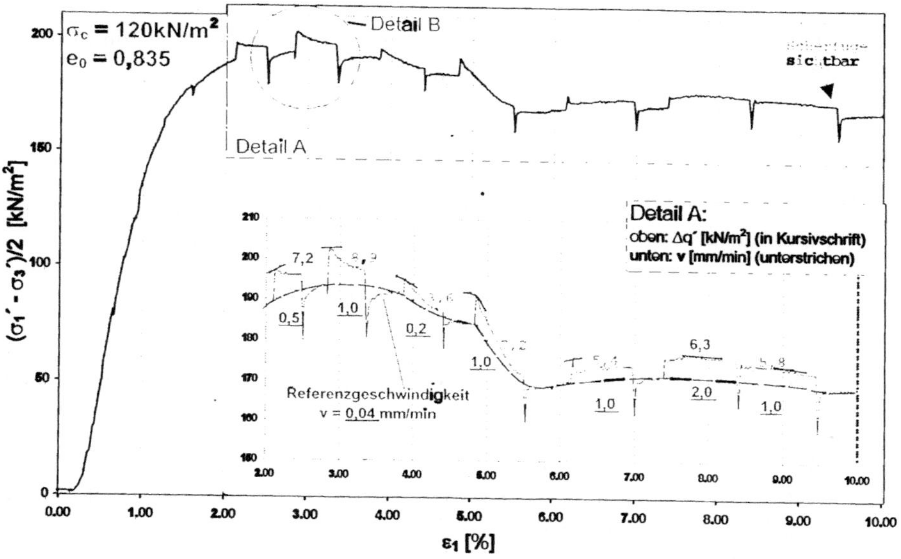

Fig. 24. Rate dependence of fine sand [13]

- How can we evaluate results obtained with relatively fast deformations (e.g. field tests with vane or cone penetration) and apply them to problems where deformation is much slower?
- Referring to radioactive waste disposals, how can we predict the behaviour of rock for the next thousands of years based on laboratory tests with a duration of a few years?
- What is the long-term behaviour of geosynthetics?

4.1 Theory of V. N. Rodionov

RODIONOV assumes that the behaviour of rock is governed by inhomogeneities of, say, spherical form, the nature of which is not further specified. He only assumes that their sizes (e.g. diameters) l_i are distributed in such a way that the material does not possess any internal length. In other words, the size l_i of the largest inhomogeneity within a sample is proportional to the size of the sample. He further assumes that in the course of loading there are formed stress fields around the inhomogeneities. The deviatoric parts of the related stresses increase with deviatoric deformation (according to a constitutive equation such as $\overset{\circ}{\mathbf{T}} = \mathbf{h}(\mathbf{T}, \mathbf{D})$. RODIONOV assumes the linear elastic relation $\dot{\mathbf{T}}^* = 2G\dot{\mathbf{E}}^*$, where $\mathbf{T}^*$ is the deviatoric part of $\mathbf{T}$ and $\dot{\mathbf{E}} = \mathbf{D}$, $\dot{\mathbf{E}}^* = \mathbf{D}^*$) and decrease due to relaxation:

$$\dot{\mathbf{T}}_i^* = 2G\mathbf{D}^* - \frac{v}{l_i}\mathbf{T}_i^* .$$

Setting $\mathbf{D}$ = const and replacing tensors with scalars (e.g. components or appropriate invariants) he obtains therefrom

$$T_i^* = 2GD^* \frac{l_i}{v}\big(1 - \exp(-\frac{v}{l_i}t)\big) .$$

Based on data from wave attenuation, RODIONOV assumes that v can be approximately considered as a universal constant for rock:

$$v \approx 2 \cdot 10^{-6} \quad \text{cm/s}.$$

To a particular time $t^\star$ that ellapsed from the begin of loading we can assign the length $l^\star = vt^\star$. We can then distinguish between large inhomogeneities ($l_i \gg l^\star$) and small ones ($l_i \ll l^\star$):

Large inhomogeneities: $\exp(-l^\star/l_i) \approx 1 - l^\star/l_i \rightsquigarrow T^* = 2GD^*t^\star$, i.e. deviatoric stress increases with time $t^\star$, a process which eventually leads to fracture (fragile or brittle behaviour).

Small inhomogeneities: $\exp(-l^\star/l_i) \approx 0 \rightsquigarrow T_i^* = 2GD^*l_i/v$, i.e. the material around small inhomogeneities behaves as a viscous material (cf. ductile behaviour).

We see thus that the material behaviour of the considered body, in particular the distinction between brittle and ductile behaviour, depends on the value of the parameter vt/l_i . As l_i correlates with a characteristic length (size) L of the body, we can consider the parameter vt/L as determining the material behaviour. If we consider a model test simulating a prototype (e.g. the convergency of a tunnel), where model and prototype consist of the same material, then the following similarity condition must be preserved according to RODIONOV:

$$\frac{t_{model}}{t_{prototype}} = \frac{L_{model}}{L_{prototype}} \quad .$$

E.g., if we consider the convergency of two cylindrical cavities with radius R_1 and R_2, respectively, observed in the same squeezing rock, then the same convergency $u_1/R_1 = u_2/R_2$ occurs after the time lapses t_1 and t_2, respectively, where $t_1/t_2 = R_1/R_2$.

With increasing time, l^* increases until it obtains the value of L. Then the body necessarily behaves in a brittle way and failure sets eventually on.

5 *Π*-theorem and dimensional analysis

It is often advantageous to group the variables of a problem into new ones. If we succeed to render the new variables dimensionless, we get the following benefits: First, the number of variables is reduced and, secondly, the variables retain their values for a whole class of similar problems. A mathematical relation between the dimensionless variables does not describe a single process but a class of similar processes. A set of variables can always be reduced to a set of dimensionless variables. This follows from the requirement that physical laws be invariant with respect to a change of units (e.g. the equation force = mass × acceleration is valid no matter if the acceleration is given in m/s^2 or ft/s^2).

According to the theorem of NOETHER, invariance is related to conservation laws. Referring to invariance with respect to units, the associated conservation is that of the dimensionless variables. It is interesting to note that scaling (or scale invariance) is *not* connected with any conservation law [10].

Let a problem be described by n significant variables $x_1, x_2, \dots x_n$. Each variable x_i can be represented as the product of a dimensionless number ξ_i and of some powers of the basic units E_i:

$$x_i = \xi_i E_1^{e_{i1}} E_2^{e_{i2}} \dots E_k^{e_{ik}}$$

k is the number of basic units which appear in this problem. Instead of the n variables $x_1, x_2, \dots x_n$ we introduce the dimensionless variables $\Pi_1, \Pi_2, \dots$. Each dimensionless variable is the product of powers of the initial variables x_i:

$$\begin{aligned} x_1^{p_1} \cdot x_2^{p_2} \cdot \ldots \cdot x_n^{p_n} = & \ \xi_1 \cdot E_1^{e_{11}p_1} \cdot E_2^{e_{12}p_1} \cdot \ldots \cdot E_k^{e_{1k}p_1} \ \times \\ & \ \xi_2 \cdot E_1^{e_{21}p_2} \cdot E_2^{e_{22}p_2} \cdot \ldots \cdot E_k^{e_{2k}p_2} \ \times \\ & \ \dots \\ & \ \xi_n \cdot E_1^{e_{n1}p_n} \cdot E_2^{e_{n2}p_n} \cdot \ldots \cdot E_k^{e_{nk}p_n} \end{aligned}$$

The new variables are dimensionless, hence

$$
\begin{aligned}
e_{11}p_1 + e_{21}p_2 + \ldots + e_{n1}p_n &= 0 \\
e_{12}p_1 + e_{22}p_2 + \ldots + e_{n2}p_n &= 0 \\
&\ldots \\
e_{1k}p_1 + e_{2k}p_2 + \ldots + e_{nk}p_n &= 0 \quad .
\end{aligned}
$$

We have a homogeneous system of k linear equations with n unknowns. Such a system possesses m linear independent solution vectors:

$$
\begin{pmatrix} p_1 \\ p_2 \\ \vdots \\ p_n \end{pmatrix}_1 , \quad \begin{pmatrix} p_1 \\ p_2 \\ \vdots \\ p_n \end{pmatrix}_2 , \quad \cdots \quad \begin{pmatrix} p_1 \\ p_2 \\ \vdots \\ p_n \end{pmatrix}_m ,
$$

Each of the solution vectors determines a dimensionless variable Π_i. There are m such dimensionless variables. The number m results from the rank r of the equation system. r is defined by the requirement that there is not any non-vanishing sub-determinant of order $r+1$. Hence $m = n - r$. In practice, the dimensionless variables Π_i are found by trial. Doing so one should take into account that there are many different sets of dimensionless variables. It is a matter of skill to find the most appropriate set such that there is a good correlation between the individual variables.

Some examples from soil mechanics are:

Permeability: The permeability $k(:= v/i)$ does not only depend on the involved materials but also on the gravity acceleration g, because the fluid pressure is expressed as pressure head (p/γ_w). The permeability k depends also on the effective grain diameter D_w (which characterizes an appropriate representative diameter of the pores) and on the viscosity η of the pore fluid: From $F(k, \gamma_w, \eta, D_w) = 0$ it follows with $\Pi = \dfrac{k\eta}{\gamma_w D_w^2}$: $\quad F_1(\Pi) = 0$. Hence Π=const or

$$
k = \text{const} \times \frac{D_w^2 \gamma_w}{\eta}
$$

Hour-glass: It is known that the discharge velocity v from a silo does not depend on the fill height h (this is why sand is appropriate for hour-glasses). The relation has the general form $F(d, v, g) = 0$, where d is the diameter of the outlet and g is the gravity acceleration. With $\Pi = \dfrac{v^2}{dg}$ we obtain $F_1(\Pi) = 0$. Hence Π=const i.e. v=const$\times\sqrt{d \cdot g}$. The discharge Q (= sand volume per time unit) reads then: $Q = \dfrac{\pi d^2}{4} \cdot v = c_1 \cdot d^{5/2}$

Consolidation time: One-dimensional consolidation is governed by a relation

$$
F(s, d, k/\gamma_w, \delta p, E_s, t) = 0 \quad ,
$$

where d is the thickness of the consolidating soil layer, s is the settlement, δp is the load, E_s is the soil stiffness and t the time. We therefrom obtain the following dimensionless variables:

$$\Pi_1 = s/d; \quad \Pi_2 = \frac{k \cdot t \cdot E_s}{\gamma_w \cdot d^2}; \quad \Pi_3 = \frac{\delta p}{E_s} \quad .$$

If the loadings of model and prototype are the same and if model and prototype consist of the same water saturated soil, we obtain $\Pi_{2model} = \Pi_{2prototype}$:

$$\left(\frac{t}{d^2}\right)_{model} = \left(\frac{t}{d^2}\right)_{prototype} \quad .$$

This confirms the general law that relaxation times decrease with the dimension of a system [12].

References

[1] D.L. Turcotte and J. Huang, Fractal distributions in Geology, In: Fractals in the Earth Sciences, Chr. Barton and P.R. La Pointe (editors), Plenum Press 1995

[2] D.L. Turcotte (1986), Fractals and Fragmentation, *J. of Geophys. Res.* **91**, 1921 - 1926

[3] L.E. Vallejo, Fractal analysis of granular materials, *Géotechnique* **45**, No. 1, 159-163 (1995)

[4] U. Gori and M. Mari, The correlation between the fractal dimension and internal friction angle of different granular materials, *Soils and Foundations* **41**, No. 6, 17-23 (2001)

[5] S.R. Brown, Dimension of self-affine fractals: Example of rough surfaces, In: Fractals in the Earth Sciences, Chr. Barton and P.R. La Pointe (editors), Plenum Press 1995

[6] C. Scavia (1996), The effect of scale on rock fracture toughness: a fractal approach. *Géotechnique* **46**, No. 4, 683 - 693,

[7] Chr. E. Krohn (1988), Sandstone Fractal and Euclidean Pore Volume Distributions. *J. of Geophys. Research*, **93**, No. B4, 3286 - 3296

[8] M.V.S. Bonala, L.N. Reddi (1999), Fractal representation of soil cohesion, *J. of Geotechn. and Geoenvironmental Eng.*, Oct. 1999, 901 - 904

[9] G.R. McDowell, A. Humphreys, Yielding of granular materials, *Granular Matter* **4** 1-8, 2002

[10] D. Sornette, Critical Phenomena in Natural Sciences – Chaos, Fractals, Selforganization and Disorder: Concepts and Tools, Springer Berlin, 2000

[11] J. Feder, Fractals, Plenum Press, New York, 1988

[12] Landau Lifschitz, Statistical Physics, Sect. 7

[13] B. Eichhorn, Der Einfluß der Schergeschwindigkeit beim Triaxialversuch, Diplomarbeit, Universitt Innsbruck, 1999
[14] J.G. Ramsay, M.I. Huber, The Techniques of Modern Structural Geology, Vol.2: Folds and Fractures, Academic Press, London, 1993
[15] Chr. A. Hecht, Appolonian Packing and Fractal slope of Grains Improving Geomechanical Properties in Engineering Geology, PAGeoPH, 157 (2000), 487-504
[16] Prandtl, L.; Ein Gedankenmodell zur kinetischen Theorie der festen Krper *ZAMM*, 8, Heft 2, April 1928, 85-106
[17] J. K. Mitchel, Shearing resistence of soils as a rate process, *Journal SMFD of ASCE,* SM1, Jan. 1964, Vol. 90, 29-61
[18] R. R. Hartley & R. P. Behringer, Logarithmic rate dependence of force networks in sheard granular materials. *Nature*, Vol 421, 27 February 2003
[19] F. Tatsuoka, et al., Time dependent deformation characteristics of stiff geomaterials in engineering practice. In: Pre-failure Deformation Characteristics of Geomaterials, Jamiolkowski et al, editors, Swets & Zeitlinger, Lisse, 2001, 1161-1262
[20] I. Herle and G. Gudehus. Determination of parameters of a hypoplastic constitutive model from properties of grain assemblies. *Mechanics of Cohesive-Frictional Materials*, 4(5):461–486, 1999.

Part II

Modelling of Boundary Value Problems

Numerical predictions and reality

Ivo Herle[1]

Institute of Geotechnical and Tunnel Engineering, University of Innsbruck
ivo.herle@uibk.ac.at

Abstract: Numerical calculations of boundary value problems must often be performed as true predictions. Unfortunately, the ability to submit a reliable prediction seems to be lacking in geotechnical engineering. Several reasons which may be responsible for this disappointing state are described.

1 Introduction

The design of geotechnical structures is mainly based on numerical calculations of boundary value problems. Contrary to standards, which usually cover only the most simple cases or give but vague recommendations, numerical simulations seem to produce impressive pictures of the overall behaviour with detailed distribution of values for all important design variables and physical quantities.

In spite of the convincing marketing of software producers and audacious projects of engineering companies, there are still many weak points in our knowledge and we still do not master numerical simulations as we would wish to do. This paper brings many examples of the unsatisfactory state of the art and tries to analyze several reasons for that.

2 Predictions

Making predictions belongs to the main tasks of civil engineering. Predictions are needed for design, for the evaluation of servicability or for the estimation of risk. They simulate *unknown* states in planned and existing structures or impact of natural phenomena. Although predictions can be done intuitively or rather empirically, nowadays it is expected to perform numerical predictions using mathematical models.

The only way of evaluating the quality of numerical predictions is to compare them with measurements and observations. It seems obvious that numerical simulations should precede the measurements, although one sometimes denotes the numerical fitting of known (measured) values as predictions as well [20]. There exist a large number of publications on such back calculations, whereas so-called *class A predictions* [20], i.e. numerical simulations done prior to the construction or experimental event, are rare. Back calculations are a usefull tool for learning and they are needed for the application of the observational method [29] but class A predictions are inevitable in many situations.

The importance of predictions was recognized already in the early years of numerical modelling but only a few competitions on geotechnical predictions have been organized in order to check the ability of the engineering forecasting. In the sequel, several of them are briefly reviewed without attempting to analyze the results in detail.

MIT trial embankment

A trial embankment was built on a normally consolidated soft clay layer (Fig. 1). Prior to the construction laboratory experiments of the subsoil were done and the first construction stage up to 12.2 m height was monitored by field measurements [20]. These data were at disposal for the prediction of deformations, pore pressures and maximum additional height of the embankment at subsequent rapid filling to failure.

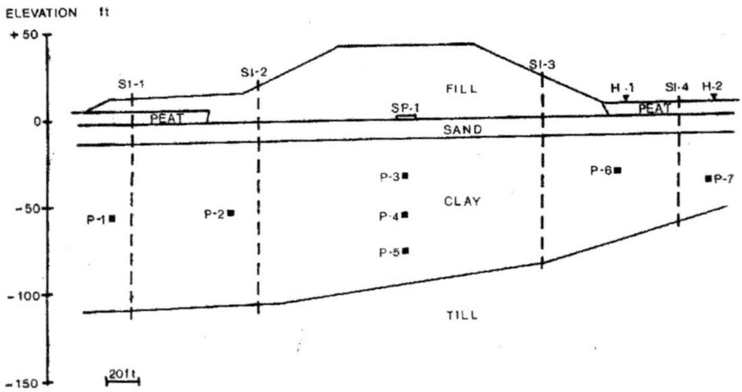

Fig. 1. MIT trial embankment, USA (1974) [47]

Predictions were submitted by ten groups. There was a large scatter of the numerical results, see Tab. 1. Although mostly linear and nonlinear elastic models were applied, one of the best predictions was based on the Modified Cam Clay model [47]. The latter was very good with respect to pore pressures but still less accurate for deformations (predicted 4 cm at SP-1 and 4.1 cm at SI-3).

Table 1. Additional defomations due to 6 ft (1.8 m) of fill, in cm

Item	Predicted	Measured
settlement of SP-1	1.9...34.8	1.7
horiz. movement of SI-3 (at -30 ft)	0.4...21.8	1.3
heave of H-1	0...12.2	-0.3

Excavation in sand

A sheet pile wall was driven into a homogeneous sand layer above the groundwater level and the task was to predict the behaviour during a 5 m deep excavation [46]. Slurry walls perpendicular to the sheet pile wall imposed plane strane conditions. For keeping the wall stable, struts were installed at 1.5 m depth. After the excavation an additional surcharge was placed at the ground surface behind the pit and the struts were loosened in order to reach the limit state.

In situ and laboratory soil investigations were performed prior to the excavation. Results from 43 predictions included horizontal displacements of the wall, vertical displacements at the ground surface, earth pressures on the wall and bending moments in the wall.

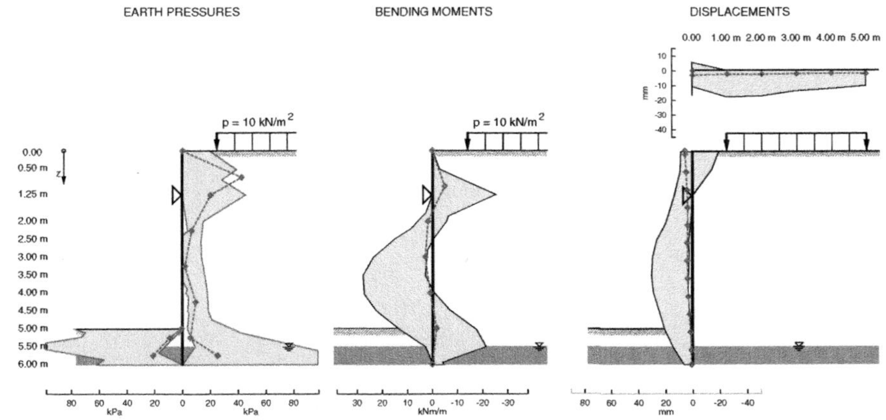

Fig. 2. Predicted (shaded range) and measured (line with points) values for the excavation in sand near Karlsruhe, Germany (1993) [46]

Most calculation methods employed finite elements or subgrade modulus models. The comparison with measured values is very disappointing, see Fig. 2 for final excavation stage with surface load, prior to the limit state. Especially worrying is that displacements have been predicted several times in the opposite direction than measured. However, even more depressing is the large scatter of predictions which becomes enormous in the limit state [46].

Excavation in clay

Inspired by the previously described excavation in sand, a similar excavation was constructed in a soft soil near Rotterdam [22]. The ground consisted of soft clay and peat with a high groundwater level. The excavation was 6.5 m deep and it was supported by struts at the top. Plane strain conditions were ensured by slurry walls at the sides. Site investigation and laboratory tests were provided.

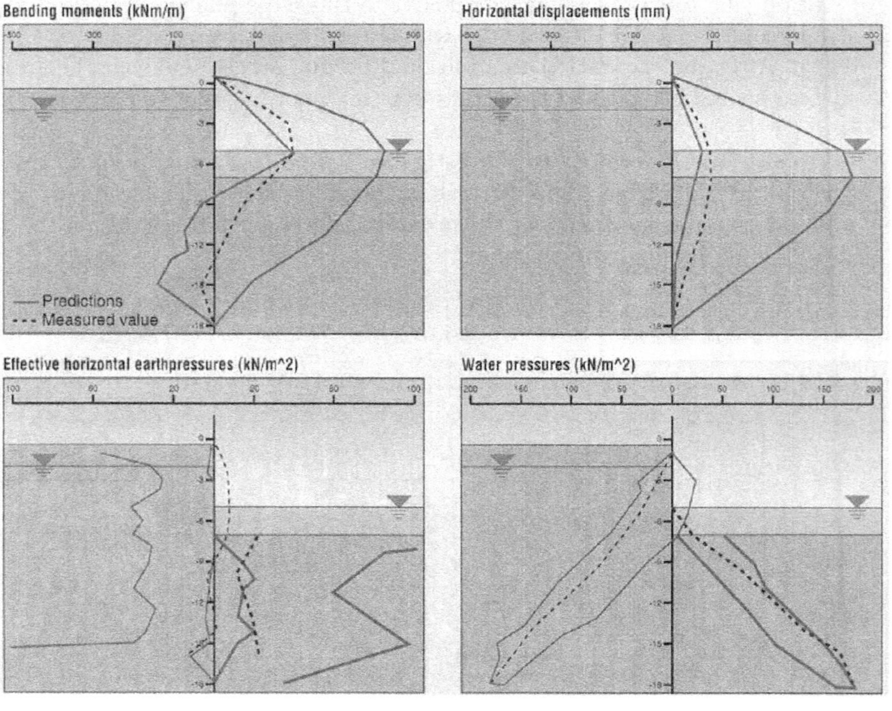

Fig. 3. Predicted (range between the full lines) and measured (dashed) values for the excavation in clay near Rotterdam, The Netherlands (1999) [22]

Although it is sometimes claimed that the behaviour of soft clay is simpler than that of sand (e.g. associated plasticity models are applicable in the first case), the predicted values displayed again a large scatter and were often far away from the measured ones (Fig. 3).

Vertically loaded model footings on sand

Spread foundations are probably the most frequent geotechnical structures but still the most difficult ones for calculations. In order to better control the subsoil conditions, a series of model experiments was prepared in Japan. Small strip footings (width b=1.0, 2.5, 5.0 and 10 cm) were placed on the surface of a dense Silver Leighton Buzzard sand and loaded vertically. Due to the small dimensions of the footings, a pronounced scale effect (stress level, ratio of the footing width to the grain size) was involved. The prediction of load-displacement curves based on laboratory (mainly plane-strain) tests was asked in advance [41].

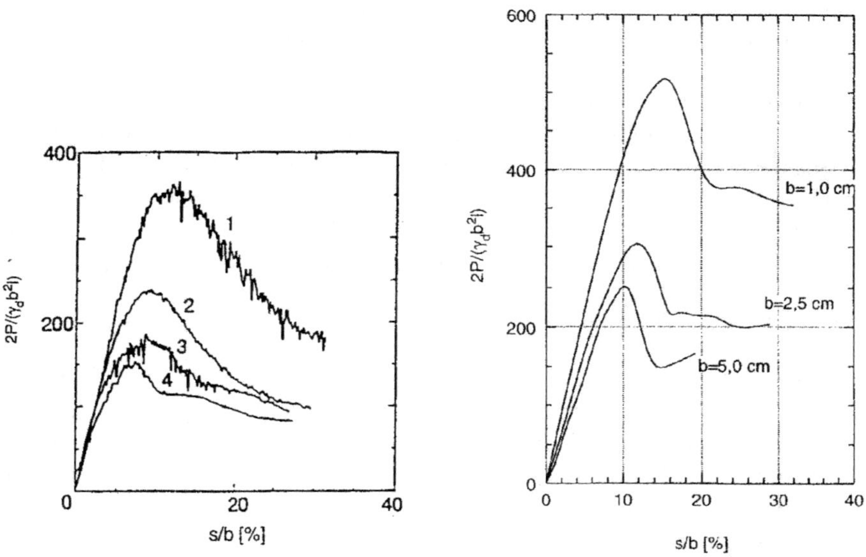

Fig. 4. Experimental [39] and predicted [16] normalized load-displacement curves of vertically loaded footings on sand (both figures are scaled identically, 1: b=1.0 cm, 2: b=2.5 cm, 3: b=5 cm)

The overall results of the prediction competition have never been published. However, a personal communication to the organizer[1] revealed that the predicted load-displacement curves based on the polar hypoplastic model [16] were the best ones. Still, although this model could reproduce very well the scale effect and the strains at the peak of the load-displacement curves, the maximum load was overpredicted by almost 40%.

[1] Personal communication from Professor F. Tatsuoka at the IS-Nagoya, 1997.

3 Benchmarks

One can object that geotechnical predictions are difficult because ground conditions can be explored only in a few discrete points, the geological history is not perfectly known and thus the initial conditions are not uniquely determined, extent of the area of interest and the boundary conditions are not precisely specified, we use different approaches to model interfaces, soil can be described by various constitutive equations, etc. This degree of freedom for engineering judgement and choice can be, however, dramatically reduced when dealing with *benchmarks*.

Benchmarks are usually designed as numerical calculations of typical boundary value problems with relatively strict input specifications without need to calibrate constitutive models on experimental outputs. Considering FE simulations, the freedom is mostly limited to numerical aspects like the choice of mesh, time step or loading procedure. Mostly we do not want to compare calculation results with the real (measured) behaviour. Rather we aim to compare the calculation results to each other, in rare cases to analytical solutions.

Benchmarking is becoming increasingly popular in recent years. Compared to prediction competions, benchmarking is faster, simpler to evaluate and cheaper. However, it can never replace prediction competitons since it covers only a limited part of the process leading to class A predictions. Moreover, a kind of benchmarking, namely parametric studies, should always accompany numerical simulations as it can discover many sensitive factors [13].

INTERCLAY II project

The project was sponsored by the European Comission during 1991-1994 as a benchmark exercise dealing with numerical predictions of the clay geomechanical behaviour, especially concerning the long term features. Several major geotechnical companies and institutes were involved (for details, see the database at [48]).

"A comparison of the theoretical capabilities of the popularly used models and computer codes and their performance on simple, somewhat hypothetical problems" [48] belonged to the main tasks of the project. One of those hypothetical problems was the advancing tunnel heading. Purely cohesive incompressible soil characterized by Tresca soil model was assumed (E=240 MPa, μ=0.499 and s_u=1.0 MPa) [31]. Isotropic initial stress conditions (2500 kPa) including pore pressures, axisymmetric analysis and rigid permeable lining were prescribed (Fig. 5 left).

The results at Fig. 5 (right) reveal a large scatter in the predicted radial convergence although the constitutive model and stress conditions seem to be rather simple.

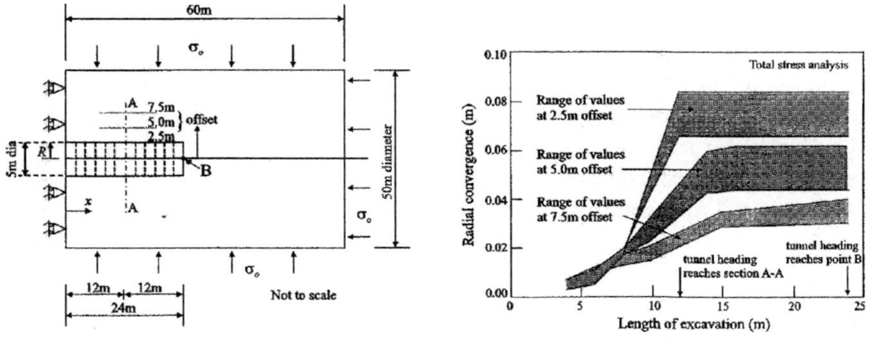

Fig. 5. Advancing tunnel heading: results from the INTERCLAY II project [31]

Tunnel excavation (DGGT)

The Working Group 1.6 of the German Society for Geotechnics (DGGT) organized in the nineties a series of benchmarks with emphasis on plane-strain simulations of tunnel and deep excavations [36]. In case of the tunnel, parameters of the linear elastic-perfectly plastic model (Mohr-Coulomb) were given and the initial state (without water) and the construction sequence were prescribed.

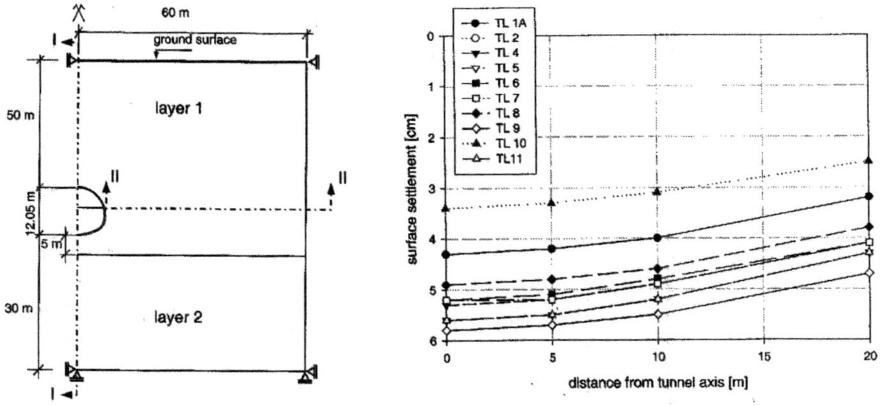

Fig. 6. Surface settlements for the full face excavation [36]

The calculated surface settlement above the tunnel axis varies between 3.3 and 5.8 cm, see Fig. 6. When leaving two predictions with the lowest settlement aside, still there is about 20% difference between the predicted values. Even larger scatter was obtained for the calculated bending moments [36].

Deep excavation (DGGT)

The second DGGT benchmark test covered a deep excavation in three steps [36], i.e. a similar task to the already described prediction of the excavation in sand. Again, parameters of the linear elastic-perfectly plastic model (Mohr-Coulomb) for soil and linear elastic model for the wall, rigid struts, the initial state and the excavation sequence were prescribed (Fig. 7). Perfect bonding between soil and wall was assumed in order to avoid problems of interfaces.

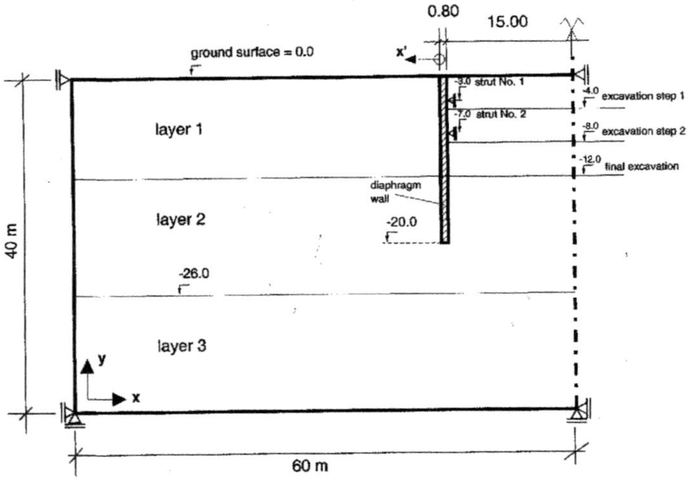

Fig. 7. Layout of the deep excavation [36]

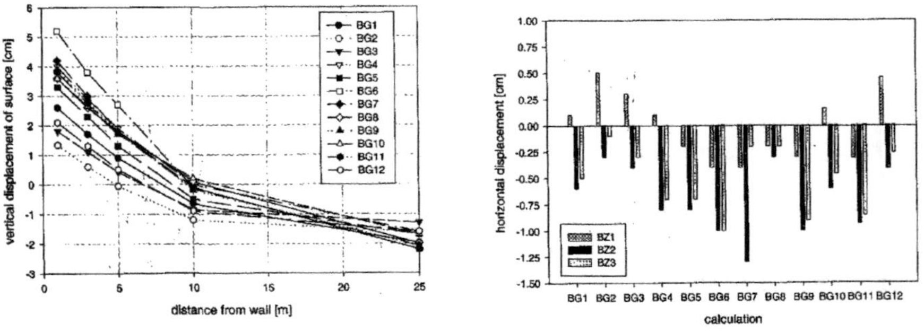

Fig. 8. Vertical (left) and horizontal (right) displacements of the surface behind the wall (BGxx is an identification of a particular prediction) [36]

The calculation results are in a narrower range than for the class A prediction (Fig. 2) but remarkable deviations remain. Taking into account the tight input specification, the scatter of the results for settlements of the surface behind the wall (Fig. 8 left) is too large. The prediction of the horizontal displacements at the top of the wall (Fig. 8 right) yields even a worse picture. There are qualitative differences in the direction of the movement for the first excavation stage (BZ1) and dramatic differences in the magnitude for the last excavation stage (BZ3).

Excavation with a tied back diaphragm wall

Although the results of the previous benchmark on excavation were not optimistic, another benchmark test for a similar problem was organized [37], see Fig. 9. The specifications were based on one particular site in Berlin and no material model was prescribed. Consequently, various elasto-plastic and hypoplastic models were involved in FE calculations using parameters derived from oedometer and triaxial curves on real samples.

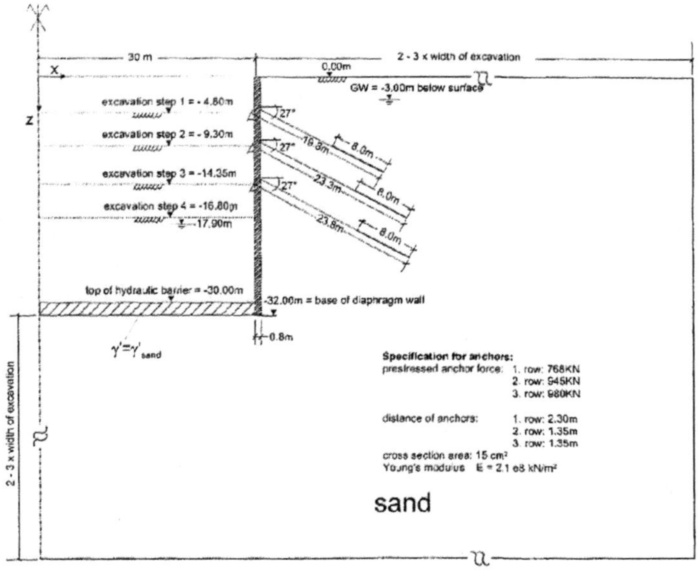

Fig. 9. Tied back diaphragm wall [37]

Bearing in mind the already shown predictions and benchmarks, it is not surprising that there is, again, a dramatic quantitative and qualitative scatter of the results (Fig. 10). Besides the arbitrarily chosen constitutive model, the construction process was rather complex with a certain degree of freedom for the modelling (e.g. lowering of the groundwater in one or more steps? how to simulate the anchors?).

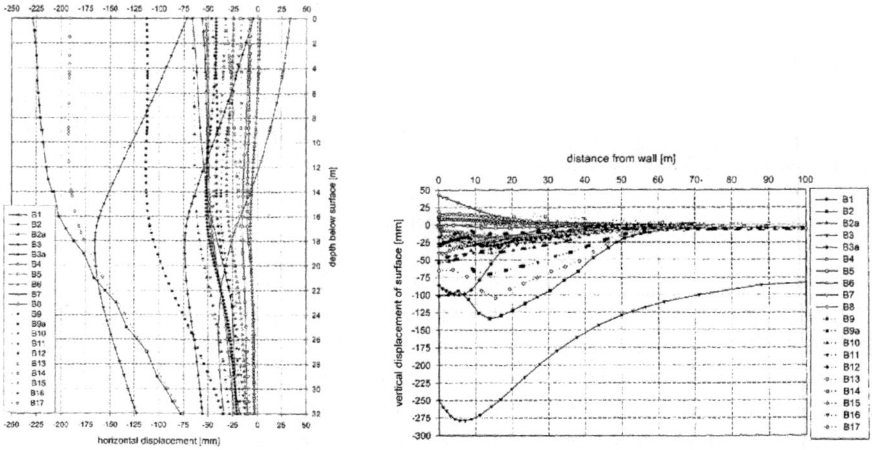

Fig. 10. Tied back diaphragm wall: wall deflection (left) and vertical surface displacements (right) [37]

Tunnel excavation (COST)

As a part of the work done by the Working Group A of the COST C7 project, a relatively simple benchmark of tunnel analysis was performed [30]. Elastic and elastic-perfectly plastic solutions of a circular tunnel in the initially isotropic stress field were to calculate (Fig. 11). Elastic parameters (G=12 MPa and ν=0.495), undrained cohesion (s_u=130 kPa or 60 kPa) and full face excavation without and with lining, respectively, were prescribed (in case of the lining the volume loss of 2% was given).

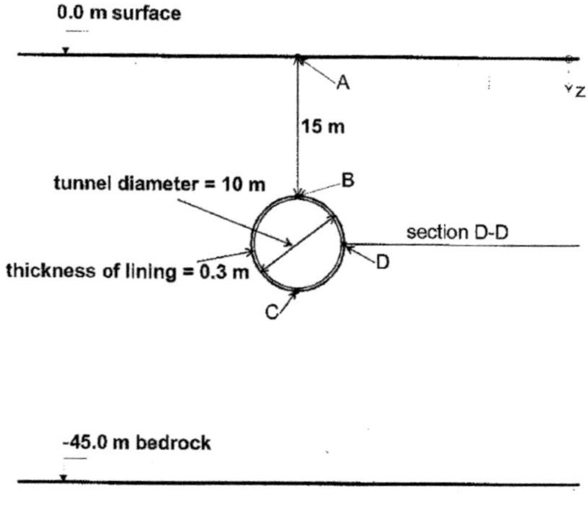

Fig. 11. Excavation of a circular tunnel [30]

Vertical settlements obtained with elastic and elastic-perfectly plastic analyses are shown in Fig. 12. Considering first the elasto-perfectly plastic calculations (Fig. 12 left) and without taking into account two results (ST4 and ST9) lying far away from the others, there is about 15% difference in the value of the maximum settlement. This means that there is no substantial improvement in the calculated results compared to the DGGT tunnel benchmark (Fig. 6). Moreover, the linear elastic calculation (Fig. 12 right) revealed a similar scatter of the results which is very disappointing since one can find even an analytical solution for this problem [43]! Unfortunately the differences cannot be attributed only to different widths of the used FE meshes.

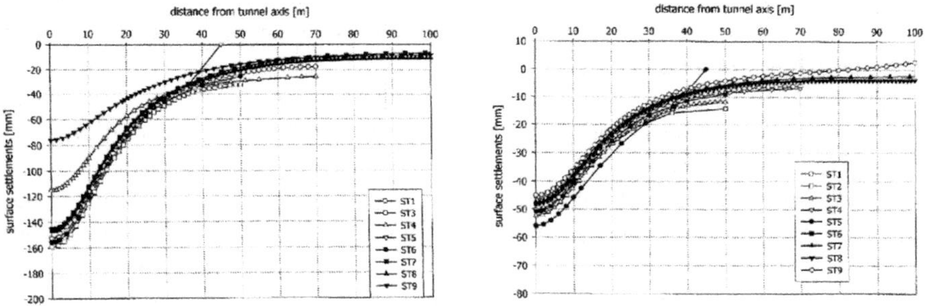

Fig. 12. Surface settlements for the tunnel excavation: elastic-perfectly plastic (left) and linear elastic (right) model [30]

Comments on benchmarks

The benchmarks show that in spite of tight specifications a large scatter of the results can be observed. This scatter becomes higher with increasing complexity of the calculation task but even for simple geometry with linear elastic model the coincidence of the results cannot be reached. It seems that the sophisticated software still leaves a lot of room for the expert's judgement. Furthermore, benchmarks participants usually belong to the academia where the theoretical background is (should be) on a higher level than in the common practice. The vulnerability of many consulting engineers to misunderstandings related to computations can be estimated even higher, also due to the time pressure.

If it is not possible to achieve a good agreement for purely calculation excercises, it cannot be expected to get reliable predictions for in situ projects from practice. However, this may not lead to the wrong conclusion that we must first improve calculation methods before solving real problems. It was shown that benchmarks are usually inadequate simplifications of the reality since they analyze only computational aspects. On the other hand, huge differences in the depicted predictions were also influenced by many additional factors out of the scope of numerical methods.

4 Making predictions

4.1 Ingredients of predictions

Geotechnical predictions are not mere numerical computations but they represent a process composed of several important modelling steps:

1. *Idealization (simplification of the reality)*
   choice of important variables, geometry, domain, boundary conditions, construction details and stages, selection of substantial aspects and disregarding minor ones
2. *Discretization*
   element type, size and density, time step, representation of loads
3. *Material models*
   appropriate constitutive equations, model for interfaces, calibration of parameters, determination of initial state, framework for calculation of strains
4. *Mathematics (incl. numerical methods)*
   type of time integration, equation solver, iteration schemes, well-posedness

All steps are equally important and it is impossible to say a priori which aspect can produce larger errors in predictions. These steps compose a chain which fails at the weakest link! Thus, for making predictions, it is necessary to have an insight in all of the mentioned topics. Computer software cannot yet replace the sound judgement based on the profound knowledge and make automatic decisions from several options interconnected with complex relations.

4.2 Sources of errors

We can distinguish many potential sources for errors in numerical calculations:

- hardware and software bugs
- application of unsuitable theory/code
- erroneous input (obvious misunderstanding/misinterpretation of data)
- lack of data (e.g. variability of geological conditions, loading scenarios)
- idealization of reality (neglecting of important aspects)
- inapropriate constitutive models (e.g. linearization of significant nonlinear effects)
- determination of material parameters
- description of the initial state (initial values of state variables)
- mathematical and numerical problems

Some of those error sources can be controlled, at least to a certain extent, by engineers. Other, like hardware/software bugs or lack of data, are independent of the qualification and contribute to the ever-present uncertainity in our predictions. Some aspects for the first kind of errors are discussed further.

4.3 Role of constitutive models

In calculations of geotechnical problems constitutive models play probably the central role. The mechanical behaviour of geomaterials is extremely complex and the material description must always look for a compromise between fitting the most important phenomena and remaining lucid and/or simple.

It is difficult to judge the suitability of constitutive models for practical applications. At the beginning, general questions arise, e.g.

- *How to define the quality of a constitutive model?*
  We know how to define the quality of a car but this is uncertain in case of constitutive models. There is no unique measure of simplicity and complexity [19].
- *What are the limitations of the selected model?*
  Publications are overflowing with diagramms showing what models can do but authors shy to disclose deffects and limitations.
- *How to check the selected model?*
  There is no clear methodology how to test constitutive models since possible loading and deformation paths in boundary value problems are too diversified. Anyway, the scatter in results of the prediction competitions on constitutive models using element tests [11, 34] resembles the scatter in results of the prediction competitions on boundary value problems.
- *How to apply/use the selected model?*
  It cannot be expected that the user (engineer) will write computer codes for novel and promising models. The vast majority of newly developed models come from the academia which should employ the approach of the information technology and give source codes with brief description at free disposal in internet. Such codes can be copyrighted e.g. by the GNU general public licence [49], in order to prevent commercial misuse by third parties.

Nevertheless, besides these general questions every predictor should be aware of several substantial aspects which are of great importance for the application of constitutive models in geomechanics. The following sections provide a few examples.

Scatter of experimental results

Many developers of constitutive models are convinced that a good model should fit measured laboratory curves of element tests as close as possible. However, laboratory experiments are subjected to many parasitic effects [2].

Even the carefully selected input data from different apparatuses used for the prediction competition on constitutive models [35] were inconsistent and according to some participants "somewhat contradictory" [28]. The range of experimental outputs for the path of a standard triaxial compression test on Hostun sand obtained in different apparatuses (standard triaxial, cubical true triaxial and hollow cylinder apparatus), see Fig. 13 [26], makes obvious that minimizing the deviation between the model and a single experimental curve is meaningless. Laboratory testing of soil elements is a fundamental part of the development and checking of constitutive model. However, instead of details one should carefully check the trends.

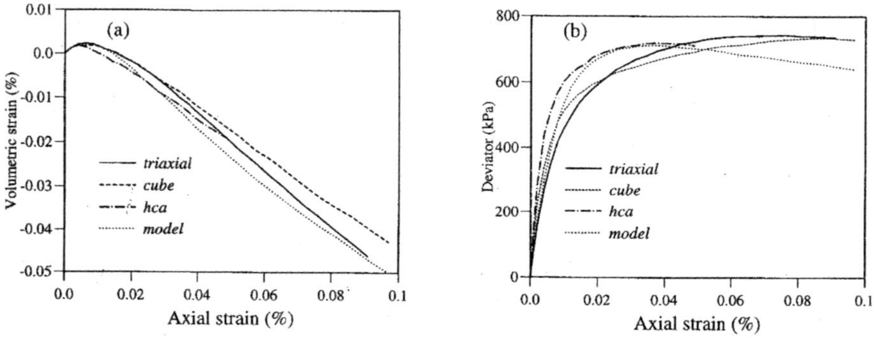

Fig. 13. Scatter of the experimental outputs for the path of a standard triaxial compression test on Hostun sand in different apparatuses [26]

Calibration

Calibration, in other words determination of material parameters (i.e. constants in the constitutive equation), is a crucial procedure for the application of any constitutive model. Model constants are uniquely related to the particular model and loose their meaning outside the model framework (thus, questions on physical meaning of model parameters are superfluous [19]).

Although the calibration procedure may be described in detail, it need not yield practicable results. Considering the most widespread constitutive equation for soils, namely the linear elastic-perfectly plastic (Mohr-Coulomb) model, the calibration underlies strong subjective decisions since the model is not particularly well suitable for soils.

Calibrating Young modulus from the initial stiffness (Fig. 14 left) results in large deviations between the experimental and model stress-strain curves although the volumetric behaviour, see Fig. 14 right, is reproduced slightly better. Note that different Young moduli were fitted to every experimental curve.

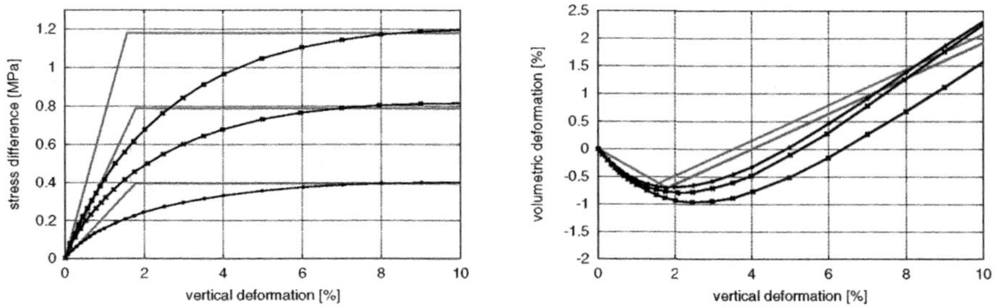

Fig. 14. Calibration of the Mohr-Coulomb model (straight lines) on standard triaxial experimental curves at different cell pressures

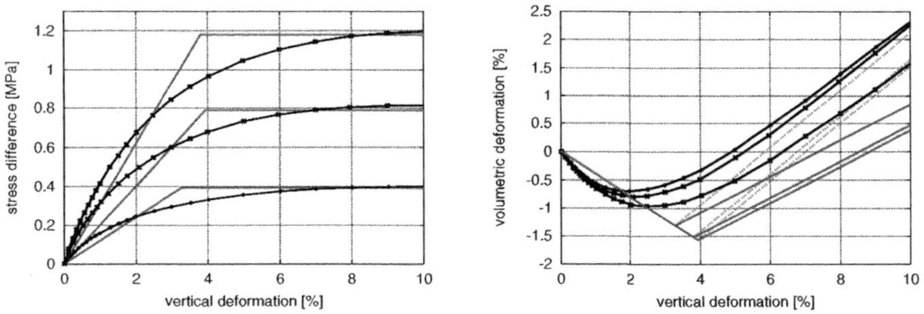

Fig. 15. Another way how to calibrate the Mohr-Coulomb model (straight lines) on standard triaxial experimental curves at different cell pressures

When trying to improve the calibration by choosing lower Young moduli, see Fig. 15 left, the secant approximations of the experimental curves appear to be a better solution than the fitting of initial stiffness. However, lower Young moduli mean larger elastic range which produces an exaggerated initial compression, see Fig. 15 right. Thus, an apparent improvement of one aspect is defeated by extra discrepancy.

Mohr-Coulomb model does not describe the behaviour realistically because its parameters depend, unfortunately, on the state of soil, i.e. they are stress- and density-dependent. We arrive to a paradox situation when material *constants* are *state-dependent* and for one and the same soil must be adjusted according to the actual conditions. However, these conditions change in the course of deformation ...

The situation of complex models is not much better. Their better agreement with measured curves is usually redeemed by a higher number of material parameters. Those parameters often represent coefficients in nonlinear functions which are interralated and cannot be determined separately. Consequently, calibration procedures are prone to yield several sets of material parameters which can fit the calibration curves equally well but diverge for complex deformation paths.

History of deformation (initial state)

The intricate behaviour of soils is manifested particularly in their memory of the previous deformation. Considering a stress increment OX from the same stress state O for several specimens (Fig. 16 left), one measures different stiffness responses (Fig. 16 right) depending on the stress path preceding the state O [33].

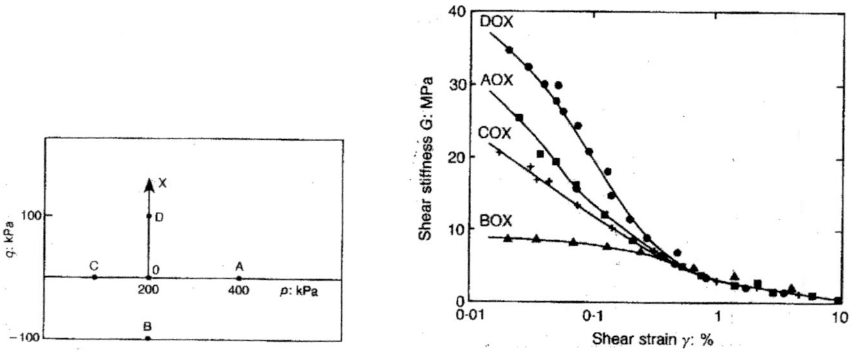

Fig. 16. Influence of the deformation history on soil stiffness [33]

Modelling phenomena of the recent deformation history requires additional state variables, often tensorial, with corresponding evolution equations. Apart from difficulties in finding such variables and equations, the major problem constitutes the determination of the initial values of those state variables. The values must be known prior to the calculation although they are mostly not measurable! Therefore, it is important that these initial values are soon replaced by actual values resulting from the course of deformation and that asymptotic values are reached for monotonous paths [27].

State variables play an eminent role in simulations of geological history, path-dependent construction processes and cyclic loading. A study of the influence of the excavation sequence during a diaphragm wall construction in Fig. 17 may serve as an example [25].

Nonlinearity of the stress-strain curve

State-dependent stiffness, which is reflected in nonlinearity of stress-strain curves, is a characteristic feature of soil behaviour even in the range of small strains [40]. Disregarding this fact may lead to qualitatively wrong predictions and dramatic mistakes.

A well-documented case is the construction of an underground car park in London close to the Big Ben tower [38]. The class A prediction was based on a linear elasticity calculation with Young modulus derived from the initial, apparently linear part of the triaxial stress-strain curve (dashed curve in

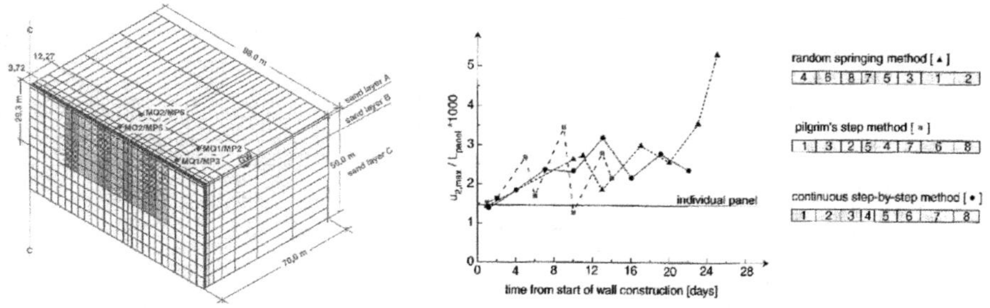

Fig. 17. Influence of the excavation method during a diaphragm wall construction [25]. The graph depicts the horizontal displacements for three different excavation sequences

Fig. 18 right). According to the predicted settlement profile behind the excavation (lower full curve in Fig. 18 left) it was expected that the tower will tilt in direction *from* the excavation. Surprisingly, leaning *towards* the excavation was measured. The unexpected behaviour could be explained using a nonlinear elastic model which yielded a very different distribution of settlements (lower dashed curve in Fig. 18 left) close to the values measured in situ. The application of the nonlinear model could be supported by advanced triaxial testing showing a clearly nonlinear stress-strain curve even in the range of small strains (full curve in Fig. 18 right).

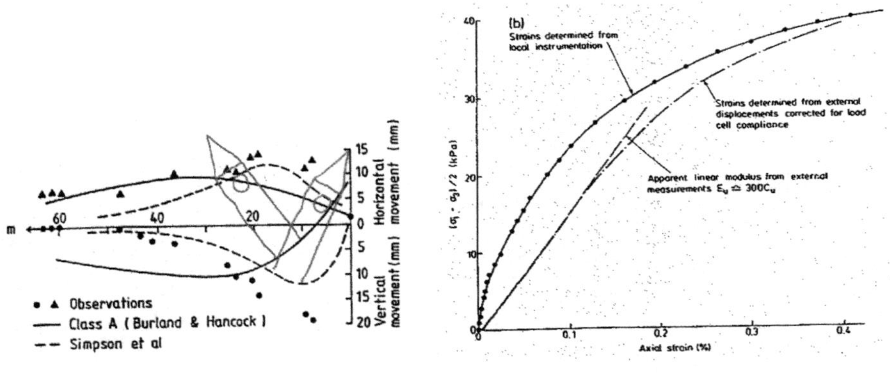

Fig. 18. Influence of nonlinearity of the stress-strain curve on the prediction of deformations [38, 8]

The influence of nonlinearity of constitutive models on results of numerical simulations of excavations was later confirmed in further works, e.g. [44].

However, is it a perfect remedy to use nonlinear elastic models? We have already seen in the section on calibration that it is misleading to look only at stress-strain curves. In Fig. 19 the numerical results of loading and un-

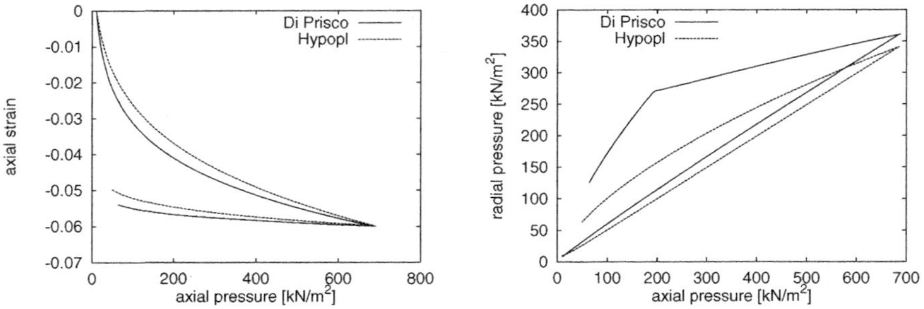

Fig. 19. Unrealistic stress path (full line) during unloading in the calculation of oedometer test [12]

loading in oedometer test are depicted for two different constitutive models (elastoplastic and hypoplastic). The calculated stress-strain curves (left) are very similar but the stress paths (right) are distinctly different during unloading. Whereas the dotted line agrees well with experimental results [3], the full line reaches unrealistic stress states. The latter case is governed by nonlinear elasticity which is unable to reproduce the observed stress path.

Nonlinearity of the stress envelope

Not only stress-strain curves are dominated by nonlinearity. The envelope of limit stresses is also often nonlinear although the popularity of friction angle φ as a typical soil *constant* survives.

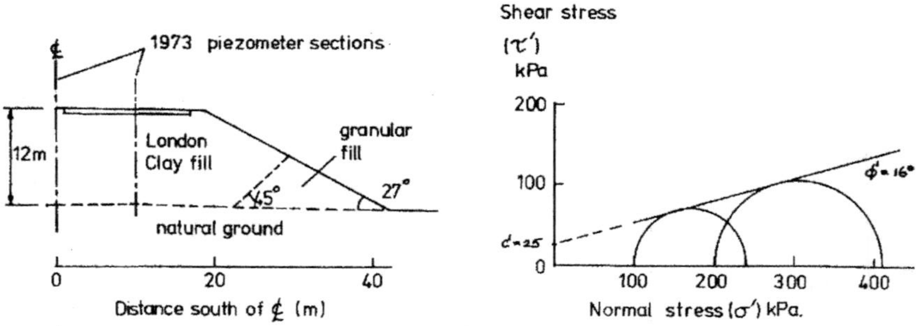

Fig. 20. Standard embankment design assuming a linear stress envelope [1]

The design of a 12 m high embankment in Fig. 20 (left) was based on standard evaluation of triaxial CID tests employing a linear stress envelope (Fig. 21 right) [1]. However, shallow landslides started to appear after finishing the construction. Therefore, an additional series of constant q tests was completed (paths E, F, G in Fig. 21 left). They revealed a strongly curved limit stress envelope in the range of low stresses (Fig. 21 right). According to the new results the range of in situ stresses was quite close to the limit stress state, although the original linear envelope suggested a high factor of safety.

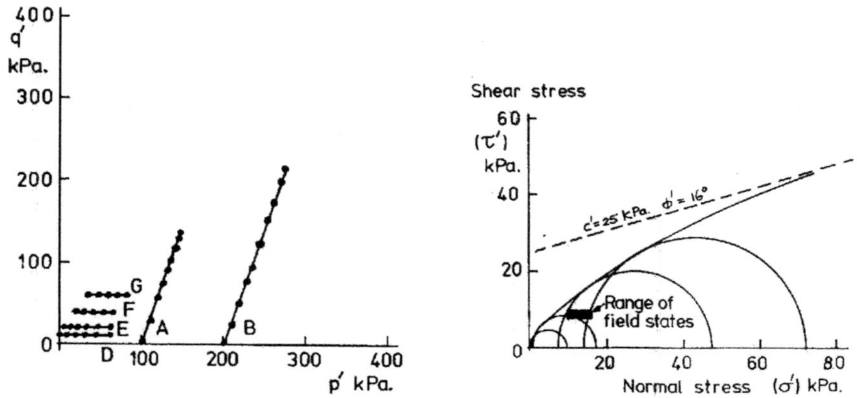

Fig. 21. Nonlinear limit stress envelope discovered by nonstandard (const q) triaxial paths [1]

Monotonic paths?

It is much simpler to describe a stress-strain curve resulting from a monotonic stress or deformation path than to take into account sharp changes in the path direction. Therefore it is often claimed that many geotechnical problems can be considered as monotonic processes and in this case the application of simple constitutive models is justified.

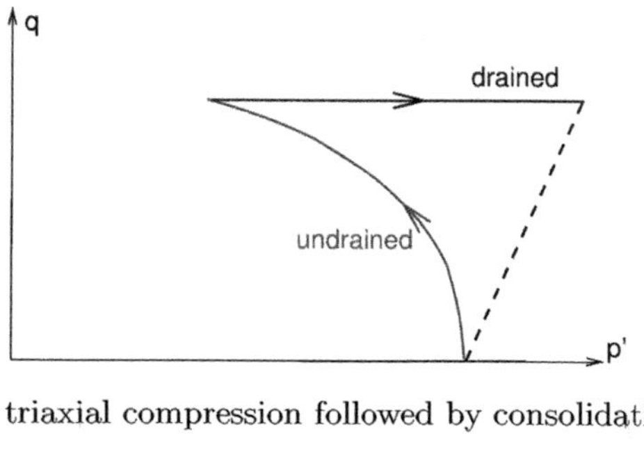

Fig. 22. Undrained triaxial compression followed by consolidation at constant vertical load

However, look at a very simple case of undrained triaxial compression followed by consolidation at constant vertical load (Fig. 22) which is a typical case for constructions on soft subsoil. Can it be denoted as "loading" or "unloading" or both? Although the process seems to be monotonic, the stress path is not!

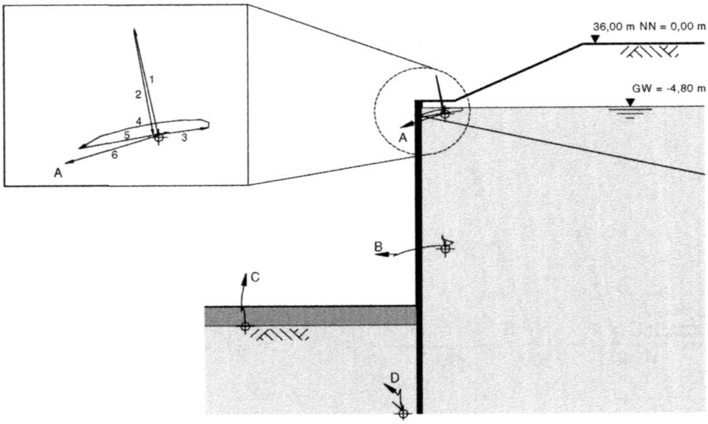

Fig. 23. Displacement paths during excavation with one anchor row [15]

Is an excavation with one anchor row a simple or a complex task? The answer provides Fig. 23 which depicts the *measured* displacement path behind the head of a sheet pile wall during excavation [15]. Simulation of such a path would be a challange for any competition on constitutive modelling.

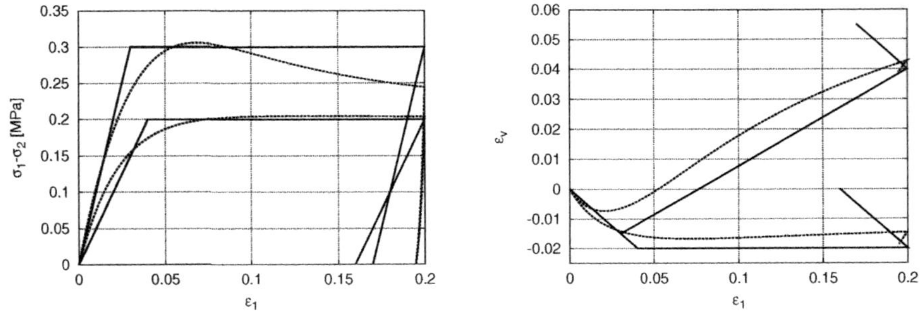

Fig. 24. Comparison of loading and unloading behaviour in standard triaxial compression at two different densities: Mohr-Coulomb model – full line, hypoplasticity – dotted line

Accepting the complexity of loading/deformation paths in most geotechnical problems, we should check the ability of widespread models in dealing with this task. Loading and unloading in standard triaxial compression is well suitable for such a study. It is known that the stiffness of Mohr-Coulomb model for unloading is the same like for loading (24 left). Consequently, the unloading behaviour is too soft and yields e.g. an unrealistic heave of the excavation bottom. This problem is often solved by introducing different Young moduli in different regions, However, the volumetric behaviour during unloading (24 right) remains qualitatively wrong independently of Young modulus. The hypoplastic model [10, 45] (dotted line) does not show this defficiency and captures both phenomena correctly in agreement with experimental results (e.g. [21]).

Calculation of strains

Constitutive models are relationships between stress and strain increments and hence the calculation of strains from displacements and vice versa is an indispensable component of numerical simulations. Nevertheless, confusement can arise from the method for calculation of strains. Whereas displacements can be measured in reality, strains are but a notion in mechanics which can be formulated in several ways.

Many computer codes and constitutive models assume tacitly "small strains", i.e. geometrical effects are negligible. However, in many cases strains can become "large", either locally in a small region or globally in a substantial domain. Especially dealing with simulation of shear zones or problems with the stiffness magnitude comparable to the stress level, the strain formulation affects the calculation results remarkably.

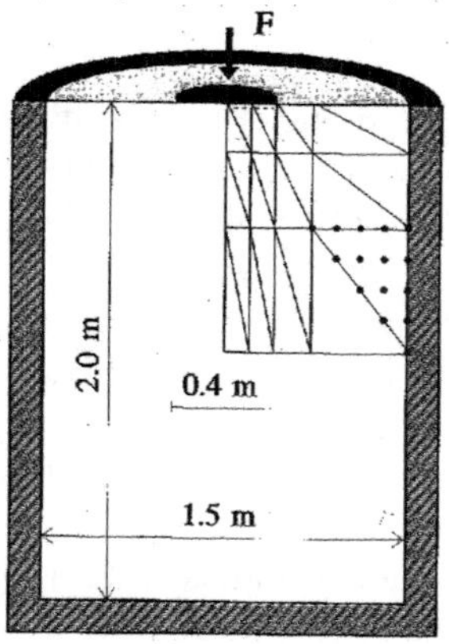

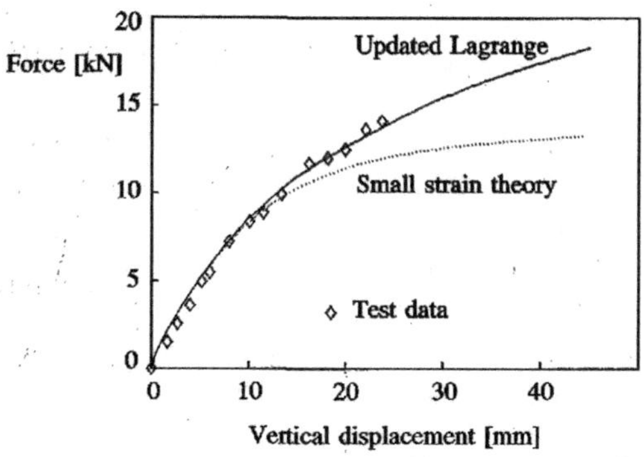

Fig. 25. Influence of the calculation of strains in a FE simulation of plate load experiment [7]

Penetration of a footing into the subsoil may be a typical example. A fitted calculation with Mohr-Coulomb model is practically independent of the strain formulation for small indentation but the divergence increases with further deformation, Fig. 25 [7]. Even larger differences can be seen in the polar hypoplastic calculation of a footing strip on the surface of a dense sand [16, 42]. The small strain formulation (curve No. 4 in Fig. 26) yields much lower bearing force P than the Updated Lagrange approach (curve No. 2 in Fig. 26).

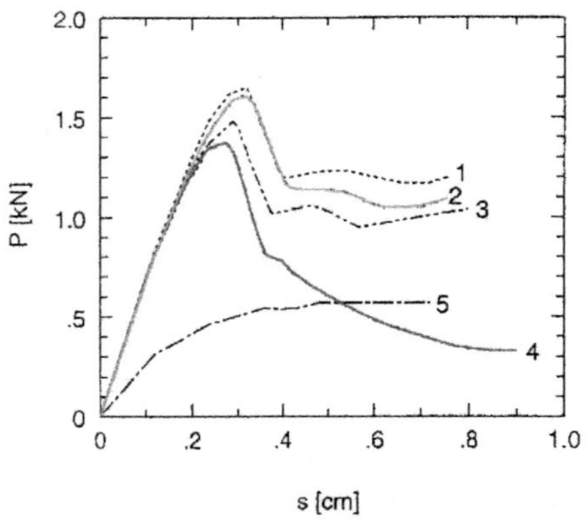

Fig. 26. Load displacement curves of a footing calculation. Compare curve No. 2 (Update Lagrange approach) with curve No. 4 (small strain approach) [16]

4.4 Idealization of reality

The right choice and application of a constitutive model is not sufficient if the boundary value problem to be simulated does not include some essential aspects of the reality. Consider an example of the settlement calculation of a reinforced embankment on soft subsoil. Taking into account plane strain conditions and the axis of symmetry of the embankment, many practitioners use one-dimensional approach ("oedometric conditions") with load provided by the embankment height (Fig. 27a). Without hardening plasticity (e.g. in case of Mohr-Coulomb model) only elastic strains develop and a single value of the maximum settlement is obtained.

Discretization of a wider domain (Fig. 27b) yields additionally the distribution of settlements beneath the embankment and, moreover, the role of plastic strains becomes important (Fig. 28 left). Differences in results are enormous and in plastic calculations the maximum settlement lies outside the symmetry axis.

As a last step one can consider a simulation of the construction of the entire embankment including reinforcement (Fig. 27c). The results change again as the reinforcement smooths the settlement differences (Fig. 28 right, for plastic calculation with stress-dependent elasticity).

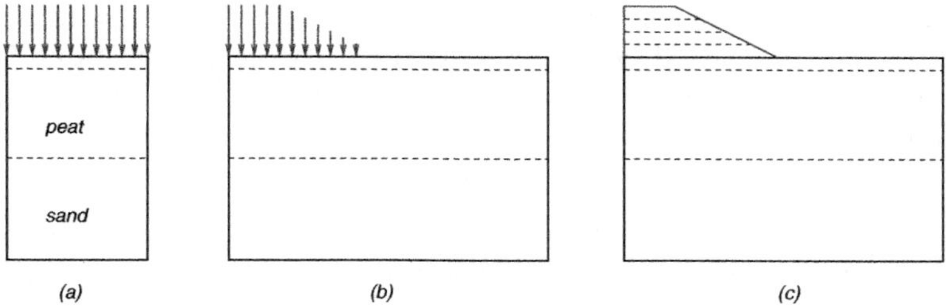

Fig. 27. Possible calculation models for a reinforeced embankment [14]

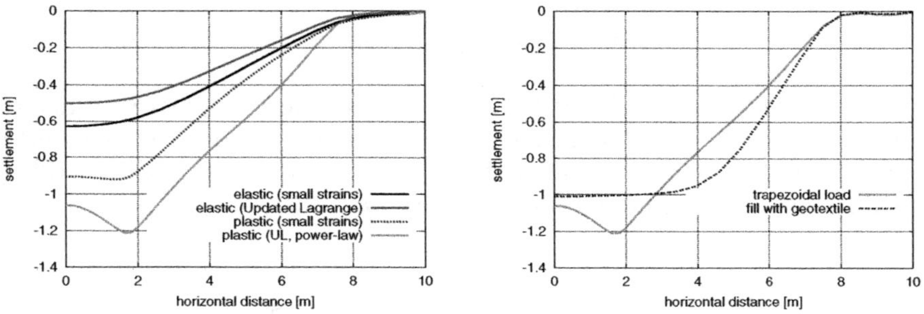

Fig. 28. Distribution of settlements beneath a reinforced embankment [14]. Plastic calculation denotes Mohr-Coulomb model and "power-law" stands for stress-dependent elasticity.

4.5 Mathematical and numerical aspects

Unfortunately, using more and more sophisticated models does not necessarily mean that we get better results. More realistic ingredients of the simulation process are inevitably connected with more complicated mathematical structures. For most geotechnical problems it is difficult, or practically impossible, to guarantee *well-posedness* which is reflected in three aspects of the mathematical solution:

1. *Existence*
   It is not difficult to imagine a set of algebraic equations without a real solution.
2. *Uniqueness*
   Even a simple quadratic equation has usually two roots.
3. *Stability*
   Small changes in input produce large changes in output.

It means that mathematically correct solutions are by no means assured. Since several years scientists realize that we live in a "chaotic" world full of bifurcations (cf. uniqueness) which can be extremely sensitive to small changes of initial and boundary conditions (cf. stability) [32].

It is rather difficult for a geotechnical engineer to abandon a fully deterministic approach although he is often confronted with instabilities. It has been proposed [17] and discussed many times (e.g. [9, 6, 18]) that the negative second order work

$$\mathring{\sigma}_{ij}\dot{\varepsilon}_{ij} < 0 \tag{1}$$

is related to the loss of uniqueness and "instability" (loss of controlability) with respect to the response of a constitutive model ($\mathring{\sigma}_{ij}$ denotes an objective stress rate and $\dot{\varepsilon}_{ij}$ deformation rate). In case of a standard triaxial test in undrained conditions, i.e. $\dot{\varepsilon}_{11} + 2\dot{\varepsilon}_{22} = 0$ and $\dot{\varepsilon}_{22} = \dot{\varepsilon}_{33}$; $\sigma_{22} = \sigma_{33}$; $\sigma_{ij} = \dot{\varepsilon}_{ij} = 0$ if $i \neq j$, this criterion reduces to:

$$\dot{\sigma}_{11}\dot{\varepsilon}_{11} + 2\dot{\sigma}_{22}\dot{\varepsilon}_{22} = 0 \quad \Longrightarrow \quad \dot{\sigma}_{11} - \dot{\sigma}_{22} \; = \; \dot{q} \; = \; 0 \tag{2}$$

Thus it can be expected that the response after the peak of stress-strain curve (Fig. 29) should cease to be unique and the response should become unstable. The real behaviour complies with this theory as may be easily verified for load-control testing in undrained conditions.

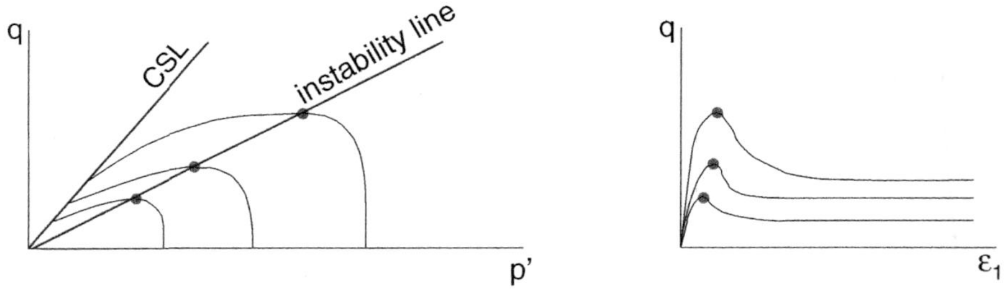

Fig. 29. Instability in undrained triaxial testing

Geotechnical calculations include many potential sources of mathematical and numerical difficulties. Material models are highly non-linear and localized deformation is often related to bifurcations manifested in shear localization and loss of controllability. Shear bands represent also discontinuity of the solution in space and introduce remarkably different scales into the problem. Consequently, it arises the question of *computability* [5]. Due to high level of mathematical difficulties one even speaks of *computer games* for problems involving geometric and material instabilities (cf. bifurcations), different resolutions (cf. shear bands) and discontinuities (in time and space) [4].

5 Concluding remarks

The reported image of the present situation in geotechnical calculations is not isolated. It is possible to find critical and warning voices throughout the whole branch of computational mechanics [5]. In car industry CAE (computer-aided engineering) one finds that "simulation science is in a state of an excessive admiration of itself" and "models are growing in size, but not in terms of physical content" [23, 24].

Prediction competitions and benchmarks in geotechnical engineering give us the lesson that our ability to submit reliable numerical prediction is very limited. Moreover, there still remain many topics related to geotechnical simulations which were not discussed in this paper, among them averaging procedures in multiphase continua (partial saturation), soil dynamics (inertial and damping effects, wave propagation) or time- and rate-dependence. They further increase the difficulty level of calculations.

The current status may seem rather controversial. On one hand we need models which involve the salient features of the soil behaviour (non-linearity, irreversibility, pressure- density- and path-dependence, ...), on the other hand increasing complexity of material models poses additional mathematical difficulties which can be hardly overcome.

Anyway, at least the necessity to use better constitutive models should be accepted. We should abandon classical soil parameters like E or φ, which are not constants for any soil. Using them implies linearity in many respects which contradicts the observed behaviour. Of course, this trend must start in education which has, probably together with standards and recommendations, the largest inertia to keep conventional way of doing.

Acknowledgement

This research was supported by a *Marie Curie Fellowship* of the European Community programme *Improving Human Research Potential* under contract number HPMF-CT-2000-01108.

References

[1] Atkinson J, Farrar D (1985) Stress path tests to measure soils strength parameters for shallow landslips. In: Proc XI ICSMFE, San Francisco 2: 983–986. Balkema

[2] Baldi G, Hight DW, Thomas GE (1988) A reevaluation of conventional triaxial test methods. In: Donaghe RT, Chaney RC, Marshall LS (eds), Advanced triaxial testing of soil and rock, ASTM STP 977, 219–263. American Society for Testing and Materials

[3] Bauer E (1992) Zum mechanischen Verhalten granularer Stoffe unter vorwiegend ödometrischer Beanspruchung. Veröffentlichungen des Institutes für Bodenmechanik und Felsmechanik der Universität Fridericiana in Karlsruhe, Heft 130

[4] Belytschko T (1996) On difficulty levels in non linear finite element analysis of solids. IACM Expressions 2:6–8

[5] Belytschko T, Mish K (2001) Computability in non-linear solid mechanics. Int J Num Meth Eng 52:3–21
[6] Bigoni D, Hueckel T (1991) Uniqueness and localization — I. Associative and non-associative elastoplasticity. Int J Solids Structures 28(2):197–213
[7] Brinkgreve RBJ (1994) Geomaterial models and numerical analysis of softening. PhD-Thesis, Delft Technical University
[8] Burland J (1989) "Small is beautiful" — the stiffness of soils at small strains. Canadian Geotechnical J 26:499–516
[9] Desrues J, Chambon R (1989) Shear band analysis for granular materials: the question of incremental non-linearity. Ingenieur Archiv 59:187–196
[10] Gudehus G (1996) A comprehensive constitutive equation for granular materials. Soils and Foundations 36(1):1–12
[11] Gudehus G, Darve F, Vardoulakis I (eds) (1984) Results of the International Workshop on Constitutive Relations for Soils, Grenoble. Balkema
[12] Herle I, Doanh T, Wu W (2000) Comparison of hypoplastic and elastoplastic modelling of undrained triaxial tests on loose sand. In: Kolymbas D (ed) Constitutive Modelling of Granular Materials, Horton 333–351. Springer
[13] Herle I, Feda J (2002) Interaction of spread footings with sandy subsoil. Part 2: Plane strain FE modelling. Engineering Mechanics 9(4):259–272
[14] Herle I, Herle V (2001) Road construction on a soft organic subsoil. In: Proc 15th Int Conf Soil Mech Geotech Eng, Istanbul 3:2081–2084. A.A.Balkema Publishers.
[15] Herle I, Mayer PM (1999) Verformungsberechnung einer Unterwasserbetonbaugrube auf der Grundlage hypoplastisch ermittelter Parameter des Berliner Sandes. Bautechnik 76(1):34–48
[16] Herle I, Tejchman J (1997) Effect of grain size and pressure level on bearing capacity of footings on sand. In: Asaoka A, Adachi T, Oka F (eds) IS-Nagoya'97: Deformation and Progressive Failure in Geomechanics 781–786. Pergamon
[17] Hill R (1958) A general theory of uniqueness and stability in elastic-plastic solids. J Mech Physics Solids 6:236–249
[18] Imposimato S, Nova R (1998) An investigation on the uniqueness of the incremental response of elastoplastic models for virgin sand. Mech Cohesive-Frictional Mater 3:65–87
[19] Kolymbas D (2000) The misery of constitutive modelling. In: Kolymbas D (ed) Constitutive Modelling of Granular Materials, Horton 11–24. Springer
[20] Lambe T (1973) Predictions in soil eingineering. Géotechnique 23(2):149–202
[21] Lanier J, Di Prisco C, Nova R (1991) Étude expérimentale et analyse théorique de l'anisotropie induite du sable d'Hostun. Rev Franç Géotech 57:59–74
[22] Lydon I (2000) Behaviour problems. Ground Engineering 33(11):31

[23] Marczyk J (1999) Recent trends in FEM. In: NAFEMS World Congress, Rhode Island

[24] Marczyk J (2000) Uncertainty management and knowladge generation in CAE. In: Europ Congress Comput Methods Appl Sci Eng, Barcelona

[25] Mayer PM (2000) Verformungen und Spannungsänderungen im Boden durch Schlitzwandherstellung und Baugrubenaushub. Veröffentlichungen des Institutes für Bodenmechanik und Felsmechanik der Universität Fridericiana in Karlsruhe, Heft 151

[26] Muir Wood D (2000) The role of models in civil engineering. In: Kolymbas D (ed) Constitutive Modelling of Granular Materials, Horton 37–55. Springer

[27] Niemunis A, Herle I (1997) Hypoplastic model for cohesionless soils with elastic strain range. Mech Cohesive-Frictional Mater 2(4):279–299

[28] Nova R (1988) 'Sinfonietta classica': An exercise on classical soil modelling. In: Saada, Bianchini (eds) Constitutive Equations for Granular Non-Cohesive Soils, 501–519. A.A.Balkema

[29] Peck R (1969) Advantages and limitations of the observational method in applied soil mechanics. Géotechnique 19(2):171–187

[30] Potts D, Axelsson K, Grande L, Schweiger H, Long M (eds) (2002) Guidelines for the use of advanced numerical analysis. COST C7 WGA report. Thomas Telford, London

[31] Potts DM, Zdravković L (2001) Finite element analysis in geotechnical engineering — Application. Thomas Telford, London

[32] Prigogine I, Stengers I (1983) Order Out of Chaos. Bantam Books, New York

[33] Richardson D (1988) Investigations of threshold effects in soil deformations. PhD-Thesis, City University, London

[34] Saada A, Bianchini G (eds) (1988) Constitutive Equations for Granular Non-Cohesive Soils, Cleveland. Balkema

[35] Saada A, Puccini P, Bianchini G (1988) Information package. In: Saada A, Bianchini G (eds) Constitutive Equations for Granular Non-Cohesive Soils, Cleveland 81–87. Balkema

[36] Schweiger HF (1998) Results from two geotechnical benchmark problems. In: Cividini A (ed) Proc 4th Europ Conf Num Meth Geotech Eng 645–654. Springer

[37] Schweiger HF (2001) Comparison of finite element results obtained for a geotechnical benchmark problem. In: Desai C, Kundu T, Harpalani S, Contractor D, Kemeny J (eds) Computer methods and advances in geomechanics — Proc 10th Int Conf, Tucson. A.A.Balkema.

[38] Simpson B, O'Riordan N, Croft D (1979) A computer model for the analysis of ground movements in London Clay. Géotechnique 29(2):149–175

[39] Tatsuoka F, Goto S, Tanaka T, Tani K, Kimura Y (1997) Particle size effects on bearing capacity of footings on granular material. In: Asaoka

A, Adachi T, Oka F (eds) IS-Nagoya'97 — Deformation and Progressive Failure in Geomechanics 133–138. Pergamon

[40] Tatsuoka F, Jardine R, Lo Presti D, Di Benedetto H, Kodaka T (1999) Characterising the pre-failure deformation properties of geomaterials. In: Proc XIV ICSMFE, Hamburg 4:2129–2164. A.A.Balkema

[41] Tatsuoka F, Siddiquee M, Yoshida T, Park C, Kamegai Y, Goto S, Kohata Y (1994) Testing methods and results of element tests and testing conditions of plane strain model bearing capacity tests using air-dried dense Silver Leighton Buzzard sand. Report prepared for class-a prediction of the bearing capacity performance of model surface footing on sand under plane strain conditions., Institute of Industrial Science, University of Tokyo

[42] Tejchman J, Herle I (1999) A "class A" prediction of the bearing capacity of plane strain footings on sand. Soils and Foundations 39(5):47–60

[43] Verruijt A, Booker JR (1996) Surface settlements due to deformation of a tunnel in an elastic half plane. Géotechnique 46(4):753–756

[44] Viggiani G, Tamagnini C (2000) Ground movements around excavations in granular soils: a few remarks on the influence of the constitutive assumptions on FE predictions. Mech Cohesive-Frictional Mater 5:399–423

[45] von Wolffersdorff PA (1996) A hypoplastic relation for granular materials with a predefined limit state surface. Mech Cohesive-Frictional Mater 1:251–271

[46] von Wolffersdorff PA (1997) Verformungsprognosen für Stützkonstruktionen. Veröffentlichungen des Institutes für Bodenmechanik und Felsmechanik der Universität Fridericiana in Karlsruhe, Heft 141

[47] Wroth CP (1977) The predicted performance of soft clay under a trial embankment loading based on the Cam-clay model. In: Gudehus G (ed) Finite Elements in Geomechanics 191–208. Wiley

[48] `http://www.cordis.lu`

[49] `http://www.gnu.org/licenses/gpl.html`

Objective integration and geometric properties of hypoplasticity

Wolfgang Fellin[1] and Alexander Ostermann[2]

[1] Institut für Geotechnik und Tunnelbau, Universität Innsbruck, wolfgang.fellin@uibk.ac.at
[2] Institut für Technische Mathematik, Geometrie und Bauinformatik, Universität Innsbruck, alexander.ostermann@uibk.ac.at

Summary. The constitutive equations of hypoplasticity are of the rate type. They can thus be considered as differential equations. In this paper, we are concerned with the time integration of such equations. The main challenge in time integration is to guarantee that the objectivity of the formulation is preserved by the numerical integrator. Among other possibilities, we discuss in detail a rotation neutralized description of the hypoplastic equations and its integration by projection methods.

In a second part of the paper, we use techniques from the theory of differential equations to analyze the hypoplastic equations without integrating particular trajectories, e.g. specific stress paths. For example, we show that the K_0-line in an oedomentric element test is exponentially attractive, and we compute possible bifurcation points of the hypoplastic equations for a triaxial element test.

1 Introduction

The constitutive equations of hypoplasticity have been developed to describe the stress-strain behaviour of granular materials such as sand. Mathematically spoken they form a system of differential equations for the unknown stresses: an objective stress rate equals a function of state variables and some constitutive parameters. In simple versions, the state variables are the stress tensor and the Eulerian stretching tensor.

The present paper is concerned with the numerical integration of such equations. More precisely, the objectives of this paper are twofold:

(i) to introduce some basic concepts for the objective time integration of hypoplastic equations;
(ii) to present methods and techniques from the numerical analysis of differential equations that are of interest in hypoplasticity.

The first aspect is important for implementing hypoplasticity correctly in a finite-element environment.

The second aspect is of interest for researchers who develop new variants of hypoplasticity. It is equally interesting in situations where one has to determine the material parameters of hypoplasticity in such a way that numerical simulations match the experimental results obtained from element tests.

The present article is organized as follows: In Section 2 we recall some basic notions from continuum mechanics. The concept of objectivity (also called frame-indifference) is briefly discussed in Section 2. We motivate, why the Lie derivative is an objective stress rate for the Cauchy stress tensor, and we introduce hypoplasticity. Further, we present an example of a constitutive equation.

In Section 4 we introduce the basic concepts of objective time stepping. We present two different approaches to objective time stepping. In the first one, the hypoplastic equations are transformed back to the reference coordinates. This corresponds to a convected description of the problem. The second approach is based on a so-called rotation neutralized description. For the here considered Jaumann-Zaremba stress rate, the rotations are basically determined by the spin tensor. This approach requires the solution of a system of differential equations on the special orthogonal group. This can be carried out with Lie group methods or with projection methods, as specified at the end of Section 4.

The second part of the paper starts with Section 5. There we show that the hypoplastic equations can (to some extend) be analyzed *without* integrating particular trajectories. We illustrate this with two examples: an oedometric and a triaxial element test. For the oedometric element test, we verify that the K_0-line is exponentially attractive in the case of loading. This shows that the considered constitutive equation fulfils Goldscheider's principle for sand. For the triaxial element test, we calculate the asymptotically stable equilibria for loading and unloading, and we comment on bifurcations. The analysis for the triaxial element test is based on a formulation of the hypoplastic equations as a differential equation on a manifold.

We further comment in Section 5 on recent developments of software for differential equations. We show that there is nowadays much more available than the forward and backward Euler method, and that is pays off to use modern methods. We emphasize that this statement also holds for low accuracy requirements, which are typical in engineering applications.

Our main conclusions are finally given in Section 6.

2 Notions from continuum mechanics

In this section, we recall the basic notions from three-dimensional continuum mechanics. For details, we refer to the monographs [5] and [12].

Let $\mathcal{B} \subset \mathbb{R}^3$ denote the *reference configuration* of the continuum body under consideration and let

$$\varphi : \mathcal{B} \times [0,T] \to \mathbb{R}^3 : \mathbf{p} \mapsto \mathbf{x} = \varphi(\mathbf{p},t) =: \varphi_t(\mathbf{p})$$

be its *motion* on the time interval $[0,T]$. The motion maps the *material coordinates* of a reference point $\mathbf{p}$ to the corresponding *spatial coordinates* $\mathbf{x}$ at time t. Since matter cannot penetrate itself, the mapping $\varphi_t : \mathcal{B} \to \mathbb{R}^3$ is one-to-one and thus invertible on its image. We denote the inverse by φ_t^{-1}. The *deformation gradient* $\mathbf{F}(\mathbf{p},t)$ of a motion is defined by

$$\mathbf{F}(\mathbf{p},t) = \nabla_{\mathbf{p}}\,\varphi(\mathbf{p},t).$$

The *spatial velocity field* is the time derivative of the motion, expressed in spatial coordinates

$$\mathbf{v}(\mathbf{x},t) = \left.\frac{\partial\varphi(\mathbf{p},t)}{\partial t}\right|_{\mathbf{p} = \varphi_t^{-1}(\mathbf{x})}. \tag{1}$$

The time derivative of the deformation gradient is the material velocity gradient. Its spatial description is called *spatial velocity gradient* and given by

$$\mathbf{L}(\mathbf{x},t) = \nabla_{\mathbf{x}}\,\mathbf{v}(\mathbf{x},t) = \left.\frac{\partial\mathbf{F}(\mathbf{p},t)}{\partial t}\mathbf{F}(\mathbf{p},t)^{-1}\right|_{\mathbf{p} = \varphi_t^{-1}(\mathbf{x})}. \tag{2}$$

The symmetric part of the spatial velocity gradient is called *stretching tensor*

$$\mathbf{D}(\mathbf{x},t) = \frac{1}{2}\Big(\mathbf{L}(\mathbf{x},t) + \mathbf{L}(\mathbf{x},t)^T\Big),$$

the skew-symmetric part is the *spin tensor*

$$\mathbf{W}(\mathbf{x},t) = \frac{1}{2}\Big(\mathbf{L}(\mathbf{x},t) - \mathbf{L}(\mathbf{x},t)^T\Big). \tag{3}$$

In the aforementioned quantities, the time t is often viewed as a parameter. In order to stress this point of view, we also write

$$\varphi_t = \varphi(\cdot,t), \quad \mathbf{F}_t = \mathbf{F}(\cdot,t), \quad \mathbf{v}_t = \mathbf{v}(\cdot,t), \quad \mathbf{D}_t = \mathbf{D}(\cdot,t), \quad \text{etc.}$$

Another advantage of this notation is that the above relations can be written more compactly. E.g., the relation for the spatial velocity field (1) becomes

$$\mathbf{v}_t = \frac{\partial\varphi_t}{\partial t} \circ \varphi_t^{-1}.$$

Here, the symbol $\circ$ denotes as usual the composition of two mappings. With these preparations, we are now in the state to formulate the constitutive equations of hypoplasticity.

3 Hypoplasticity

We first consider, how the above tensors transform under a change of the observer. Let

$$\widetilde{\mathbf{x}} = \widetilde{\varphi}(\mathbf{p},t) = \mathbf{q}(t) + \mathbf{Q}(t)\varphi(\mathbf{p},t) = \mathbf{q}(t) + \mathbf{Q}(t)\,\mathbf{x}$$

be a superposed rigid body motion. Here $\mathbf{q}(t)$ is a translation and $\mathbf{Q}(t)$ a rotation. Using (2) and $\widetilde{\mathbf{F}}_t = \mathbf{Q}(t)\mathbf{F}_t$ shows at once that the stretching tensor transforms objectively, whereas the spin tensor does *not*

$$\widetilde{\mathbf{D}}_t = \mathbf{Q}(t)\mathbf{D}_t\mathbf{Q}(t)^T, \qquad \widetilde{\mathbf{W}}_t = \mathbf{Q}(t)\mathbf{W}_t\mathbf{Q}(t)^T + \dot{\mathbf{Q}}(t)\mathbf{Q}(t)^T.$$

The material time derivative of the transformed Cauchy stress tensor

$$\frac{\mathrm{d}}{\mathrm{d}t}\Big(\mathbf{Q}(t)\mathbf{T}_t\mathbf{Q}(t)^T\Big) = \mathbf{Q}(t)\dot{\mathbf{T}}_t\mathbf{Q}(t)^T + \dot{\mathbf{Q}}(t)\mathbf{Q}(t)^T\mathbf{T}_t - \mathbf{T}_t\dot{\mathbf{Q}}(t)\mathbf{Q}(t)^T$$

is *not* an objective quantity either. The geometric object that is behind an objective stress rate in the Eulerian description is the *Lie derivative with respect to the vector field* $\mathbf{v}_t$. The Lie derivative of the Cauchy stress tensor is given by

$$\mathfrak{L}_{\mathbf{v}_t}\mathbf{T}_t := \left(\mathbf{F}_t\frac{\mathrm{d}}{\mathrm{d}t}\Big(\mathbf{F}_t^{-1}(\mathbf{T}_t\circ\varphi_t)\mathbf{F}_t^{-T}\Big)\mathbf{F}_t^T\right)\circ\varphi_t^{-1}.$$

Recall that the composition of $\mathbf{T}_t$ and φ_t is defined by $(\mathbf{T}_t\circ\varphi_t)(p) = \mathbf{T}_t\big(\varphi_t(p)\big) = \mathbf{T}(\mathbf{x},t)$. The Lie derivative corresponds to a *pull-back* to the reference configuration, followed by a differentiation in the convected description and a *push-forward* to the spatial description, see e.g. Abraham and Marsden [1].

Constitutive equations in hypoplasticity are usually formulated with the *Jaumann-Zaremba* stress rate

$$\overset{\circ}{\mathbf{T}} := \dot{\mathbf{T}} - \mathbf{W}\mathbf{T} + \mathbf{T}\mathbf{W},$$

which differs from the Lie derivative by an objective quantity

$$\overset{\circ}{\mathbf{T}} = \mathfrak{L}_{\mathbf{v}}\mathbf{T} + \mathbf{D}\mathbf{T} + \mathbf{T}\mathbf{D}.$$

The constitutive equations in hypoplasticity are of the form

$$\overset{\circ}{\mathbf{T}} = \mathbf{H}(\mathbf{T},\mathbf{D}) \tag{4}$$

and represent an evolution equation for the Cauchy stress tensor. Note that the right-hand side of (4) is objective, i.e. for any rotation $\mathbf{Q}$ it holds

$$\mathbf{Q}^T\mathbf{H}(\mathbf{T},\mathbf{D})\,\mathbf{Q} = \mathbf{H}\big(\mathbf{Q}\mathbf{T}\mathbf{Q}^T,\mathbf{Q}\mathbf{D}\mathbf{Q}^T\big).$$

Therefore, the change from the Jaumann-Zaremba stress rate to any objective stress rate just corresponds to an objective modification of the right-hand side of (4). For a thorough introduction to hypoplasticity, we refer to the monograph of Kolymbas [10].

Modern hypoplastic equations depend further on additional state variables, e.g. the void ratio. For simplicity, we will base our computations in Section 5 on the following version of (4) by Wu and Kolymbas [15]

$$\overset{\circ}{\mathbf{T}} = C_1(\operatorname{tr}\mathbf{T})\mathbf{D} + C_2\frac{\operatorname{tr}\mathbf{TD}}{\operatorname{tr}\mathbf{T}}\mathbf{T} + C_3\frac{\mathbf{T}^2}{\operatorname{tr}\mathbf{T}}\sqrt{\operatorname{tr}\mathbf{D}^2} + C_4\frac{\mathbf{T}^{*2}}{\operatorname{tr}\mathbf{T}}\sqrt{\operatorname{tr}\mathbf{D}^2}. \quad (5)$$

Here, $\mathbf{T}^* = \mathbf{T} - \frac{1}{3}(\operatorname{tr}\mathbf{T})\,\mathbf{I}$ denotes the *deviatoric stress*, and C_1, C_2, C_3 and C_4 are material parameters.

4 Objective time stepping

In this section, we discuss some strategies for solving the hypoplastic equations numerically. Our main emphasis will lie on algorithms that preserve the objectivity of the constitutive equations. We assume that the motion φ_t is known for all $t \in [0,T]$. An excellent introduction to *incrementally objective time stepping* where the motion is only known for a set of discrete points t_n is given in the monograph by Simo and Hughes [12].

We are concerned with the numerical solution of

$$\overset{\circ}{\mathbf{T}} = \dot{\mathbf{T}} - \mathbf{WT} + \mathbf{TW} = \mathbf{H}(\mathbf{T},\mathbf{D}).$$

Note that standard discretizations of

$$\dot{\mathbf{T}} = \mathbf{WT} - \mathbf{TW} + \mathbf{H}(\mathbf{T},\mathbf{D})$$

are, in general, *not* objective. We will basically discuss two different approaches for objective time stepping. In the first approach, the hypoplastic equations are transformed back to the reference configuration (convected description). The second approach uses a rotation neutralized description of the spatial hypoplastic equations and is related to the solution of differential equations on Lie groups.

4.1 Convected description

We have seen in Section 2 that it is sufficient to consider numerical methods for

$$\mathcal{L}_{\mathbf{v}}\mathbf{T} = \mathbf{H}(\mathbf{T},\mathbf{D}) - \mathbf{DT} - \mathbf{TD} = \widetilde{\mathbf{H}}(\mathbf{T},\mathbf{D}).$$

The main idea of the convected description is to transform this evolution problem back to the reference configuration. Let

$$\mathbf{S}(\mathbf{p},t) = \mathbf{F}(\mathbf{p},t)^{-1}\,\mathbf{T}\big(\varphi(\mathbf{p},t),t\big)\,\mathbf{F}(\mathbf{p},t)^{-T},$$

denote the pull-back of the Eulerian stress tensor to the reference configuration. For short, we write this relation as

$$\mathbf{S}_t = \mathbf{F}_t^{-1}(\mathbf{T}_t \circ \varphi_t)\,\mathbf{F}_t^{-T}.$$

Then $\mathbf{S}$ satisfies in Lagrangian coordinates the ordinary differential equation

$$\frac{\mathrm{d}}{\mathrm{d}t}\mathbf{S}_t = \mathbf{F}_t^{-1}\,\widetilde{\mathbf{H}}\Big(\mathbf{F}_t\,\mathbf{S}_t\,\mathbf{F}_t^T,\,\mathbf{D}_t \circ \varphi_t\Big)\mathbf{F}_t^{-T}. \tag{6}$$

Since the Lagrangian coordinates are not affected by rotations, this differential equation can be discretized by *any* numerical method, as long as the motion φ_t is known for all $t \in [0,T]$. For example, the forward Euler discretization of (6) is given by

$$\frac{\mathbf{S}_{n+1}-\mathbf{S}_n}{h} = \mathbf{F}_n^{-1}\,\widetilde{\mathbf{H}}\Big(\mathbf{F}_n\,\mathbf{S}_n\,\mathbf{F}_n^T,\mathbf{D}_n \circ \varphi_n\Big)\,\mathbf{F}_n^{-T}.$$

The push-forward to the spatial description

$$\mathbf{T}_{n+1} = \Big(\mathbf{F}_{n+1}\mathbf{S}_{n+1}\mathbf{F}_{n+1}^T\Big) \circ \varphi_{n+1}^{-1}$$

gives then an objective discretization of the hypoplastic equations.

4.2 Rotation neutralized description

An alternative way to obtain objective algorithms is based on the concepts of Lie groups and Lie algebras. For an introduction to this subject, see Olver [11].

The main idea is to consider an auxiliary matrix-valued initial value problem

$$\frac{\mathrm{d}}{\mathrm{d}t}\mathbf{Y}(t) = \mathbf{A}(t)\mathbf{Y}(t), \qquad \mathbf{Y}(0) = \mathbf{I} \tag{7}$$

with skew-symmetric $\mathbf{A}(t)$. The set of skew-symmetric matrices

$$\mathrm{so}(3) = \{3 \times 3 \text{ matrices } \mathbf{A} \mid \mathbf{A} + \mathbf{A}^T = \mathbf{0}\}$$

forms a vector space, and a *Lie algebra* with the matrix commutator

$$[\mathbf{A},\mathbf{B}] = \mathbf{AB} - \mathbf{BA}$$

as the bracket. On the other hand, the set of orthogonal matrices with determinant one forms a matrix group, the so-called *special orthogonal group*

$$\mathrm{SO}(3) = \{3 \times 3 \text{ matrices } \mathbf{Y} \mid \mathbf{Y}^T\mathbf{Y} = \mathbf{I},\ \det \mathbf{Y} = 1\}$$

which is the corresponding *Lie group*. Lie algebra and Lie group are related by the matrix exponential

$$\exp : \mathrm{so}(3) \to \mathrm{SO}(3) : \mathbf{A} \mapsto \exp(\mathbf{A}) = \sum_{k=0}^{\infty} \frac{\mathbf{A}^k}{k!}$$

which maps skew-symmetric matrices to orthogonal ones.

Differentiating the expression $\mathbf{Y}^T\mathbf{Y}$ with respect to t and inserting the differential equation (7) yields

$$\frac{\mathrm{d}}{\mathrm{d}t}\left(\mathbf{Y}^T\mathbf{Y}\right) = \frac{\mathrm{d}\mathbf{Y}^T}{\mathrm{d}t}\mathbf{Y} + \mathbf{Y}^T\frac{\mathrm{d}\mathbf{Y}}{\mathrm{d}t} = (\mathbf{AY})^T\mathbf{Y} + \mathbf{Y}^T\mathbf{AY} = \mathbf{Y}^T(\mathbf{A}^T + \mathbf{A})\mathbf{Y} = \mathbf{0}.$$

This shows that $\mathbf{Y}^T(t)\mathbf{Y}(t)$ is a *first integral* of (7), i.e. it remains constant along solutions. Since the initial value $\mathbf{Y}(0) = \mathbf{I}$ is orthogonal, we have that $\mathbf{Y}(t)$ remains orthogonal for all t. Equation (7) is thus a differential equation on the Lie group SO(3).

Having the Jaumann-Zaremba rate in mind, we choose $\mathbf{A}(t) = \mathbf{W}(\varphi_t, t)$, where $\mathbf{W}$ is the spin tensor (3) in Eulerian coordinates. Let

$$\mathbf{S}(t) := \mathbf{Y}^T(t)\,\mathbf{T}(\varphi_t, t)\mathbf{Y}(t)$$

be the rotated stress tensor. Differentiating $\mathbf{S}(t)$ with respect to t and inserting (7) gives

$$\frac{\mathrm{d}}{\mathrm{d}t}\mathbf{S}(t) = \mathbf{Y}^T(t)\,\mathring{\mathbf{T}}(\varphi_t, t)\,\mathbf{Y}(t) = \mathbf{H}\Big(\mathbf{S}(t), \mathbf{Y}^T(t)\,\mathbf{D}(\varphi_t, t)\,\mathbf{Y}(t)\Big). \qquad (8)$$

This equation has to be solved simultaneously with (7). Since (8) is an ordinary differential equation in the reference configuration, *any* numerical method can be used to solve it. For the solution of (7), however, particular methods are necessary which will be described in the following two subsections.

Lie group methods

We still have to solve the auxiliary differential equation (7) on the Lie group SO(3). Following an approach of Magnus (cf. [6, Section IV.7]), we search a numerical solution $\mathbf{Y}_n \approx \mathbf{Y}(nh)$ of the form

$$\mathbf{Y}_{n+1} = \exp\big(h\,\boldsymbol{\Omega}_n\big)\mathbf{Y}_n.$$

Since the matrix exponential maps skew-symmetric matrices to orthogonal ones, the method stays on SO(3) as long as $\boldsymbol{\Omega}_n$ is a skew-symmetric matrix.

A simple yet interesting method is the implicit midpoint rule, given by

$$\mathbf{Y}_{n+1} = \exp\Big(h\,\mathbf{A}\big(nh + h/2\big)\Big)\mathbf{Y}_n, \qquad \mathbf{Y}_0 = \mathbf{I}.$$

It has order 2 and can be evaluated easily with the help of the Rodrigues formula, which in our case is just a polynomial degree 2, see [12, Section 8.3].

Methods of higher order exist as well, but they are more complicated due to the involved matrix commutators, see [6].

Projection methods

In order to solve (7) on the Lie group SO(3), we consider SO(3) as a submanifold of $\mathbb{R}^9$. A natural approach for solving differential equations on manifolds are projection methods. We describe here a method originally proposed by Higham [9], see also [6, Example IV.4.6].

Starting from the numerical approximation $\mathbf{Y}_n \approx \mathbf{Y}(nh)$ on SO(3), we first compute a numerical approximation $\widetilde{\mathbf{Y}}_{n+1}$ to the exact solution at time $t = (n+1)h$ by *any* numerical method. An improved approximation $\mathbf{Y}_{n+1}$ is then computed by projecting $\widetilde{\mathbf{Y}}_{n+1}$ back onto the manifold SO(3). This projection is achieved by

$$\text{minimize } \|\mathbf{Y}_{n+1} - \widetilde{\mathbf{Y}}_{n+1}\|_F \quad \text{subject to} \quad \mathbf{Y}_{n+1}^T \mathbf{Y}_{n+1} = \mathbf{I},$$

where $\|\mathbf{M}\|_F = \sqrt{\operatorname{tr}(\mathbf{M}^T\mathbf{M})}$ denotes the Frobenius norm of $\mathbf{M}$. Therefore, the searched projection $\mathbf{Y}_{n+1}$ is just the first factor of the *polar decomposition*

$$\widetilde{\mathbf{Y}}_{n+1} = \mathbf{Y}_{n+1}\,\mathbf{P}$$

where $\mathbf{Y}_{n+1}$ is orthogonal, and $\mathbf{P}$ is symmetric and positive definite. Note that for symmetric tensors the Frobenius norm is $\|\mathbf{D}\|_F = \sqrt{\operatorname{tr}(\mathbf{D}^2)} = |\mathbf{D}|$.

The main advantage of this projection approach is that all existing numerical integrators can be used. Equations (8) and (7) can be solved simultaneously with the same numerical method, and the solution of (7) is projected back to SO(3) after each successful step of the integrator.

5 Geometric theory for element tests

The *geometric* or *qualitative theory* of differential equations is concerned with the properties of the corresponding dynamical system. For our purpose, a dynamical system is a mapping that assigns the initial values of a differential equation to the corresponding solutions of the equation at time t. In that sense the hypoplastic equations form a dynamical system. The objects of interest in dynamical systems are equilibria, bifurcation points, etc., as well as their stability properties.

Some knowledge of the qualitative properties of the hypoplastic equations is important for choosing an appropriate numerical integrator. For example, it is well-known that nonstiff and stiff problems need different integrators. But also problems with particular features often need integrators that are specially tailored to the problem.

Element tests are essential for finding the appropriate constitutive equations. They are further indispensable for the calibration of the material parameters. In the following two subsections we illustrate with two simple examples, how useful information can be drawn from the hypoplastic equations, *without* integrating particular trajectories, e.g. specific stress paths.

5.1 Oedometric element test

The oedometric test simulates a one-dimensional deformation. Since no lateral deformations are allowed and due to axial symmetry, we have

$$\mathbf{D} = \begin{bmatrix} D_1 & 0 & 0 \\ 0 & 0 & 0 \\ 0 & 0 & 0 \end{bmatrix} \qquad \text{and} \qquad \mathbf{T} = \begin{bmatrix} T_1 & 0 & 0 \\ 0 & T_2 & 0 \\ 0 & 0 & T_2 \end{bmatrix} .$$

Here, D_1 is considered as parameter. The unknowns of the problem are thus T_1 and T_2. Inserting these relations into the hypoplastic equations (5) results in the following two-dimensional dynamical system

$$\begin{aligned} \frac{\mathrm{d}}{\mathrm{d}t} T_1 = {} & C_1 (T_1 + 2T_2) D_1 + C_2 \frac{T_1 D_1}{T_1 + 2T_2} T_1 + C_3 \frac{T_1^2 \, |D_1|}{T_1 + 2T_2} \\ & + \frac{4C_4}{9} (T_1 - T_2)^2 \frac{|D_1|}{T_1 + 2T_2} , \end{aligned} \tag{9a}$$

$$\frac{\mathrm{d}}{\mathrm{d}t} T_2 = C_2 \frac{T_1 D_1}{T_1 + 2T_2} T_2 + C_3 \frac{T_2^2 \, |D_1|}{T_1 + 2T_2} + \frac{C_4}{9} (T_1 - T_2)^2 \frac{|D_1|}{T_1 + 2T_2} . \tag{9b}$$

Since the right-hand side of (9) is positively homogeneous in D_1, we can restrict our attention to $D_1 = \mp 1$ (loading/unloading). This reduction is achieved by an appropriate scaling of time.

We study the right-hand side of (9) with the help of the symbolic package `Maple` for $T_1 \leq 0$, $T_2 \leq 0$. In our simulations, we use the parameters for loose Karlsruhe sand (due to Bauer [2])

$$C_1 = -69.23, \quad C_2 = -670.72, \quad C_3 = -653.26, \quad C_4 = 690.9 . \tag{10}$$

The thin straight lines in the phase portrait in Fig. 1 are *invariant* under the dynamics of the differential equation. This means that for initial values in this set, the solution stays in this set for all times. In the theory of dynamical systems, such sets are called *invariant manifolds*.

For loading, the invariant manifold is given by $T_2 = 0.51\, T_1$. This corresponds to the *earth pressure coefficient* at rest $K_0 = T_2/T_1 = 0.51$ that can be calculated directly from (9). The thick lines in Fig. 1 are the limit states for triaxial conditions. They will be calculated in the following subsection. Note that, in principle, only stress states in between these lines are admissible.

The Jacobian of the right-hand side of (9), restricted to the invariant manifold (K_0-line), has the eigenvalues

$$\lambda_1 = 184.99, \qquad \lambda_2 = -100.34 .$$

The positive eigenvalue corresponds to the evolution on the invariant manifold $\dot{z}(t) = 184.99\, z(t)$, whereas the negative eigenvalue *proves* that the K_0-line

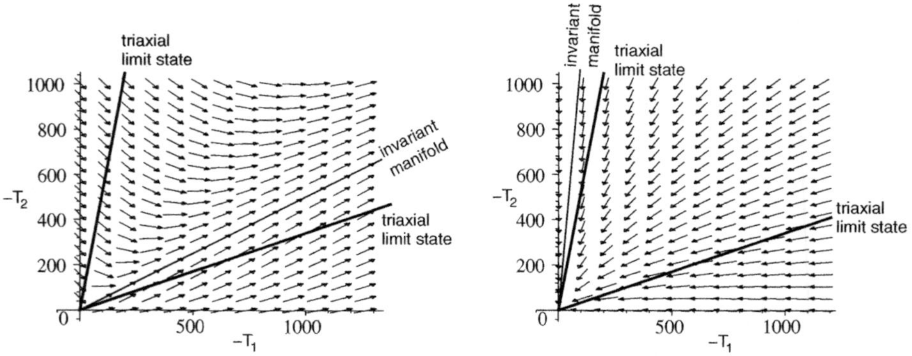

Fig. 1: Oedometric element test for loading (left) and unloading (right). Phase portrait of (9) with `Maple`.

is asymptotically stable and attracts all its neighbouring solutions with an exponential speed. This shows that (9) fulfils Goldscheider's principle [4] for sand, which is also seen from Fig. 1, left.

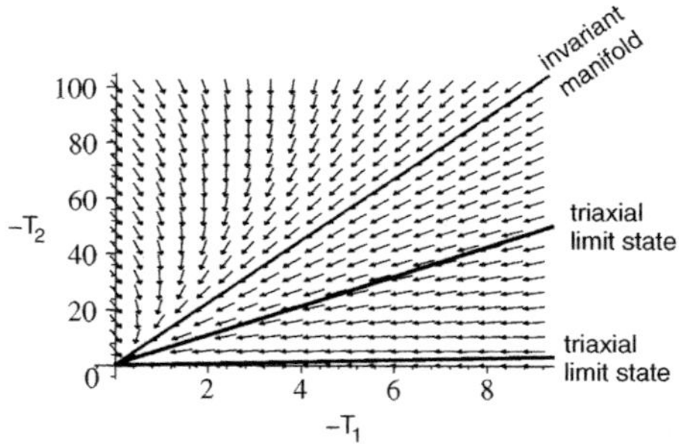

Fig. 2: Oedometric element test for unloading. Zoomed and distorted phase portrait of (9) with `Maple`.

For unloading, the invariant manifold is given by $T_2 = 11.23\,T_1$, see Fig. 2. It lies only slightly outside the limit state, see Fig. 1. This situation is related to the so-called bound surface, which lies outside the failure surface for this version of hypoplasticity, cf. [13].

Accuracy

Next, we study the achieved accuracy for three loading and unloading cycles. We compare the forward Euler method with an explicit Runge-Kutta method.

The aim of this comparison is to show that high-order methods with automatic error and stepsize control are generally preferable to the forward Euler method.

time	D_1	T_1	T_2
0.000	−1	−10	−10
0.020	1	−407	−208
0.025	−1	−15	−22
0.045	1	−660	−338
0.051	−1	−13	−24
0.071	1	−641	−328
0.076		−23	−35

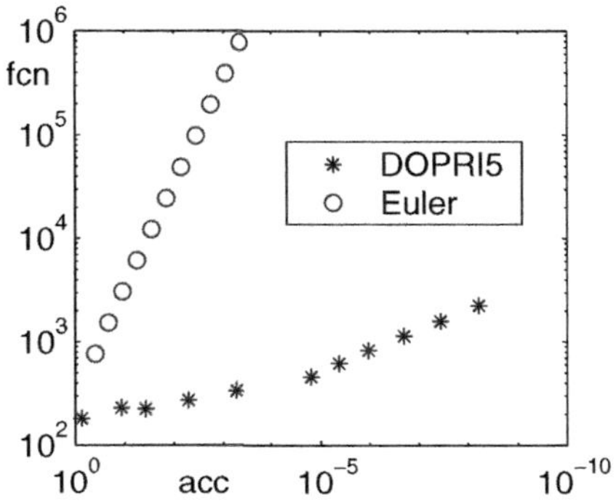

Fig. 3: Numerical work (number of right-hand side evaluations fcn) as a function of the achieved accuracy (acc). Comparison of the forward Euler method with the explicit Runge-Kutta method DOPRI5. The relevant data *time* and D_1 for the simulation are given in the table on the left.

DOPRI5 is an explicit Runge-Kutta method of order 5, based on a coefficient set by Dormand and Prince. We used the FORTRAN-implementation by E. Hairer, see [7]. The method needs seven function evaluations per step. One step with DOPRI5 is therefore seven times more expensive than a forward Euler step. However, DOPRI5 is very efficient since it achieves high accuracy with large steps. This is obvious from Fig. 3, which shows that DOPRI5 is *always* superior to the forward Euler method. For three digits of accuracy, e.g., it is about 50 times faster than the forward Euler method.

Further features of DOPRI5 include the following. The program supplies a continuous output of the solution. This can be used for graphical purposes or to stop the integration at a prescribed stress. It has an automatic error and stepsize control which allows adaptivity. We deem this to be indispensable for serious simulations. DOPRI5 further has good stability properties and an automatic stiffness detection. It is our favorite code for nonstiff problems.

5.2 Triaxial element test

Next, we perform a triaxial compression test. Let

$$\mathbf{D} = \begin{bmatrix} D_1 & 0 & 0 \\ 0 & D_2 & 0 \\ 0 & 0 & D_2 \end{bmatrix} \quad \text{and} \quad \mathbf{T} = \begin{bmatrix} T_1 & 0 & 0 \\ 0 & T_2 & 0 \\ 0 & 0 & T_2 \end{bmatrix} .$$

Here, D_1 is considered as parameter and T_2 is kept constant. The unknowns of the problem are thus T_1 and D_2. Inserting the above relations into the hypoplastic equations (5) results in the following two-dimensional dynamical system

$$
\begin{aligned}
\frac{\mathrm{d}}{\mathrm{d}t}T_1 = C_1(T_1+2T_2)D_1 + C_2\frac{T_1D_1+2T_2D_2}{T_1+2T_2}T_1 + C_3\frac{T_1^2\sqrt{D_1^2+2D_2^2}}{T_1+2T_2} \\
+ \frac{4C_4}{9}(T_1-T_2)^2\frac{\sqrt{D_1^2+2D_2^2}}{T_1+2T_2},
\end{aligned}
\tag{11a}
$$

$$
\begin{aligned}
0 = C_1(T_1+2T_2)D_2 + C_2\frac{T_1D_1+2T_2D_2}{T_1+2T_2}T_2 + C_3\frac{T_2^2\sqrt{D_1^2+2D_2^2}}{T_1+2T_2} \\
+ \frac{C_4}{9}(T_1-T_2)^2\frac{\sqrt{D_1^2+2D_2^2}}{T_1+2T_2}.
\end{aligned}
\tag{11b}
$$

We study the system for $T_1 \le 0$ and $D_1 = \pm 1$, again with the parameters for loose Karlsruhe sand (10).

The underlying mathematical structure of this triaxial compression test is a so-called *differential equation on a manifold*

$$\frac{\mathrm{d}y}{\mathrm{d}t} = f(y,z), \qquad 0 = g(y,z). \tag{12}$$

We are thus given a differential equation for $y = T_1$ and a nonlinear equation for $z = D_2$. Differentiation of the constraining manifold $0 = g(y,z)$ with respect to t gives

$$0 = \frac{\partial g}{\partial y}(y,z)\,\frac{\mathrm{d}y}{\mathrm{d}t} + \frac{\partial g}{\partial z}(y,z)\,\frac{\mathrm{d}z}{\mathrm{d}t}. \tag{13}$$

If the derivative of g with respect to z is invertible (*index-1 condition*), we can formally solve (13) for the derivative of z. Inserting (12), we finally get a differential equation for z

$$\frac{\mathrm{d}z}{\mathrm{d}t} = -\left(\frac{\partial g}{\partial z}(y,z)\right)^{-1}\frac{\partial g}{\partial y}(y,z)\,f(y,z). \tag{14}$$

Equations (12) and (14) form a system of differential equations and therefore posses a locally unique solution (for Lipschitz continuous right-hand sides). When the index-1 condition is violated, the uniqueness of the solution can be lost and bifurcations occur.

A simple example where bifurcation occurs is the following dynamical system

$$\frac{\mathrm{d}y}{\mathrm{d}t} = z, \qquad y^2 + z^2 = 1.$$

For the (consistent) initial values $y(0) = 0$ and $z(0) = 1$, the index-1 condition is satisfied. The locally unique solution

$$y(t) = \sin t, \qquad z(t) = \cos t$$

exists for all t. At $t = \pi/2$, however, the index-1 condition

$$\frac{\partial g}{\partial z} = 2z = 2\cos(\pi/2) = 0$$

is first violated and bifurcation takes place. A second solution through $(y, z) = (1, 0)$ is given by

$$y \equiv 1, \qquad z \equiv 0.$$

Implicit Runge-Kutta methods like RADAU5 can solve the index-1 problem (12) directly *without* performing the reduction to (14). This is achieved by introducing an appropriate mass matrix, as explained in [8]. Note that the user nevertheless has to supply the code with *consistent* initial values y_0 and z_0, i.e. values on the manifold $g(y_0, z_0) = 0$, in order to commence the integration of (12) properly.

The FORTRAN-code RADAU5 is an implicit Runge-Kutta method of order 5, based on the Radau IIA formulas. It has been implemented by Hairer and Wanner and is described in their monograph [8]. The method needs three function evaluations per step and is in general very efficient due to the employed modified Newton iterations. Its excellent stability properties make it the ideal code for stiff problems. Further features of RADAU5 include a continuous output of the solution and an automatic error and stepsize control. RADAU5 is our favorite code for stiff problems.

Geometric properties

In this subsection, we determine some interesting features of the dynamical system (11) *without* integrating the equations. This gives an important connection to critical state soil mechanics.

We first determine equilibrium points of the dynamical system (11). Such points must be zeros of the right-hand side of (11). In our example, the zeros of the right-hand side of (11a) form a one-dimensional manifold of candidates for an equilibrium. In Fig. 4, we display this manifold together with the constraining manifold (11b) for the data $T_2 = T_3 = -100$ and $D_1 = \mp 1$. The intersection of these two manifolds defines the equilibrium points.

For loading, we find an asymptotically stable equilibrium point at $\sigma_1 = -T_1 = 293.02$ and $D_2 = 0.50$. This is the limit state for loading, see Fig. 5. The linearized differential equation at the equilibrium point is given by

$$\frac{\mathrm{d}}{\mathrm{d}t} w(t) = -177.65\,\big(w(t) - 293.02\big).$$

This problem is stiff with stiffness $\lambda = -177.65$. Since the forward Euler method has the stability barrier $-2 \leq h\lambda \leq 0$, the stepsize h has to be chosen less than $h_{\max} = 2/|\lambda| \approx 0.01$. A similar stability bound applies to *all*

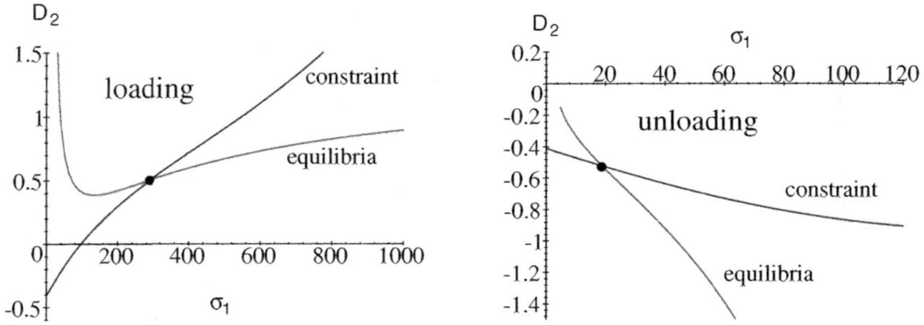

Fig. 4: Zeros of (11a) (manifold of equilibria) and of (11b) (constraining manifold) in the triaxial compression test. Their intersection points give the equilibria for loading and unloading, respectively.

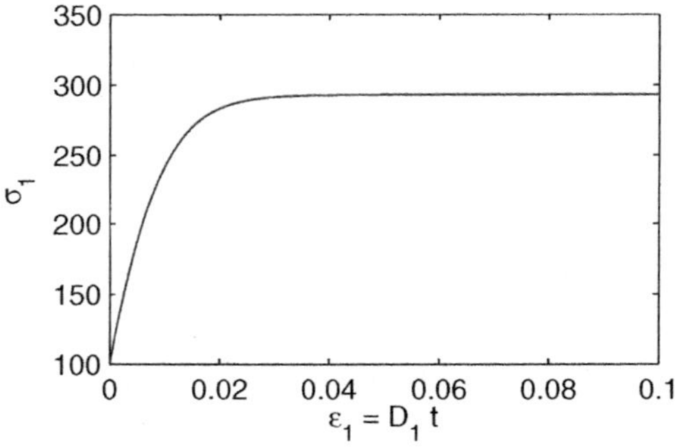

Fig. 5: Stress-strain curve for the triaxial element test.

explicit Runge-Kutta methods. The implicit method RADAU5, on the other hand, has no stepsize restriction and allows *large* stepsizes in a neighbourhood of the equilibrium.

For unloading, we find an asymptotically stable equilibrium point at $\sigma_1 = -T_1 = 18.82$ and $D_2 = -0.52$. Note that in this version of hypoplasticity loading and unloading results in different mobilized friction angles, cf. [14]. The linearized problem at the equilibrium point is again stiff with stiffness $\lambda = -250.65$.

Finally, we look for possible bifurcation points in system (11). For a bifurcation point, the partial derivative of (11b) with respect to D_2 must vanish. This gives in our example a one-dimensional manifold of candidates for bifurcation points. In Fig. 6, we display this manifold together with the constraining manifold (11b) for the data $T_2 = T_3 = -100$ and $D_1 = \mp 1$. The intersection of these two manifolds defines the bifurcation points. There is no bifurcation in the case of loading, but one bifurcation point for unloading at

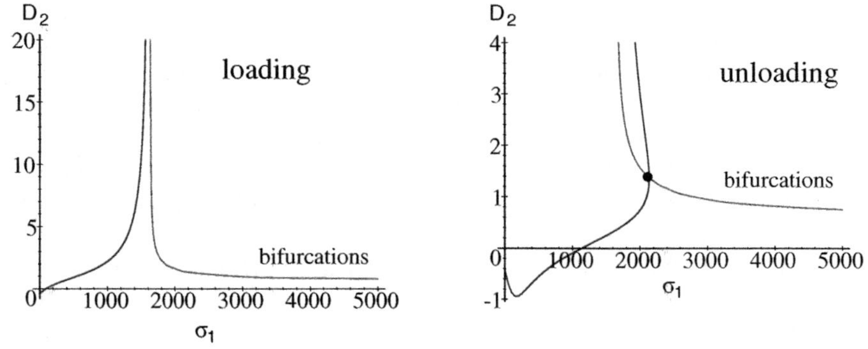

Fig. 6: Constraining manifold and manifold of bifurcation points in the triaxial compression test for loading and unloading, respectively.

$\sigma_1 = -T_1 = 2133.81$ and $D_2 = 1.36$. However this point can never be reached by triaxial path.

6 Conclusions

In this article, we discussed the time integration of the constitutive equations in hypoplasticity. Among different approaches for objective time integration, the rotation neutralized description in combination with projection methods turned out to be particularly attractive.

Based on techniques from the qualitative theory of differential equations, we have further shown, how interesting features of the constitutive equations can be determined without integrating the equations.

Moreover, we have demonstrated that high-order methods generally pay. Instead of using Euler's method with fixed step sizes, we recommend the use of the explicit Runge-Kutta method DOPRI5 and the implicit Runge-Kutta method RADAU5, depending on whether the problem is non-stiff or stiff.

References

[1] Abraham R, Marsden J E (1978) Foundations of Mechanics. Benjamin-Cummings, Reading, Mass.

[2] Bauer E (1992) Zum mechanischen Verhalten granularer Stoffe unter vorwiegend ödometrischer Beanspruchung. Institut für Bodenmechanik und Felsmechanik der Universität Karlsruhe, Karlsruhe

[3] Fellin W, Ostermann A (2002) Int. J. Numer. Anal. Meth. Geomech. 26:1213–1233

[4] Goldscheider M (1976) Mech. Res. Comm. 3, 463–468

[5] Gurtin M E (1981) An Introduction to Continuum Mechanics. Academic Press, San Diego New York
[6] Hairer E, Lubich C, Wanner G (2002) Geometric Numerical Integration. Structure-Preserving Algorithms for Ordinary Differential Equations. Springer, Berlin Heidelberg New York
[7] Hairer E, Nørsett S P, Wanner G (1993) Solving Ordinary Differential Equations I. Nonstiff Problems. 2nd rev. ed., Springer, Berlin Heidelberg New York
[8] Hairer E, Wanner G (1996) Solving Ordinary Differential Equations II. Stiff and Differential-Algebraic Problems. 2nd rev. ed., Springer, Berlin Heidelberg New York
[9] Higham D J (1997) BIT 37:24–36
[10] Kolymbas D (2000) Introduction to Hypoplasticity. Balkema, Rotterdam
[11] Olver P J (1993) Applications of Lie Groups to Differential Equations. 2nd ed., Springer, Berlin Heidelberg New York
[12] Simo J C, Hughes T J R (1998) Computational Inelasticity. Springer, Berlin Heidelberg New York
[13] Wu W (1997) Int. J. Numer. Anal. Meth. Geomech. 21:153–174
[14] Wu W (1992) Hypoplastizität als mathematisches Modell zum mechanischen Verhalten granularer Stoffe. Institut für Bodenmechanik und Felsmechanik der Universität Karlsruhe, Karlsruhe
[15] Wu W, Kolymbas D (1990) Mechanics of Materials 9:245–253

Using constitutive models of the rate type in implicit finite-element calculations: error-controlled stress update and consistent tangent operator

Wolfgang Fellin[1] and Alexander Ostermann[2]

[1] Institut für Geotechnik und Tunnelbau,Universität Innsbruck, wolfgang.fellin@uibk.ac.at
[2] Institut für Technische Mathematik, Geometrie und Bauinformatik, Universität Innsbruck, alexander.ostermann@uibk.ac.at

Summary. The consistent tangent operator is important for efficient finite-element calculations. By means of a one-dimensional model problem, we explain how this operator can be calculated. Based on this, a general approach for obtaining the consistent tangent operator for constitutive equations of the rate type is given. In order to achieve reliable results in the finite-element calculations, an error controlled time integration of the constitutive law should be provided. Here an adaptive time integration method is used which is based on a control of the local error. The effects of the numerical time integration scheme are studied in detail with the help of a one-dimensional hypoplastic example. Further a finite-element example is given to show the performance of the presented general approach. Quadratic convergence of the equilibrium iteration is shown.

1 Introduction

For an efficient finite-element analysis with non-linear constitutive laws, the consistent tangent operator is of crucial importance [11]. Any inconsistency with the stress-update algorithm of the constitutive law will spoil the quadratic convergence of the equilibrium iterations based on Newton's method. A consequence will be computational inefficiency.

The first goal of the present paper is to exemplify the consistent tangent operator and to point out its importance for an implicit finite-element analysis. Then we show how an efficient, accurate and robust approximation of the consistent tangent operator can be obtained for constitutive equations of the rate type. Our approach is quite general and can be used in combination with any time integrator. In the here presented implementation we use an adaptive step size strategy that is based on the control of the local integration error. In our opinion such an error control is indispensable for complex problems.

We present our ideas by means of a hypoplastic constitutive model. Since we do not assume that the reader is familiar with hypoplasticity, we give a short introduction to this model. We emphasize, however, that our approach is sufficiently general to cover all constitutive laws of the rate type such as the CloE-model [2].

The article is organized as follows. In Section 2 we give a short outline of hypoplasticity. In the following Section 3 we give a one-dimensional illustration of the consistent tangent operator. In Section 4 we explain how the consistent tangent operator is obtained in general, and we discuss our time integration method. Examples are presented in Section 5. In Section 6 we show the result of a finite-element calculation. We consider the deformation of a vertically compressed soil specimen that develops a shear band due to an initial imperfection. Our main conclusions are finally given in Section 7.

2 A short introduction to hypoplasticity

In this section, a simple one-dimensional hypoplastic law is developed which simulates the behaviour of a soil sample in the confined compression test. As usual in the theory of hypoplasticity, we denote the stress by T. Tension and stretching are positive.

2.1 A confined compression test

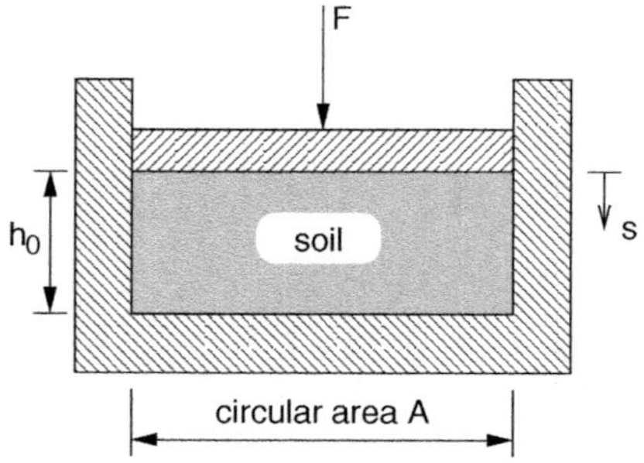

Fig. 1: Confined compression test: schematic experimental setup

The cylindrical soil sample in Fig. 1 is axially compressed with confined lateral strains. The vertical stress $T = -F/A$ as a function of the vertical strain $\varepsilon = -s/h_0$ is plotted in Fig. 2 for loading and unloading of a sand sample.

The curved lines indicate a non-linear behaviour. The different branches for loading and unloading signify inelastic behaviour. Our goal is to find a mathematical formulation as simple as possible, but nevertheless representing a good approximation to this behaviour.

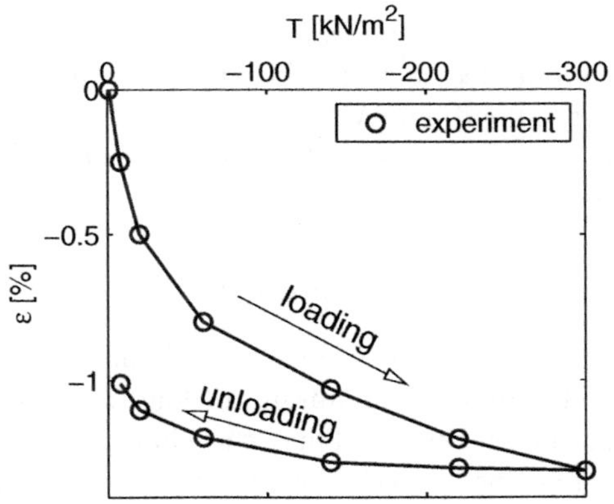

Fig. 2: Confined compression test with loose sand: stress-strain relation

2.2 A simple mathematical model

The simplest approximation of the stress-strain curve of a confined compression test are two straight lines (Fig. 3). This is called a linear inelastic constitutive law. A naive way to formulate this law is the following relation between stress and strain:

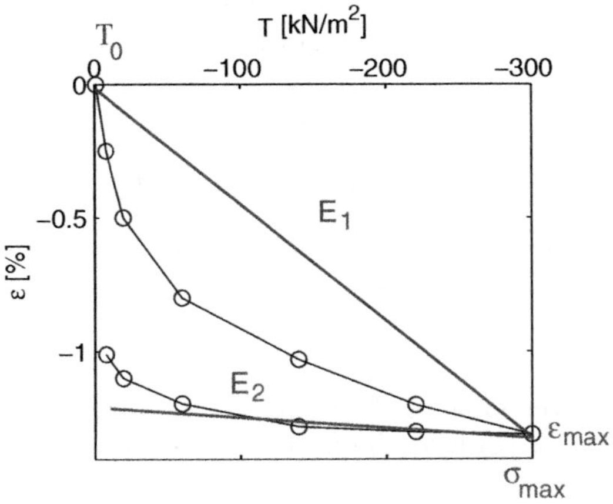

Fig. 3: Linear inelastic approximation

$$\text{loading:} \quad T = T_0 + E_1 \varepsilon \,, \tag{1}$$

$$\text{unloading:} \quad T = T_{\max} + E_2 (\varepsilon - \varepsilon_{\max}) \,. \tag{2}$$

But the maximum stress $T_{\max}$ is usually not known a priori, it is a result of the loading history. Thus the equation for unloading (2) is not handy defined. In addition we cannot decide from the current value of the strain whether there is loading or unloading. We have to regard rather the change of the strain. This leads directly to the following incremental formulation, which is based on the strain rate.

2.3 A constitutive law of the rate type

A more clever formulation of the non-linear inelastic behaviour in Fig. 2 is possible using a rate equation. Thinking of loading as a process in time, we introduce the parameter time t. Then the change of the stress is regarded as a function of time for a given strain rate. In one-dimensional problems, this strain rate is equal to the Eulerian stretching D and is given as the derivative of the strain with respect to the time: $\dot{\varepsilon}(t) = \mathrm{d}\varepsilon(t)/\mathrm{d}t$. The strain rate is negative for loading (compression) $\dot{\varepsilon} < 0$ and positive for unloading (expansion) $\dot{\varepsilon} > 0$.

A simple constitutive law of the rate type for modeling the inelastic behaviour looks then like:

$$\text{loading } \dot{\varepsilon} < 0: \quad \dot{T} = E_1 \dot{\varepsilon}\,, \tag{3}$$

$$\text{unloading } \dot{\varepsilon} > 0: \quad \dot{T} = E_2 \dot{\varepsilon}\,. \tag{4}$$

In order to obtain the stress-strain relation, we have to integrate this rate equation over the time. Time integration of (3) for loading yields:

$$\int_0^t \dot{T}\,\mathrm{d}t = \int_0^t E_1 \dot{\varepsilon}\,\mathrm{d}t$$

$$T(t) - T(0) = E_1 \varepsilon(t) - E_1 \varepsilon(0)\,.$$

The initial values are $\varepsilon(0) = 0$ and $T(0) = T_0$. Thus the stress as a function of the strain during loading is

$$T = T_0 + E_1 \varepsilon\,,$$

which is exactly the same as (1). If loading is changed to unloading at time t_1, time integration of (4) with the initial values $\varepsilon(t_1) = \varepsilon_{\max}$ and $T(t_1) = T_{\max}$ gives

$$T = T_{\max} + E_2(\varepsilon - \varepsilon_{\max})\,,$$

which coincides with (2).

We can combine equations (3) and (4) to get

$$\dot{T} = \frac{E_1 + E_2}{2}\dot{\varepsilon} + \frac{E_2 - E_1}{2}|\dot{\varepsilon}|\,. \tag{5}$$

The absolute value of the strain rate provides different stiffnesses for loading and unloading.

2.4 An improved model

Since the linear stress-strain relation is rather unsatisfactory, we try now to obtain a curved line. The stiffness of soil is often assumed to be proportional to the stress. Thus the stress rate $\dot{T}$ should be a function of the actual stress. We can find a proper mathematical formulation by introducing the stress as a state variable at the right-hand side of (5)

$$\dot{T} = K_1 T\dot{\varepsilon} + K_2 T|\dot{\varepsilon}| = K_1 TD + K_2 T|D| = h(T, D) . \tag{6}$$

Here we used two new material constants K_1 and K_2 as additional parameters.

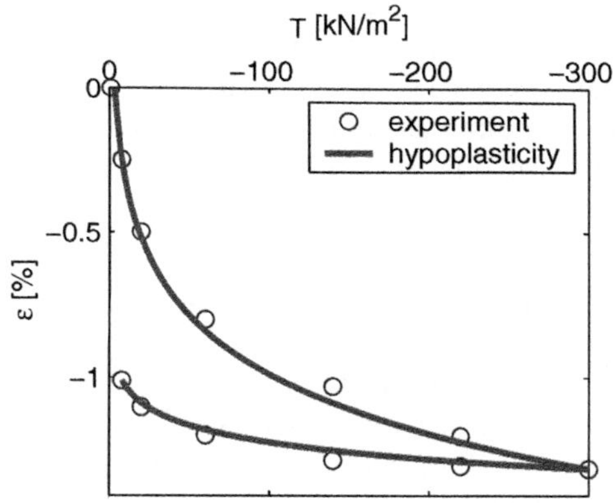

Fig. 4: Hypoplastic approximation with (6), $K_1 = -775$, $K_2 = -433$ ($T_0 = -3.4$ kN/m^2)

Time integration of (6) for loading ($D = \dot{\varepsilon} < 0$) yields

$$\ln \frac{T}{T_0} = (K_1 - K_2)(\varepsilon - \varepsilon_0) ,$$

the well-known logarithmic stress-strain relation in the confined compression test. The curve is plotted for loading and unloading in Fig. 4 and compared with the experimental values.

2.5 Three-dimensional hypoplasticity

The hypoplastic law represents the objective stress rate [12] of the effective Cauchy stress $\mathring{\mathbf{T}}$ as a function of the effective Cauchy stress $\mathbf{T}$, the Eulerian stretching $\mathbf{D}$ and some additional state variables

$$\mathring{\mathbf{T}} = \mathbf{h}(\mathbf{T}, \mathbf{D}, e) . \tag{7}$$

In the version of von Wolffersdorff [13] used here, the additional state variable is the void ratio e of the soil. Since the mass is assumed to stay constant, the evolution of the void ratio e is described by

$$\dot{e} = (1+e)\,\mathrm{tr}\,\mathbf{D} = k(\mathbf{T},\mathbf{D},e)\ . \tag{8}$$

For a more thorough discussion, we refer to [9] and [3].

3 A one-dimensional illustration of the consistent tangent operator for hypoplasticity

The concept of the consistent tangent operator is not commonly known. To set the stage for our general treatment of consistent tangent operators, we first outline the basic structure of the initial-boundary value problem within the framework of a one-dimensional problem. Using the one-dimensional hypoplastic law of the previous section, we then show how to calculate the basic ingredients of the consistent tangent operator.

3.1 One-dimensional finite-element notation

In this section, we follow closely the presentation of Simo [11]. We consider a one-dimensional body $\mathcal{B}$ with length L, see Fig. 5. This bar is fixed at the left end and loaded with the stress $\overline{T}(t)$ at the right end. Thus the boundary of the body $\partial\mathcal{B}$ has two parts, a part with prescribed deformations $\partial_u\mathcal{B}$ (left end) and a part with prescribed stress $\partial_T\mathcal{B}$ (right end).

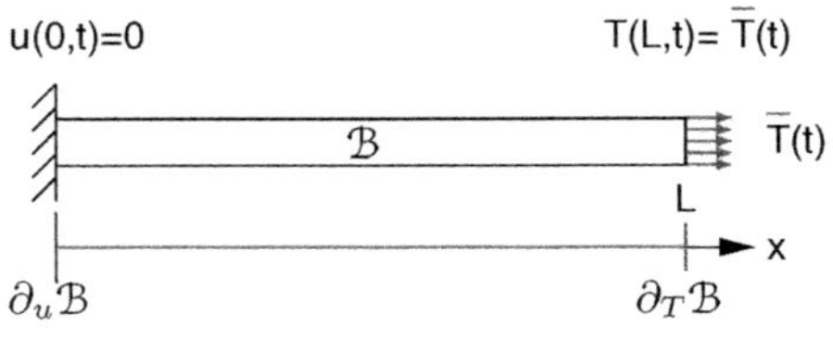

Fig. 5: One-dimensional example

We denote by $u(x,t)$ the displacement, and by

$$\varepsilon(x,t) := \frac{\partial u(x,t)}{\partial x} \tag{9}$$

the strain field.

The local momentum equation

The balance of momentum for the static case yields the equilibrium equation

$$\frac{\partial T}{\partial x} = 0 \, . \tag{10}$$

Equations (9) and (10) are supplied with the boundary conditions

$$u\big|_{\partial_u \mathcal{B}} = u(0,t) = 0 \, , \tag{11}$$

$$T\big|_{\partial_T \mathcal{B}} = T(L,t) = \overline{T}(t) \, . \tag{12}$$

To solve the problem, i.e., to calculate the deformation at the right end of the bar, we need in addition the constitutive law of the material, which is a relation between stress and strain. The simplest case is a linear elastic material, described by

$$T = E\varepsilon = E\frac{\partial u}{\partial x} \, . \tag{13}$$

In our case we use a non-linear constitutive law of the rate type

$$\dot{T} = h(T,\dot{\varepsilon}) = h(T,D) \, . \tag{14}$$

Equations (9), (10) and (14) together with the boundary conditions (11), (12) and the initial conditions

$$T(x,t_0) = T_0(x) \, , \qquad u(x,t_0) = u_0(x) = 0 \, , \tag{15}$$

define our initial-boundary value problem.

Note that the stress rate is here given as a function of the actual stress and the strain rate. This brings additional difficulties in the solution strategy.

The weak formulation

The basis for a finite-element calculation is a weak formulation of the initial-boundary value problem. To obtain it, we multiply the equilibrium equation (10) with an arbitrary test function $\eta \in \mathbb{V}$ of the linear space of admissible test functions

$$\mathbb{V} = \{\eta : \mathcal{B} \to \mathbb{R} \; ; \; \eta\big|_{\partial_u \mathcal{B}} = 0\} \tag{16}$$

and integrate over the body to get

$$\int\limits_{\mathcal{B}} \frac{\partial T}{\partial x} \eta \, \mathrm{d}x = 0 \, . \tag{17}$$

Using an integration by parts, we end up with the weak form

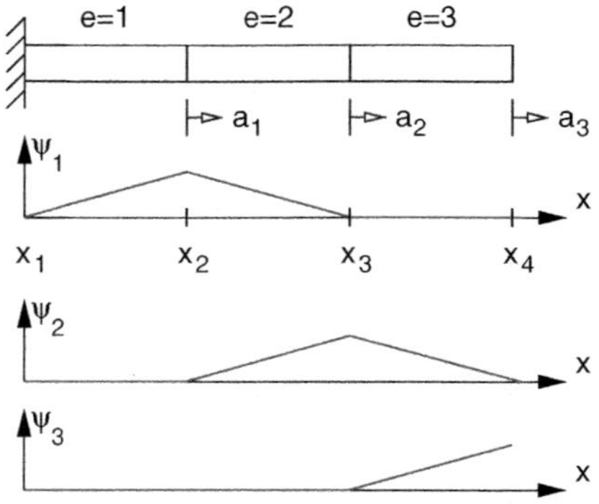

Fig. 6: Basis of the test functions for $n_{el} = 3$

$$(T\eta)\big|_{\partial\mathcal{B}} - \int\limits_{\mathcal{B}} T\frac{\partial\eta}{\partial x}\,\mathrm{d}x = 0\,. \tag{18}$$

We discretize our body in n_{el} elements (Fig. 6)

$$\mathcal{B} = \bigcup_{e=1}^{n_{el}} \mathcal{B}_e \qquad \text{with} \qquad \mathcal{B}_e = [x_e, x_{e+1}]\,, \tag{19}$$

and use test functions $v \in \mathbb{V}^h \subset \mathbb{V}$, which are a linear combination of the virtual nodal displacements a_i of the elements (Galerkin method)

$$v(x) = \sum_{i=1}^{n_{el}} \psi_i(x)\, a_i\,, \tag{20}$$

where ψ_i are a basis of $\mathbb{V}^h$.

The test function v fulfils the kinematical boundary conditions

$$v\big|_{\partial_u\mathcal{B}} = 0\,. \tag{21}$$

Using (21) and the static boundary condition (12) the weak form (18) yields

$$\overline{T}v\big|_{\partial_T\mathcal{B}} - \int\limits_{\mathcal{B}} T\frac{\partial v}{\partial x}\,\mathrm{d}x = 0\,. \tag{22}$$

From local (element) to global (system)

Stress updating in finite-element calculations happens separately in each element. Therefore we need to transfer the informations from the elements (local formulation) to the system (global formulation), see Fig. 7.

To describe the test function v locally, we define the linear shape functions

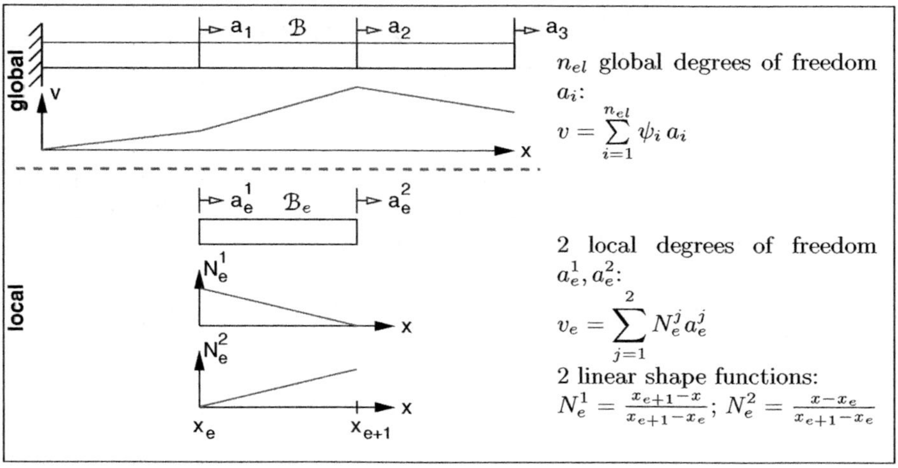

Fig. 7: Discretization in finite-elements

$$N_e^1 = \frac{x_{e+1} - x}{x_{e+1} - x_e} \qquad \text{and} \qquad N_e^2 = \frac{x - x_e}{x_{e+1} - x_e}\,, \tag{23}$$

and write

$$v_e = \sum_{j=1}^{2} N_e^j a_e^j = \left[N_e^1, N_e^2\right] \begin{bmatrix} a_e^1 \\ a_e^2 \end{bmatrix} = \mathbf{N}_e \mathbf{a}_e = \mathbf{a}_e^T \mathbf{N}_e^T \,. \tag{24}$$

We further denote

$$\mathbf{B}_e = \left[\frac{\partial}{\partial x} N_e^1, \frac{\partial}{\partial x} N_e^2\right] \,. \tag{25}$$

The transition from the local nodal displacements a_e^1 and a_e^2 to the global displacements a_i works via the incidence or location matrix $\mathbf{Z}_e$, which reflects the topology of the discretization

$$\begin{bmatrix} a_e^1 \\ a_e^2 \end{bmatrix} = \mathbf{a}_e = \mathbf{Z}_e \mathbf{a} = \mathbf{Z}_e \begin{bmatrix} a_1 \\ a_2 \\ \vdots \\ a_{n_{el}} \end{bmatrix} \,. \tag{26}$$

The weak form (22) can be calculated as sum over the elements

$$\sum_{e=1}^{n_{el}} \left(\overline{T} v_e \big|_{\partial_T \mathcal{B}} - \int_{\mathcal{B}_e} T \frac{\partial v_e}{\partial x} \, \mathrm{d}x \right) = 0 \,. \tag{27}$$

Inserting (24) and (26) we obtain

$$\mathbf{a}^T \sum_{e=1}^{n_{el}} \left(\mathbf{Z}_e^T \underbrace{\overline{T} \mathbf{N}_e^T |_{\partial_T \mathcal{B}}}_{\mathbf{f}_e^{\text{ext}}} - \mathbf{Z}_e^T \underbrace{\int\limits_{\mathcal{B}_e} \mathbf{B}_e^T T \,\mathrm{d}x}_{\mathbf{f}_e^{\text{int}}} \right) = 0 \,. \tag{28}$$

This is the equilibrium condition expressed with the element internal force vector $\mathbf{f}_e^{\text{int}}$ and the element external load vector $\mathbf{f}_e^{\text{ext}}$. Since (28) must hold for any test function in $\mathbb{V}^h$ and thus for any $\mathbf{a}$, we get

$$\sum_{e=1}^{n_{el}} \left(\mathbf{Z}_e^T \int\limits_{\mathcal{B}_e} \mathbf{B}_e^T T \,\mathrm{d}x - \mathbf{Z}_e^T \overline{T} \mathbf{N}_e^T |_{\partial_T \mathcal{B}} \right) = \sum_{e=1}^{n_{el}} \left(\mathbf{Z}_e^T \mathbf{f}_e^{\text{int}} - \mathbf{Z}_e^T \mathbf{f}_e^{\text{ext}} \right) = 0 \,. \tag{29}$$

We can also write this in terms of global force vectors

$$F^{\text{int}} - F^{\text{ext}} = 0 \tag{30}$$

as it is common in finite-element literature, wherein the external force vector is a function of the load $F^{\text{ext}}(\overline{T})$ and the internal force vector is a function of the internal stress $F^{\text{int}}(T)$.

3.2 The consistent tangent operator

In the following we will see what the consistent tangent operator is, and for what purpose it is needed. For sake of simplicity, we illustrate this with one element only, i.e., $n_{el} = e = 1$. This means that we have only one global degree of freedom, thus $\mathbf{a} = a$ is the virtual nodal displacement at the end of the bar, see Fig. 8.

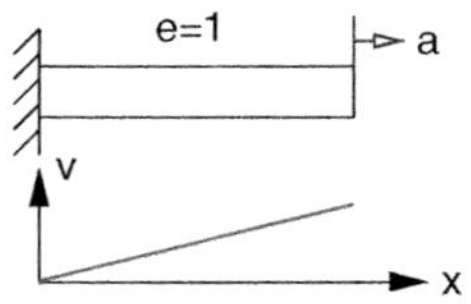

Fig. 8: Discretization with one finite element

The incidence matrix is then given by

$$\begin{bmatrix} a_e^1 \\ a_e^2 \end{bmatrix} = \mathbf{a}_e = \mathbf{Z}_e \, a = \begin{bmatrix} 0 \\ 1 \end{bmatrix} a \,. \tag{31}$$

Incremental loading

In non-linear finite-element calculations the load is applied in time increments $t_{n+1} = t_n + \Delta t_n$. Let us start with a equilibrated body at the time $t = t_n$ with given stress field T_n. Thus

$$F^{\text{int}}(T_n) - F^{\text{ext}}(\overline{T}_n) = \mathbf{Z}_e^T \int\limits_{\mathcal{B}_e} \mathbf{B}_e^T T_n \, \mathrm{d}x - \mathbf{Z}_e^T \overline{T}_n \mathbf{N}_e^T \big|_{\partial_T \mathcal{B}} = 0 \tag{32}$$

holds.

During the time increment the load is changed from $\overline{T}_n$ by the load increment to $\overline{T}_{n+1} = \overline{T}_n + \Delta\overline{T}_n$. Our task is to find the updated displacement field $u_{n+1} = u_n + \Delta u_n$ and a stress field T_{n+1} such that:

- the body is equilibrated at time t_{n+1}

$$\mathbf{Z}_e^T \int\limits_{\mathcal{B}_e} \mathbf{B}_e^T T_{n+1} \, \mathrm{d}x - \mathbf{Z}_e^T \overline{T}_{n+1} \mathbf{N}_e^T \big|_{\partial_T \mathcal{B}} = 0 \,, \tag{33}$$

- and the stress update is compatible with the constitutive law.

The standard solution strategy is an iterative procedure: the equilibrium equation is solved with the help of a finite-element package, and the constitutive law by a solver for ordinary differential equations. The relevant constitutive information is passed between these two processes.

The equilibrium iteration

Within the framework of finite elements, the unknown displacement $u(x)$ is searched as a linear combination of some ansatz functions. In our case we take

$$u(x) = \sum_{i=1}^{n_{el}} \psi_i(x) \, d_i \,, \tag{34}$$

where d_i are the unknown nodal displacements. In particular, for one element the displacement is $u(x) = \psi d$, where d is the nodal displacement at the right end of the bar, see Fig. 8. Locally the displacements are given by

$$u_e = \sum_{j=1}^{2} N_e^j d_e^j = \mathbf{N}_e \mathbf{d}_e \tag{35}$$

with $\mathbf{N}_e$ as in (23).

We choose a displacement increment Δd_n. Starting from the equilibrated body with the known nodal displacement d_n, we calculate the nodal displacement at the end of the increment

$$d_{n+1} = d_n + \Delta d_n \,. \tag{36}$$

The strain in the element at the end of the increment is

$$\varepsilon_{n+1}\big|_{\mathcal{B}_e} \overset{(9)}{=} \frac{\partial u_{e,n+1}}{\partial x} \overset{(24)}{=} \frac{\partial \mathbf{N}_e}{\partial x} \mathbf{d}_{e,n+1} \overset{(25)}{=} \mathbf{B}_e \mathbf{d}_{e,n+1} \overset{(31)}{=} \mathbf{B}_e \mathbf{Z}_e d_{n+1} \,. \tag{37}$$

The element is deformed by the strain increment

$$\Delta\varepsilon_n = \varepsilon_{n+1} - \varepsilon_n \,, \tag{38}$$

wherein ε_n is known. The stress increment ΔT_n follows from a time integration of the constitutive law $\dot{T}_n = h(T_n, \dot{\varepsilon}_n)$ with the strain rate $\dot{\varepsilon}_n = \Delta\varepsilon_n/\Delta t_n$. The new stress is given by

$$T_{n+1} = T_n + \Delta T_n \,. \tag{39}$$

Inserting this new stress in the equilibrium equation (33) generally results in a non-zero right-hand side, the so called residual R

$$\mathbf{Z}_e^T \int\limits_{\mathcal{B}_e} \mathbf{B}_e^T T_{n+1}\, \mathrm{d}x - \mathbf{Z}_e^T\, \overline{T}_{n+1}\, \mathbf{N}_e^T \big|_{\partial_T \mathcal{B}} = R(d_{n+1}) \,. \tag{40}$$

Note that T_{n+1} is a non-linear function of d_{n+1} due to the non-linear constitutive law. Therefore the internal force vector is also a non-linear function $F^{\mathrm{int}}(d_{n+1})$. The external force vector F^{ext} is known for the given load $\overline{T}_n$

$$F^{\mathrm{int}}(d_{n+1}) - F^{\mathrm{ext}} = R(d_{n+1}) \,. \tag{41}$$

Thus $R(d_{n+1})$ is also a non-linear function of d_{n+1}. We will use Newton's method to find the correct displacement increment, i.e., a zero of R.

Newton's method

Let us denote the first guess of the new displacement with d_{n+1}^0. This displacement will cause a new stress field T_{n+1}^0 with the help of (37–39). Eq. (40) will give the first residual $R(d_{n+1}^0)$, see Fig. 9.

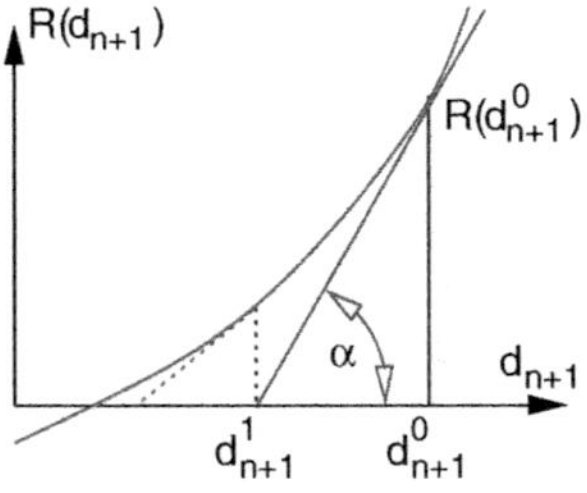

Fig. 9: Newton's method

The derivative of the residual with respect to the displacement at d_{n+1}^0 is the slope of the line with angle α in Fig. 9

$$R'(d_{n+1}^0) = \tan\alpha \,. \tag{42}$$

The next iteration d_{n+1}^1 is a zero of this line. The angle α is determined by the triangle d_{n+1}^0, d_{n+1}^1 and $R(d_{n+1}^0)$

$$\tan\alpha = \frac{R(d_{n+1}^0)}{d_{n+1}^0 - d_{n+1}^1} . \tag{43}$$

Thus we can find the next iteration

$$d_{n+1}^1 = d_{n+1}^0 - R'(d_{n+1}^0)^{-1} R(d_{n+1}^0) . \tag{44}$$

The derivative of the residual $R'(d_{n+1}^k)$ is called *consistent tangent (operator)* of the k'th equilibrium iteration, because it is consistent with Newton's method.

If any other gradient is used in the Newton iteration, the quadratic convergence of the method is lost.

Consistent tangent operator

To calculate the consistent tangent operator in the k'th equilibrium iteration, we have to differentiate the equilibrium equation (40) with respect to d_{n+1}, knowing that the external force $F^{\text{ext}} = \mathbf{Z}_e^T \,\overline{T}_{n+1}\, \mathbf{N}_e^T\big|_{\partial_T \mathcal{B}}$ is independent of d_{n+1}. We evaluate this derivative at d_{n+1}^k and get

$$\left.\frac{\mathrm{d}R(d_{n+1})}{\mathrm{d}d_{n+1}}\right|_{d_{n+1}^k} = R'(d_{n+1}^k) = \mathbf{Z}_e^T \int\limits_{\mathcal{B}_e} \mathbf{B}_e^T \,\frac{\partial T_{n+1}^k}{\partial \varepsilon_{n+1}^k}\,\frac{\partial \varepsilon_{n+1}^k}{\partial d_{n+1}^k}\,\mathrm{d}x . \tag{45}$$

Knowing from (37) that $\varepsilon_{n+1} = \mathbf{B}_e \mathbf{Z}_e d_{n+1}$, we can reformulate this to get the consistent tangent operator (global tangent stiffness)

$$R'(d_{n+1}^k) = \mathbf{Z}_e^T \underbrace{\left(\int\limits_{\mathcal{B}_e} \mathbf{B}_e^T \,\frac{\partial T_{n+1}^k}{\partial \varepsilon_{n+1}^k}\,\mathbf{B}_e\,\mathrm{d}x \right)}_{\text{element tangent stiffness}} \mathbf{Z}_e . \tag{46}$$

3.3 The Jacobian

The material information on the element level to build the consistent tangent operator (46) is the so called Jacobian

$$\frac{\partial T_{n+1}}{\partial \varepsilon_{n+1}} = \frac{\partial (T_{n+1} - T_n)}{\partial (\varepsilon_{n+1} - \varepsilon_n)} = \frac{\partial \Delta T}{\partial \Delta \varepsilon} . \tag{47}$$

The Jacobian has to be determined by the subroutine that supplies the constitutive law for the implicit finite-element code.

We consider a one-dimensional compression test with the one-dimensional hypoplastic law (6) for loading

$$\dot{T} = h(T, D) = KTD = KT\dot{\varepsilon} \tag{48}$$

to explain, how the Jacobian can be calculated for a constitutive law of the rate type.

Analytical solution

In this simple example the Jacobian can be found analytically by differentiation. Starting from an initial stress $T(0) = T_0$ and an initial strain $\varepsilon(0) = 0$, time integration of (48) yields

$$T(t) = T_0 e^{KDt} \,. \tag{49}$$

The stress increment for a given strain increment $\Delta\varepsilon = D\Delta t$ is

$$\Delta T = T(\Delta t) - T_0 = T_0 \left(e^{K\Delta\varepsilon} - 1\right) \,. \tag{50}$$

The analytic Jacobian is the derivative of the stress increment with respect to the strain increment

$$\frac{\partial \Delta T}{\partial \Delta\varepsilon} = KT_0 e^{K\Delta\varepsilon} = KT_0 e^{KD\Delta t} \,. \tag{51}$$

Numerical time integration

A numerical time integration of (48) with one forward Euler step results in a stress increment $\Delta T^{\mathrm{num}} = KT_0\Delta\varepsilon$ (straight line with gradient KT_0 in Fig. 10), which is different from the analytic solution (49) (solid curved line in Fig. 10). Calculating the tangent of the constitutive law at the end of the time step will result in the gradient of the dashed line $K(T_0 + \Delta T^{\mathrm{num}})$. The Jacobian required by the finite-element code is the gradient of the numerical scheme (straight line) $C = \frac{\partial \Delta T}{\partial \Delta\varepsilon} = KT_0$. We see clearly that the Jacobian and the constitutive tangent are different, even in one-dimensional cases!

Variational equation

The Jacobian depends on the numerical time integration scheme. Therefore we search for a calculation method which takes into account the time integration method. This can be performed with the help of the variational equation of the constitutive law.

We differentiate the constitutive law $\dot{T} = h(T, D)$ with respect to the stretching D using the chain rule

$$\frac{\mathrm{d}}{\mathrm{d}t}\frac{\partial T}{\partial D} = \frac{\partial h}{\partial T}\frac{\partial T}{\partial D} + \frac{\partial h}{\partial D} \,. \tag{52}$$

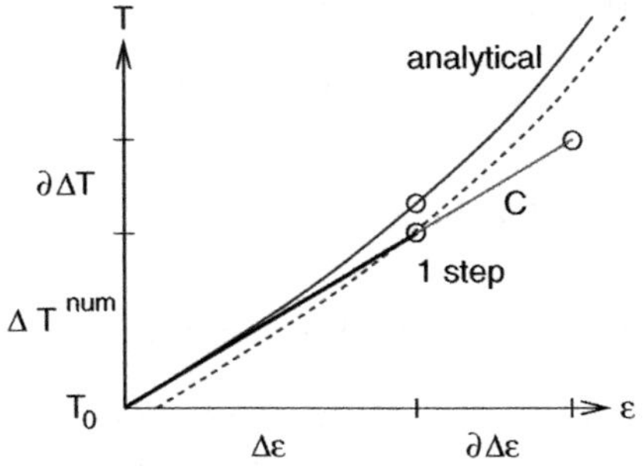

Fig. 10: Jacobian with numerical scheme

Replacing the right-hand side by (48) we arrive at

$$\frac{\mathrm{d}}{\mathrm{d}t}\frac{\partial T}{\partial D} = K\frac{\partial T}{\partial D}D + KT\,. \tag{53}$$

Next we denote $\partial T/\partial D$ by C

$$\frac{\mathrm{d}}{\mathrm{d}t}C = KCD + KT \tag{54}$$

and end up with a differential equation for C.

Equations (48) and (54) form a coupled system of differential equations. They have to be solved together with the initial conditions

$$T(0) = T_0 \tag{55}$$

$$C(0) = \frac{\partial T}{\partial D}(0) = \frac{\partial T_0}{\partial D} = 0\,. \tag{56}$$

Note that $C(0) = 0$, since $T(0)$ is the initial condition and therefore independent of D.

Analytic solution of stress update and Jacobian

In this simple case, we can solve (48) and (54) – (56) analytically

$$T(t) = T_0 e^{KDt}\,, \tag{57}$$

$$C(t) = KT_0 t e^{KDt}\,. \tag{58}$$

At the end of the time increment Δt we get the stress update

$$T(\Delta t) = T_0 e^{KD\Delta t} = T_0 e^{K\Delta\varepsilon} = \Delta T + T_0 \tag{59}$$

and

$$C(\Delta t) = KT_0 \Delta t e^{KD\Delta t} = \Delta t K T_0 e^{K\Delta\varepsilon}\,. \tag{60}$$

Comparing this with (51), we see that

$$\frac{C(\Delta t)}{\Delta t} = \frac{\partial \Delta T}{\partial \Delta \varepsilon}\,, \tag{61}$$

i.e., we have found the Jacobian.

Numerical approximation of the Jacobian

Working out the Jacobian analytically for a complex constitutive law can be a tedious task or sometimes even not feasible. Thus we want to use the following numerical approximation B. Using a small variation ϑ we define

$$\frac{\mathrm{d}}{\mathrm{d}t} B = \frac{1}{\vartheta}\Big(h(T + \vartheta B, D + \vartheta) - h(T, D) \Big)\,. \tag{62}$$

Substituting (48) yields

$$\frac{\mathrm{d}}{\mathrm{d}t} B = \frac{1}{\vartheta}\Big(KT \cdot (D + \vartheta) + K\vartheta B \cdot (D + \vartheta) - KTD \Big) \tag{63}$$

$$= KB \cdot (D + \vartheta) + KT\,. \tag{64}$$

Comparing (64) with (54) we see that $B = C$ for $\vartheta \to 0$. If ϑ is sufficiently small, $B(\Delta t)/\Delta t$ will be thus a good approximation to the Jacobian.

4 A general approach for consistent tangent operators

The ideas of the previous section can easily be generalized to two and three dimensional problems. This is explained in detail in our paper [5]. Here, we will briefly summarize the strategy and comment on the numerical algorithm in order to make the present article self-contained.

As we have seen in the previous section, the equilibrium equation together with the constitutive law form a coupled system consisting of an initial-boundary value problem and an ordinary differential equation. To distinguish between the stress history in the overall initial-boundary value problem and in the time integration of the constitutive law, we denote the stresses with $\boldsymbol{\sigma}$ and $\mathbf{T}$, respectively.

4.1 Consistent tangent operator, Jacobian

We start from an equilibrium at time t_n. For the given initial stress tensor $\mathbf{T}(0) = \boldsymbol{\sigma}(t_n)$ and the strain increment $\Delta\varepsilon$, the constitutive subroutine has to provide the new stress tensor $\boldsymbol{\sigma}(t_n + \Delta t) = \mathbf{T}(\Delta t)$ at time $t_n + \Delta t$ as well as its derivative with respect to the strain increment

$$\frac{\partial \Delta \boldsymbol{\sigma}}{\partial \Delta \varepsilon} = \frac{\partial \boldsymbol{\sigma}(t_n + \Delta t)}{\partial \Delta \varepsilon}. \tag{65}$$

Due to the incremental solution procedure, the temporal rate of the strain tensor is not known as a function of time. Only its mean value over the time-window Δt

$$\mathbf{D} = \frac{\Delta\varepsilon}{\Delta t} \tag{66}$$

is available for use in the constitutive law. Assuming that the finite-element program only needs the co-rotational parts, e.g. [1], we have to solve the following system of differential equations for $0 \le t \le \Delta t$

$$\begin{aligned} \frac{\mathrm{d}}{\mathrm{d}t}\mathbf{T} &= \mathbf{h}(\mathbf{T}, \mathbf{D}, \mathbf{Q}) , \qquad & \mathbf{T}(0) &= \boldsymbol{\sigma}(t_n) , \\ \frac{\mathrm{d}}{\mathrm{d}t}\mathbf{Q} &= \mathbf{k}(\mathbf{T}, \mathbf{D}, \mathbf{Q}) , \qquad & \mathbf{Q}(0) &= \mathbf{Q}_0 . \end{aligned} \tag{67}$$

Here, $\mathbf{Q}$ denotes the additional state variables, and $\mathbf{Q}_0$ are their values at time t_n. The hypoplastic law considered in this paper has only one additional state variable, denoted by e. In this case the function $\mathbf{k}$ is given by (8).

Differentiation of (67) with respect to $\mathbf{D}$ yields the variational equations

$$\begin{aligned} \frac{\mathrm{d}}{\mathrm{d}t}\frac{\partial \mathbf{T}}{\partial \mathbf{D}} &= \frac{\partial \mathbf{h}}{\partial \mathbf{T}} \cdot \frac{\partial \mathbf{T}}{\partial \mathbf{D}} + \frac{\partial \mathbf{h}}{\partial \mathbf{Q}} \cdot \frac{\partial \mathbf{Q}}{\partial \mathbf{D}} + \frac{\partial \mathbf{h}}{\partial \mathbf{D}} , \qquad & \frac{\partial \mathbf{T}}{\partial \mathbf{D}}(0) &= 0 , \\ \frac{\mathrm{d}}{\mathrm{d}t}\frac{\partial \mathbf{Q}}{\partial \mathbf{D}} &= \frac{\partial \mathbf{k}}{\partial \mathbf{T}} \cdot \frac{\partial \mathbf{T}}{\partial \mathbf{D}} + \frac{\partial \mathbf{k}}{\partial \mathbf{Q}} \cdot \frac{\partial \mathbf{Q}}{\partial \mathbf{D}} + \frac{\partial \mathbf{k}}{\partial \mathbf{D}} , \qquad & \frac{\partial \mathbf{Q}}{\partial \mathbf{D}}(0) &= 0 . \end{aligned} \tag{68}$$

Let $\Delta\boldsymbol{\sigma} = \boldsymbol{\sigma}(t_n + \Delta t) - \boldsymbol{\sigma}(t_n)$. In order to get

$$\frac{\partial \Delta\boldsymbol{\sigma}}{\partial \Delta\varepsilon} = \frac{\partial \boldsymbol{\sigma}(t_n + \Delta t)}{\partial \Delta\varepsilon} = \frac{1}{\Delta t} \cdot \frac{\partial \mathbf{T}}{\partial \mathbf{D}}(\Delta t) , \tag{69}$$

system (68) has to be solved simultaneously with system (67). Due to the complicated structure of our constitutive law, the calculation (and implementation) of the expressions appearing on the right-hand side of (68) seems to be a tedious task. We therefore strongly recommend to replace (68) by the following approximation which is obtained by numerical differentiation

$$\begin{aligned} \frac{\mathrm{d}}{\mathrm{d}t}\mathbf{B}_{ij} &= \frac{1}{\vartheta}\Big(\mathbf{h}(\mathbf{T} + \vartheta\mathbf{B}_{ij}, \mathbf{D} + \vartheta\mathbf{V}_{ij}, \mathbf{Q} + \vartheta\mathbf{G}_{ij}) - \mathbf{h}(\mathbf{T}, \mathbf{D}, \mathbf{Q})\Big) , \\ \frac{\mathrm{d}}{\mathrm{d}t}\mathbf{G}_{ij} &= \frac{1}{\vartheta}\Big(\mathbf{k}(\mathbf{T} + \vartheta\mathbf{B}_{ij}, \mathbf{D} + \vartheta\mathbf{V}_{ij}, \mathbf{Q} + \vartheta\mathbf{G}_{ij}) - \mathbf{k}(\mathbf{T}, \mathbf{D}, \mathbf{Q})\Big) , \end{aligned} \tag{70}$$

with $\mathbf{B}_{ij}(0) = 0$ and $\mathbf{G}_{ij}(0) = 0$ for $1 \le i \le j \le 3$, see (62). Here, $\mathbf{V}_{ij}$ denotes the standard basis tensor

$$\mathbf{V}_{ij} = (\delta_{ik}\delta_{j\ell})_{k,\ell=1}^{3} \tag{71}$$

with the $\delta_{ik} = 1$ if $i = k$ and $\delta_{ik} = 0$ else. A Taylor series expansion of the right-hand side of (70) shows that

$$\mathbf{B}_{ij} = \frac{\partial \mathbf{T}}{\partial D_{ij}} + \mathcal{O}(\vartheta)\,. \tag{72}$$

Thus the six tensors $\mathbf{B}_{ij}$ are good approximations to the Jacobian for ϑ suitably chosen, see [5]. We propose to solve (67) and (70) simultaneously with the same numerical method (as described subsequently). This guarantees the consistency of the derivatives.

4.2 Adaptive time integration

In this subsection we will explain how to solve (67) and (70) with a second-order Runge-Kutta method based on the forward Euler scheme. Collecting all the variables of our problem in a super-vector

$$\begin{aligned}\mathbf{y} = \big[&T_{11}, T_{22}, T_{33}, T_{12}, T_{13}, T_{23},\\ &(B_{11})_{11}, (B_{11})_{22}, (B_{11})_{33}, (B_{11})_{12}, (B_{11})_{13}, (B_{11})_{23},\\ &(B_{22})_{11}, ..., (B_{22})_{23}, (B_{33})_{11}, ..., (B_{23})_{23},\\ &Q_1, ..., Q_m, (G_{11})_1, ..., (G_{11})_m, (G_{22})_1, ..., (G_{23})_m\big]^T\end{aligned} \tag{73}$$

and denoting the right-hand sides of (67) and (70) by $\mathbf{F}$, we have to solve the initial value problem

$$\frac{\mathrm{d}}{\mathrm{d}t}\mathbf{y}(t) = \mathbf{F}(\mathbf{y}(t))\,, \qquad \mathbf{y}(0) = \mathbf{y}_0 \quad \text{given}\,. \tag{74}$$

We choose an initial step size τ satisfying $0 \le \tau \le \Delta t$ where Δt is the given time increment and calculate the two approximations

$$\mathbf{v} = \mathbf{y}_0 + \tau \mathbf{F}(\mathbf{y}_0) \tag{75}$$

and

$$\mathbf{w} = \mathbf{y}_0 + \frac{\tau}{2}\mathbf{F}(\mathbf{y}_0) + \frac{\tau}{2}\mathbf{F}\big(\mathbf{y}_0 + \frac{\tau}{2}\mathbf{F}(\mathbf{y}_0)\big) \tag{76}$$

$$= \mathbf{y}_0 + \tau \mathbf{F}(\mathbf{y}_0) + \frac{\tau^2}{4}\frac{\partial \mathbf{F}}{\partial \mathbf{y}}(\mathbf{y}_0)\,\mathbf{F}(\mathbf{y}_0) + \mathcal{O}(\tau^3). \tag{77}$$

Note that $\mathbf{v}$ is the result of the forward Euler method with step size τ, whereas $\mathbf{w}$ is the result of two Euler steps of size $\tau/2$.

Our error and step size control is based on Richardson extrapolation. It is robust and very reliable. To explain this approach, we study the error of $\mathbf{v}$ and $\mathbf{w}$ after one step. For this we expand the exact solution into a Taylor series

$$\mathbf{y}(\tau) = \mathbf{y}(0) + \tau\frac{\mathrm{d}\mathbf{y}}{\mathrm{d}t}(0) + \frac{\tau^2}{2}\frac{\mathrm{d}^2\mathbf{y}}{\mathrm{d}t^2}(0) + \mathcal{O}(\tau^3) \tag{78}$$

and insert the identities

$$\frac{\mathrm{d}\mathbf{y}}{\mathrm{d}t}(0) = \mathbf{F}(\mathbf{y}_0) \qquad \text{and} \qquad \frac{\mathrm{d}^2\mathbf{y}}{\mathrm{d}t^2}(0) = \frac{\partial \mathbf{F}}{\partial \mathbf{y}}(\mathbf{y}_0)\frac{\mathrm{d}\mathbf{y}}{\mathrm{d}t}(0) = \frac{\partial \mathbf{F}}{\partial \mathbf{y}}(\mathbf{y}_0)\mathbf{F}(\mathbf{y}_0) \quad (79)$$

which are obtained from (74) with the help of the chain rule. This yields

$$\mathbf{y}(\tau) = \mathbf{y}_0 + \tau \mathbf{F}(\mathbf{y}_0) + \frac{\tau^2}{2}\frac{\partial \mathbf{F}}{\partial \mathbf{y}}(\mathbf{y}_0)\,\mathbf{F}(\mathbf{y}_0) + \mathcal{O}(\tau^3)\,. \quad (80)$$

and shows at once

$$\mathbf{y}(\tau) - \mathbf{v} = \frac{\tau^2}{2}\frac{\partial \mathbf{F}}{\partial \mathbf{y}}(\mathbf{y}_0)\,\mathbf{F}(\mathbf{y}_0) + \mathcal{O}(\tau^3) \quad (81a)$$

and

$$\mathbf{y}(\tau) - \mathbf{w} = \frac{\tau^2}{4}\frac{\partial \mathbf{F}}{\partial \mathbf{y}}(\mathbf{y}_0)\,\mathbf{F}(\mathbf{y}_0) + \mathcal{O}(\tau^3)\,. \quad (81b)$$

Therefore, the difference

$$\mathbf{w} - \mathbf{v} = \frac{\tau^2}{4}\frac{\partial \mathbf{F}}{\partial \mathbf{y}}(\mathbf{y}_0)\,\mathbf{F}(\mathbf{y}_0) + \mathcal{O}(\tau^3) \quad (82)$$

is an asymptotically correct estimate for the error of the more accurate numerical solution $\mathbf{w}$. It can thus can be used to control the error. We set

$$\mathrm{EST} = \|\mathbf{w} - \mathbf{v}\| \quad (83)$$

and employ the following step size strategy which is standard in this field, see [6].

If the estimated error EST is below the user-supplied tolerance TOL, the step is accepted and we use the extrapolated value

$$\mathbf{y}_{\mathrm{new}} = 2\mathbf{w} - \mathbf{v} \quad (84)$$

which due to (81) is a second-order approximation to the solution. Further, we enlarge our step size. We then continue our integration until we reach Δt. If the estimated error EST is larger than TOL, we have to reject the step and to redo it with a smaller step size. For more details on the step size control, we refer to [4] and [5].

5 An illustrative academic example

In this section, we treat a simple but illustrative example to study some numerical aspects of our implementation. We assume small strains and use a simplified one-dimensional hypoplastic law (6) to simulate loading in a one-dimensional compression test

$$\dot{T} = KT\dot{\varepsilon}\,, \quad (85)$$

with the constant $K = -2000$ and the initial stress $T_0 = -100$ N/m^2. The loading is $\overline{T} = -1000$ N/m^2. We are searching an ε such that (85) gives an internal stress T which equals the external load $\overline{T}$. It is therefore convenient we to consider T also as a function of ε. The initial-boundary value problem to be solved reduces to

$$T(\varepsilon) - \overline{T} = 0\,. \tag{86}$$

Note that the analytic solution of problem (86), (85) is given by

$$\varepsilon^{\text{an}} = \frac{1}{K} \ln \frac{\overline{T}}{T_0} = -1.1512 \times 10^{-3}\,. \tag{87}$$

Equation (86) was solved numerically with the standard Newton method

$$\varepsilon^{i+1} = \varepsilon^{i} - \left(\frac{\mathrm{d}T^{i}}{\mathrm{d}\varepsilon^{i}}\right)^{-1} R^{i}\,, \tag{88}$$

with the residual R in each iteration i

$$R^{i} = T(\varepsilon^{i}) - \overline{T}\,. \tag{89}$$

The consistent tangent $\mathrm{d}T^i/\mathrm{d}\varepsilon^i$ is calculated together with the numerical stress $T^i = T(\varepsilon^i)$ during the integration of (85) with the above described scheme. As starting value, $\varepsilon^0 = 10^{-3}$ was chosen.

Several calculations with different time integration schemes were done in IEEE quadruple precision. To compare the numerical results ε^i with the analytical solution (87), the relative error

$$\text{err}\ \varepsilon = \frac{\varepsilon^{i} - \varepsilon^{\text{an}}}{\varepsilon^{\text{an}}} \tag{90}$$

was used.

5.1 Accuracy of the numerical time integration

The results of the numerical time integration in the last equilibrium iteration i are plotted in Fig. 11. The Newton iterations guarantees that the system is in equilibrium. Thus all curves end at $T(\varepsilon^i) = \overline{T} = -1000$ N/m^2.

The different numerical time integration schemes result in different strains in the last Newton iteration. The error (90) is quite large for one forward Euler step (err $\varepsilon = -2.908$) and for 10 steps (err $\varepsilon = -0.124$). Only with 1000 steps the error err $\varepsilon = -1.152 \times 10^{-3}$ is sufficiently small. Note that also one backward Euler step gives a large error.

Finite-element programs do not check the result of the time integration $\Delta T(\Delta\varepsilon)$. Thus the errors have to be *controlled* by the user. This can be done using the extrapolated forward Euler method (second-order Runge-Kutta)

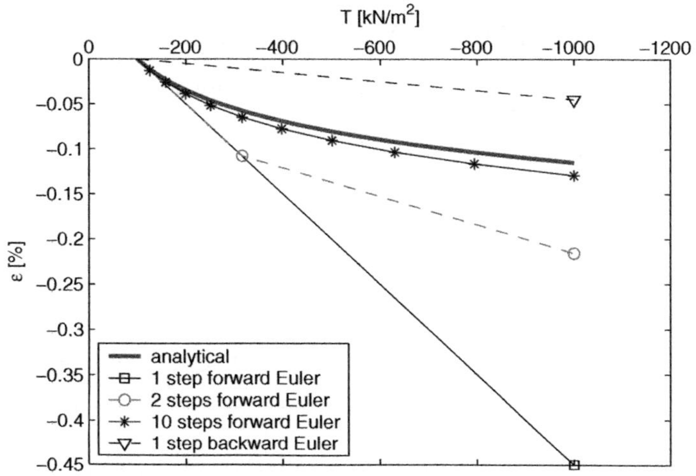

Fig. 11: Numerical time integration in the last Newton iteration

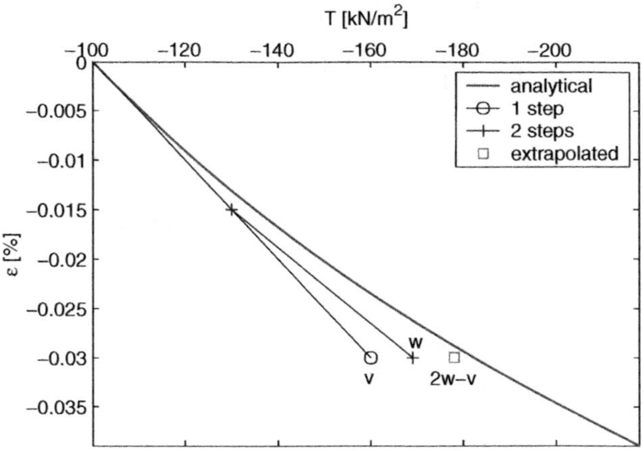

Fig. 12: Extrapolated forward Euler: one step v, two steps w, extrapolated solution $2w - v$

presented in Sect. 4.2. We calculate the stress increment with a forward Euler time integration using one step, $\Delta T^{1\ \mathrm{step}}(\Delta\varepsilon) = v$, and two steps, $\Delta T^{2\ \mathrm{steps}}(\Delta\varepsilon) = w$, see Fig. 12. The error estimate is $\mathrm{EST} = |w - v|/(|w| + \mathrm{ATOL})$. If this error is smaller than the user defined tolerance TOL, we calculate the extrapolated solution $\Delta T^{\mathrm{extr.}}(\Delta\varepsilon) = 2w - v$ which is very close to the analytical one. The limit ATOL (absolute error) is necessary for solutions near zero.

Using the adaptive time integration scheme of Sect. 4.2 we obtain a close approximation to the analytical solution, see Fig. 13. The size of the steps are

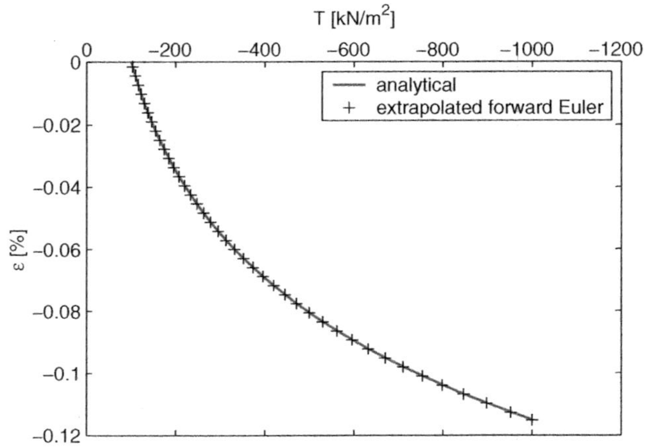

Fig. 13: Time integration with extrapolated forward Euler: EST $<$ TOL $= 10^{-3}$, ATOL $= 10^{-3}$, 41 steps

automatically increased when the curvature decreases. Note that our computation needs only 41 steps and thus requires the same work as a computation with 82 forward Euler steps.

5.2 Influence of the time integration on the Newton scheme

The residual R is a function of the strain ε depending on the time integration method, see Fig. 14.

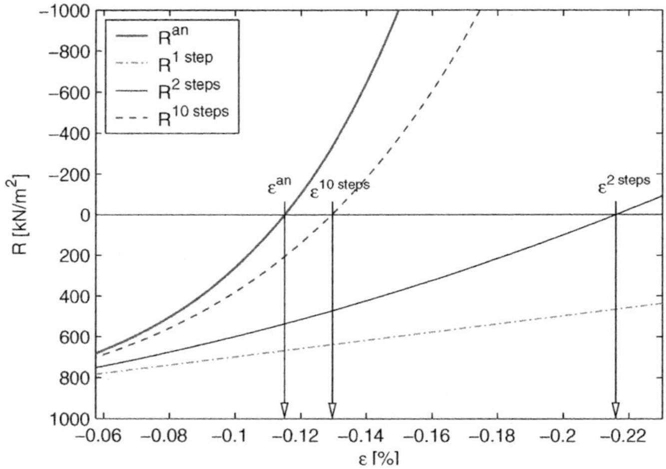

Fig. 14: Dependence of R^{num} on the step size sequence

The residual obtained with analytical time integration is

$$R^{\mathrm{an}} = T^{\mathrm{an}}(\varepsilon) - \overline{T} = T_0 e^{K\varepsilon} - \overline{T} \,. \tag{91}$$

The residual calculated with the numerical time integration is a composition of forward Euler steps, e.g., for one and two steps

$$R^{1\ \mathrm{step}} = T_0 + T_0 K\varepsilon - \overline{T} \,, \tag{92}$$

$$R^{2\ \mathrm{steps}} = T_0 + T_0 K\varepsilon/2 + (T_0 + T_0 K\varepsilon/2)K\varepsilon/2 - \overline{T} \,. \tag{93}$$

If the step size sequence in the time integration is kept fixed for all Newton iterations, we stay on the same numerical approximation R^{num} and thus quadratic convergence to the zero of R^{num} is achieved, see Fig. 15.

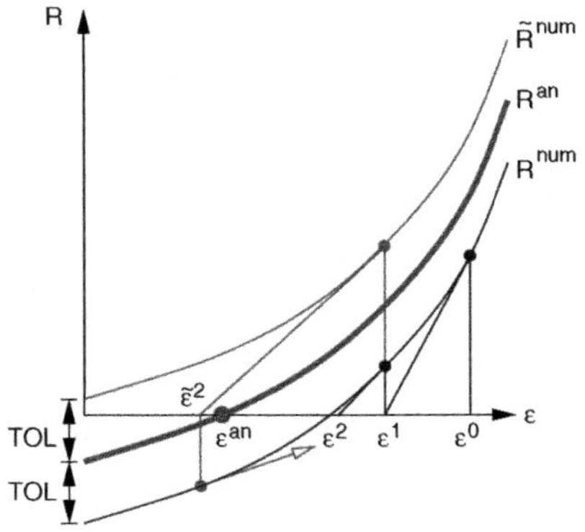

Fig. 15: Newton iterations with varying time integration scheme

However, we change the problem, i.e. switch from R^{num} to a neighbouring $\widetilde{R}^{\mathrm{num}}$, whenever changing the step size sequence from one Newton iteration to the other due to the error control of the time integrator. Such a behavior is illustrated in Fig. 15 by the sequence ε^0, ε^1, $\tilde{\varepsilon}^2$, etc. The iteration might thus become quite irregular whenever the numerical solution is very close to the exact one. However, this is *no problem at all* since such a numerical solution is accurate enough anyway and should therefore be accepted.

Nevertheless, the solution with fixed step size sequence is not better than the adaptive one, since it is only a zero of R^{num} which in general is different from $\varepsilon^{\mathrm{an}}$, compare Fig. 14 and Fig. 11.

We study this in detail for the Newton iterations in our example, see Fig. 16. The analytic function of the residual (91) appears as a straight line R^{an} in this double logarithmic plot. The composition of 10 Euler steps results in the curve $R^{10\ \mathrm{steps}}$. The circles on this curve denote the results of each Newton iteration (R^i, 10 Steps). The convergence is quadratic (the vertical distance between the circles is approximately doubled in each iteration), but the error is large, compare Tab. 1.

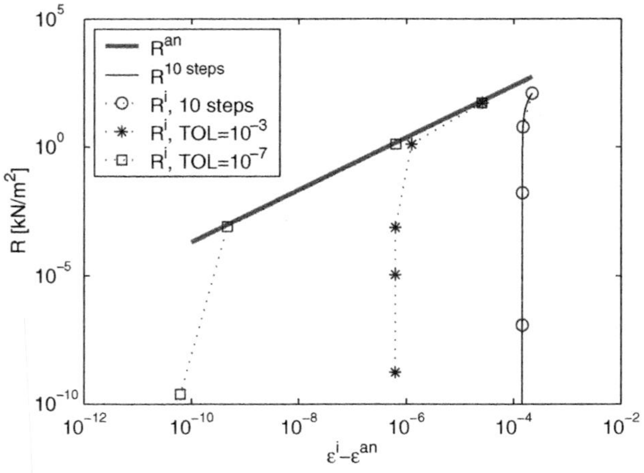

Fig. 16: Loss of quadratic convergence using different step size sequence

Using an adaptive time integration scheme, we end up with a much more accurate solution, see Tab. 1 and Fig. 16: R^i, TOL $= 10^{-3}$ and R^i, TOL $= 10^{-7}$. The Newton iteration shows quadratic convergence when the numerical residual R^i is near the analytic residual R^{an}. We loose the quadratic convergence when the residual is of the same order as the tolerance of the time integration. However, the solution ε is then near the analytic one and the iteration can be stopped.

Table 1. Newton iteration of (86) with numerical time integration of (85).

10 Euler steps	
It.No.	R^i
1	380.8
2	−124.6
3	−6.040
4	-1.629×10^{-2}
5	-1.195×10^{-7}
6	-5.573×10^{-18}
err ε	−0.124

Extrapolated Euler, TOL $= 10^{-10}$	
It.No.	R^i
1	261.1
2	−52.07
3	−1.267
4	-8.024×10^{-4}
5	-3.218×10^{-10}
6	4.629×10^{-20}
err ε	-5.400×10^{-11}

6 Finite-element Example

The performance of the proposed time integration scheme is shown with a shear band calculation. The computations were performed on a Silicon Graphics o200 with ABAQUS 5.8-10. We use the default convergence criteria and load incrementation of ABAQUS, and we set the error tolerances in the constitutive time integration to TOL $= 10^{-3}$ (relative error) and ATOL $= 10^{-3}$ (absolute error).

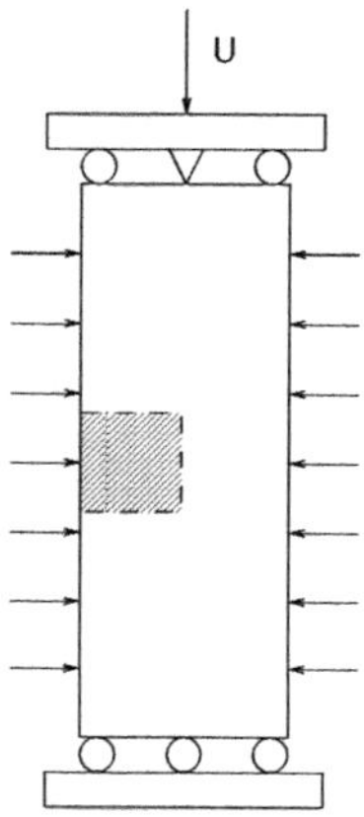

Fig. 17: Biaxial test.

The results are compared with an implementation of the von Wolffersdorff hypoplastic law by Roddeman [10], which is very robust and provides a relatively cheap but only rough approximation of the Jacobian. Therefore, the convergence of the global system is quite slow.

The biaxial example was taken from Hügel [8]. A soil specimen of 0.04 m width and 0.14 m height is laterally compressed with a constant stress of size 4×10^5 N/m^2. The specimen is compressed vertically by a prescribed displacement $u = 0.01$ m. The material in the dashed area of Fig. 17 is given an initially higher void ratio of 0.45, whereas the void ratio is 0.40 elsewhere. In this way, an initial imperfection in the dashed area is simulated. The dashed area is of the size 0.02 m by 0.02 m. The specific weight of the sample was $\gamma = 17$ kN/m^3.

The biaxial test is modeled with 8 by 28 linear plane strain elements. The calculation accounts for large strains and large rotations, so the lateral pressure remains perpendicular to the edge of the specimen.

Fig. 18 is a filled contour plot for the void ratio at the end of the test. This is plotted in the deformed piece of material. Note that the deformations are not scaled. It is clear that the void ratio did increase in the shear band which is formed during the test.

A comparison of the computational costs in Table 2 shows that the proposed implementation of the hypoplastic law is also more efficient.

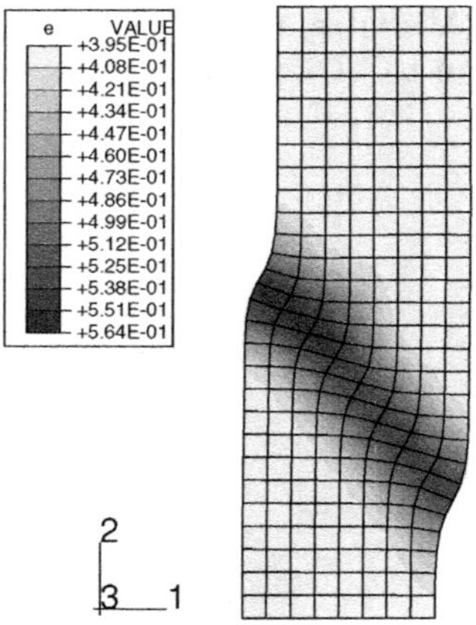

Fig. 18: Biaxial compression test with gravity: void ratio; solution with the proposed implementation of the hypoplastic law.

Table 2. Comparison of computational costs in the biaxial test without gravity. ABAQUS: automatic load incrementation with $\Delta t_{\text{start}} = 0.125$, $\Delta t_{\max} = 0.125$, and default convergence criteria for equilibrium iteration $R_n^\alpha = 5\times 10^{-3}$

implementation	No. of Inc.	No. of It.	CPU [s]
proposed	28	67	224.84
Roddeman	69	706	385.76

7 Conclusions

The consistent tangent operator is one of the main ingredients of any implicit non-linear finite-element calculation which solves the equations of motion and the non-linear constitutive law separately in each loading increment. The integration of the constitutive law in each element is usually done in a subroutine which has to provide the stresses at the end of the increment and the derivatives of the stresses with respect to the strain increments, the so called Jacobian. With this information the finite-element program performs the equilibrium iterations.

There are two crucial points which have to be considered carefully in the integration subroutine:

- An error control of the stress update has to be provided.
- The Jacobian has to be calculated accurately enough, so that the equilibrium iterations converges fast.

The above proposed scheme is able to handle both tasks. We achieved quadratic convergence of the equilibrium Newton iteration in implicit finite-element calculations using the incrementally non-linear hypoplastic constitutive law. The error in the stress update can be controlled by the user. Our approach is even sufficiently general to cover all constitutive laws of the rate type.

References

[1] Abaqus (1998) User's Manual, Version 5.8, Volume 3. HKS Inc., Hibbit, Karlson & Sorenson, Rhode Island, U.S.A.

[2] Desrues J, Chambon R (1993) A new rate type constitutive model for geomaterials. In: Kolymbas D (ed) Modern Approaches to Plasticity. Elsevier

[3] Fellin W, Kolymbas D (2002) Bautechnik 12:830–841

[4] Fellin W, Ostermann A (2001) A hypoplasticity routine for ABAQUS with consistent tangent operator and error control http://geotechnik.uibk.ac.at/res/FEhypo.html

[5] Fellin W, Ostermann A (2002) International Journal for Numerical and Analytical Methods in Geomechanik 26:1213–1233

[6] Hairer E, Nørsett S P, Wanner G (1993) Solving Ordinary Differential Equations I. Nonstiff Problems. Springer, Berlin

[7] Heeres O M, de Borst R (2000) Implicit integration of hypoplastic models. In: Kolymbas D (ed) Constitutive Modelling of Granular Materials. Springer

[8] Hügel H (1995) Prognose von Bodenverformungen. Veröffentlichung des Institutes für Bodenmechanik und Felsmechanik der Universität Fridericiana in Karlsruhe, Heft 136

[9] Kolymbas D (2000) Introduction to Hypoplasticity. Number 1 in Advances in Geotechnical Engineering and Tunnelling. Balkema

[10] Roddeman D (1997) FEM-implementation of hypoplasticity. European Union Project (ERBCHBGCT 940554), Institut für Geotechnik und Tunnelbau, Baufakultät, Universität Innsbruck

[11] Simo J C, Hughes T J R (1998) Computational Inelasticity, Interdisciplinary Applied Mathematics, Volume 7. Springer

[12] Truesdell C, Noll W (1965) The Non-Linear Field Theories of Mechanics. Springer

[13] von Wolffersdorff P A (1996) Mechanics of Cohesive-Frictional Materials 1:251–271

Efficient and Reliable Nonlocal Damage Models *

Antonio Huerta, Antonio Rodríguez-Ferran, and Irene Morata

Laboratori de Càlcul Numèric (LaCàN)
Edifici C2, Campus Nord, Universitat Politècnica de Catalunya
Barcelona, Spain.
e-mails: {antonio.huerta,antonio.rodriguez-ferran,irene.morata}@upc.es

Summary. We present an efficient and reliable approach for the numerical modelling of failure with nonlocal damage models. The two major numerical challenges – the strongly nonlinear, highly localized and parameter-dependent structural response of quasi-brittle materials, and the interaction between non-adjacent finite elements associated to nonlocality – are addressed in detail. Efficiency is achieved with a suitable combination of load-stepping control technique and nonlinear solver for equilibrium equations. Reliability of the numerical results is ensured by an h-adaptive strategy based on error estimation. We use a residual-type error estimator for nonlinear FE analysis based on local computations, which, at the same time, accounts for the nonlocality of the damage model. The proposed approach is illustrated by means of three application examples: the three-point bending test, the single-edge notched beam test and the Brazilian test.

Keywords: quasi-brittle materials; nonlocal damage models; adaptivity; error estimation; consistent tangent matrix; quadratic convergence; nonlocal displacements

1 Introduction

Damage models are nowadays a common choice in the numerical modelling of failure of quasi-brittle materials [19]. To avoid the pathological mesh dependence exhibited by local damage models, one may use either gradient damage models or nonlocal damage models. These two related strategies regularize the problem and ensure mesh objectivity. In gradient damage models, strain derivatives are incorporated into the constitutive equation [8]. In nonlocal damage models [25, 3, 21], strain (or, rather, a strain-related state variable)

* Contract grant sponsor: Ministerio de Ciencia y Tecnología; Contract grant number: DPI2001-2204

is smoothed by means of an integral average in the vicinity – associated to a characteristic length – of each point. The latter approach is considered in this paper.

Another clear trend in computational mechanics is the quest for reliable computations. The quality of the results must be guaranteed in an quantifiable, objective manner. This has led to adaptive finite element analysis based on error estimation.

The price to pay for reliable results is a large number of degrees of freedom, especially in nonlinear solid mechanics. This means that one needs computationally efficient numerical methods.

Of course, efficiency and reliability is a general concern in all the fields of computational mechanics. However, when modelling quasi-brittle failure with nonlocal damage models, we face some specific challenges:

1. Due to brittleness, the structural response is strongly nonlinear, very localized and highly dependent (at the quantitative and even qualitative level) on the value of the material parameters.
2. Due to nonlocality, there is interaction between non-adjacent finite elements. This poses several difficulties. The consistent tangent stiffness matrix, for instance (needed for quadratic convergence in Newton iterations), cannot be assembled from elementary contributions solely.
3. Many error estimators are based on local (element-wise) computations. This fact must be conciliated with the nonlocal nature of the damage model: adaptivity typically leads to element sizes smaller than characteristic length.

1.1 Objectives

In this context, the main goal of this paper is to present an efficient and reliable approach for the numerical modelling of failure with nonlocal damage models. The key ingredients are:

1. A residual-type error estimator based on element-wise computations which, at the same time, accounts for the nonlocality of the constitutive model [28].
2. An h-adaptive strategy driven by the error estimator which yields numerical results with the desired accuracy. The FE discretization errors are kept under control and, thus, the physical significance of the computations is guaranteed [29, 28].
3. Advanced arc-length control techniques, adapted to the highly localized failure patterns.
4. A flexible approach to achieve quadratic convergence in Newton iterations. The element-to-element stiffness matrices can either be assembled into the global tangent stiffness matrix [17] or accounted for in the right-hand-side vector to prevent fill-in.

This paper also addresses, in a more exploratory manner, another issue: The use of nonlocal *displacements* as the basis of the nonlocal damage model. The standard approach is to define the nonlocal state variable as the nonlocal average (NLA) of the (strain-related) local state variable. Other approaches have been proposed in the literature (see [16] for a comparative analysis), based, for instance, on nonlocal strains or nonlocal damage. An alternative approach is explored here: to use nonlocal displacements, obtained as the NLA of local displacements, to drive the evolution of damage. According to our preliminary studies, the resulting model exhibits a satisfactory behaviour and it is very attractive from the computational point of view.

1.2 Outline of Paper

The rest of the paper is organized as follows. Section 2 deals with material modelling. It starts by reviewing the main ingredients of nonlocal damage models in Sect. 2.1. Then, the proposed model based on nonlocal displacements is sketched in Sect. 2.2.

Numerical aspects are discussed in Sect. 3. Although they are clearly interrelated, the issues of efficiency and reliability are covered separately for expository purposes. Regarding efficiency, see Sect. 3.1, the key features are adequate control strategies (for load-stepping) and iterative solvers (for iterations within the load-step). Reliability, see Sect. 3.2, refers to the adaptive strategy based on error estimation.

Section 4 contains some illustrative applications. Finally, the concluding remarks of Sect. 5 close the paper.

2 Material Modelling

2.1 Overview of Nonlocal Damage Models

For simplicity, only elastic-scalar damage models are considered here. However, many of the ideas, methods and algorithms can be extended to more complex damage models incorporating, for instance, anisotropy or plasticity [21, 2].

A generic nonlocal model of such type consists of the following equations, summarized in Table 1:

- A relation between Cauchy stresses $\boldsymbol{\sigma}$ and small strains $\boldsymbol{\varepsilon}$, where the loss of stiffness (from elastic stiffness $\mathbf{C}$ to zero stiffness) is described by means of a scalar damage parameter D which ranges from 0 to 1, (1);
- The definition of a local state variable Y as a function of strain $\boldsymbol{\varepsilon}$, (2);
- The definition of the nonlocal state variable $\widetilde{Y}$ as the average of the local state variable Y, (3);

- A weighting function α which depends on the distance r between two points and contains a characteristic length l_c as a parameter, (4);
- A damage evolution law, where the nonlocal state variable $\widetilde{Y}$ drives the evolution of the non-decreasing damage parameter D, (5).

Table 1. General expression of an elastic-damage model

Stress-strain relationship	$\boldsymbol{\sigma}(\mathbf{x},t) = \big(1 - D(\mathbf{x},t)\big)\mathbf{C} : \boldsymbol{\varepsilon}(\mathbf{x},t)$	(1)
Local state variable	$Y(\mathbf{x},t) = Y\big(\boldsymbol{\varepsilon}(\mathbf{x},t)\big)$	(2)
Nonlocal state variable	$\widetilde{Y}(\mathbf{x},t) = \dfrac{\int_V \alpha(\mathbf{x}-\mathbf{z})Y(\mathbf{z},t)\mathrm{d}V}{\int_V \alpha(\mathbf{x}-\mathbf{z})\mathrm{d}V}$	(3)
Weighting function	$\alpha(\mathbf{x}-\mathbf{z}) = \alpha(r; l_c)$ with $r = \Vert\mathbf{x}-\mathbf{z}\Vert$	(4)
Damage evolution	$D(\mathbf{x},t) = D\big(\max_{\tau \leq t} \widetilde{Y}(\mathbf{x},\tau)\big)$	(5)

Many nonlocal damage models encountered in the literature can be accommodated with little or no modification into the general framework of Table 1. The most common choices for (2), (4) and (5) are reviewed next.

Local State Variable

The local state variable Y is a suitable scalar measure of the strain tensor $\boldsymbol{\varepsilon}$. Three common definitions are the energy release rate [19, 15]

$$Y = \frac{1}{2}\boldsymbol{\varepsilon} : \mathbf{C} : \boldsymbol{\varepsilon}\,, \tag{6}$$

the average of positive principal strains ε_i used in the Mazars model [20]

$$Y = \sqrt{\sum_i \left[\max(0,\varepsilon_i)\right]^2}\,, \tag{7}$$

and a function of strain invariants used in the modified von Mises model [9]

$$Y = \frac{k-1}{2k(1-2\nu)}I_1 + \frac{1}{2k}\sqrt{\left(\frac{k-1}{1-2\nu}I_1\right)^2 + \frac{12k}{(1+\nu)^2}J_2}\,. \tag{8}$$

In (8), I_1 and J_2 are the first and second invariants of the strain and deviatoric strain tensors respectively, and k is the ratio of compressive to tensile strength.

Weighting Function

The weighting function α is typically the Gaussian function [22, 24, 29]

$$\alpha(r; l_c) = \exp\left[-\left(\frac{2r}{l_c}\right)^2\right] . \tag{9}$$

As a remark, is is worth noting that this function is sometimes written as [16, 1]

$$\alpha(r; l_c) = \exp\left[-\left(\frac{r}{\sqrt{2}l_c}\right)^2\right] . \tag{10}$$

Note that the characteristic lengths in (9) and (10) differ by a factor of $2\sqrt{2}$.

For computational efficiency, the infinite support of the Gaussian function is truncated for the nonlocal averaging. Another possibility is to use a parabolic function with compact support, see [16].

Damage Evolution Law

Two typical choices to describe the evolution of damage are the exponential law [20]

$$D = 1 - \frac{Y_0(1-A)}{\widetilde{Y}} - A\exp\left[-B\left(\widetilde{Y} - Y_0\right)\right] \tag{11}$$

and the polynomial law [26, 1]

$$D = 1 - \frac{1}{1 + B(\widetilde{Y} - Y_0) + A(\widetilde{Y} - Y_0)^2} . \tag{12}$$

In both expressions, parameter A is associated to residual strength and parameter B controls the slope of the softening branch at the peak (i.e. at $\widetilde{Y} = Y_0$), see [29].

In Mazars model, damage D is expressed as a combination of tensile damage D_t and compressive damage D_c [20]. Each of these two components evolves according to an exponential law (11), with the corresponding parameters A_c and B_c for compression and A_t and B_t for tension.

2.2 A Nonlocal Damage Model Based on Nonlocal Displacements

As Table 1 reflects, the standard approach is to define a scalar local state variable Y (as a function of strains) and then to average it into the nonlocal state variable $\widetilde{Y}$, which drives the evolution of damage, see Table 1.

However, other variables can be selected for averaging. In fact, a number a proposals can be found in the literature. Either scalar (for instance: damage D) or tensorial (for instance: strain ε) Gauss-point quantities may be averaged into the corresponding nonlocal quantities ($\widetilde{D}$ and $\widetilde{\varepsilon}$ in the two examples mentioned). The existing approaches are compared in [16] by means of a simple 1D numerical test (bar under uniaxial tension).

A new proposal is made here: to compute nonlocal displacements $\widetilde{\mathbf{u}}$ by averaging the local (i.e. standard) displacements $\mathbf{u}$. These nonlocal displacements $\widetilde{\mathbf{u}}$ drive the evolution of damage, see Table 2.

Table 2. Standard approach (nonlocal state variable) vs. alternative approach (nonlocal displacements). Subscript NL denotes quantities with nonlocal information but computed locally. The tilde is reserved to truly nonlocal quantities (i.e. computed via nonlocal average, NLA, of a local quantity)

Standard approach	**Alternative approach**
Local displacement $\mathbf{u}$	Local displacement $\mathbf{u}$
Local strain $\boldsymbol{\varepsilon} = \nabla^s \mathbf{u}$	Nonlocal displacement $\widetilde{\mathbf{u}} = \mathrm{NLA}(\mathbf{u})$
Local state var. $Y = Y(\boldsymbol{\varepsilon})$	Nonlocal strain $\boldsymbol{\varepsilon}_{\mathrm{NL}} = \nabla^s \widetilde{\mathbf{u}}$
Nonlocal state var. $\widetilde{Y} = \mathrm{NLA}(Y)$	Nonlocal state var. $Y_{\mathrm{NL}} = Y(\boldsymbol{\varepsilon}_{\mathrm{NL}})$
Damage evolution $D = D(\widetilde{Y})$	Damage evolution $D = D(Y_{\mathrm{NL}})$
	Local strain $\boldsymbol{\varepsilon} = \nabla^s \mathbf{u}$
Stress-strain law $\boldsymbol{\sigma} = (1-D)\mathbf{C} : \boldsymbol{\varepsilon}$	Stress-strain law $\boldsymbol{\sigma} = (1-D)\mathbf{C} : \boldsymbol{\varepsilon}$

From the numerical point of view, this model has very attractive properties. The consistent tangent matrix is quite simpler to compute than in the standard case described in [17], because now the nonlocal average is performed completely "upstream" in the constitutive equation (with displacements, the primal unknowns in the FE analysis).

From the viewpoint of physical modelling, our preliminary numerical results are also encouraging. For the 1D tension test, the model exhibits a satisfactory behaviour: no pathological mesh dependence and width of damaged zone controlled by the characteristic length.

3 Numerical Aspects

As discussed in the introduction, *reliability* and *efficiency* are the two major issues in virtually any branch of computational (solid and fluid) mechanics. When modelling quasi-brittle materials, the development of reliable and efficient algorithms must take into account that the structural response is typically (1) very nonlinear, (2) localized, (3) highly dependent on the value of material parameters and (4) geometrically complex (curved cracks, primary and secondary failure mechanisms, etc.).

3.1 Efficiency

Due to material nonlinearity, FE discretization of the equilibrium equation leads to a nonlinear system of algebraic equations, which is solved in an incremental-iterative fashion [7, 5]. Thus, the two key ingredients for an efficient numerical model are:

1. An adequate control strategy to define the increments (i.e. time- or load-stepping).
2. A suitable nonlinear solver for the iterations within each increment, including the appropriate (secant or tangent) stiffness matrices.

Control Strategy (for Increments)

The softening behaviour of the stress-strain law leads to structural softening of the snap-through or snap-back type, as shown in Figs. 6, 15 and 21 of Sect. 4. As a consequence, force control is not a suitable control strategy; either displacement control (for snap-through) or arc-length control (for snap-through or snap-back) are required.

In any case, the localized nature of quasi-brittle failure must be accounted for. For arc-length control, for instance, the classical constraint [7]

$$\sqrt{\Delta\mathbf{u}^T\Delta\mathbf{u} + (\Delta\mathbf{f}^{\text{ext}})^T\Delta\mathbf{f}^{\text{ext}}} = \Delta s\,, \qquad (13)$$

which prescribes the norm of the increment of solution (displacement and external forces) to a given arc-length Δs is not appropriate, because $\Delta\mathbf{u}$ is a global quantity that does not reflect localization.

Alternative constraints are required. Two possibilities are [29]

$$\max|\Delta\varepsilon_{ij}| = \Delta s \qquad (14)$$

and

$$|\Delta u_{\text{characteristic}}| = \Delta s\,, \qquad (15)$$

where $u_{\text{characteristic}}$ is one (or a combination of a few) characteristic degree(s) of freedom. For notched specimens, for instance, one can use the CMOD

(Crack-Mouth Opening Displacement) or CMSD (Crack-Mouth Sliding Displacement) as the arc-length parameter.

Another important issue is the proper selection of the prescribed arc-length Δs. For complex computations, it is not practical to use a constant value, so automatic load-stepping is required to adapt the "load-step" Δs along the analysis. A common choice is to use ad-hoc formulas based on comparing the desired number of iterations per step with the actual number of iterations in the last step [7].

A more rigorous approach is to use adaptive time-stepping strategies based on techniques for the numerical solution of ordinary differential equations. The time error is controlled such that it is at least one order of magnitude smaller than the space error. Thus, the effect of the time integration in the space error may be considered negligible. This approach is developed in [11] for initial boundary value problems (i.e. evolutionary problems involving, for instance, viscoelasticity or coupled chemo-mechanical behaviour of concrete). This idea makes even more sense for the boundary value problem under consideration here, where inertia effects are neglected (quasistatic problem), so physical time does not appear in the governing equations. If load steps (that is, pseudo-time increments) are properly chosen, then the error associated to load-stepping is negligible with respect to the space error for each load level.

Nonlinear Solver and Stiffness Matrix (for Iterations)

Within each increment, the equilibrium equation remains nonlinear and demands an iterative solution. A nonlinear solver amounts basically to the selection of a particular stiffness matrix.

One possibility [26, 24, 29, 28] is to work with the secant stiffness matrix, computed from the damaged elastic moduli $(1 - D)\mathbf{C}$. The main advantage of this approach is that the secant matrix is symmetric positive definite and very simple to compute (the factor $(1 - D)$ at each Gauss point is the only difference with respect to the elastic stiffness matrix). The main drawback is that it must be supplemented with convergence acceleration and, even so, convergence is only linear.

If quadratic convergence is desired (full Newton-Raphson method), the consistent tangent matrix is required [5]. For nonlocal damage models, this poses a substantial difficulty: due to nonlocality, there is interaction between non-adjacent nodes, and the consistent tangent matrix exhibits a larger bandwidth (with respect to the sparsity pattern of the elastic or secant matrices) [26, 17] Moreover, symmetry is lost.

To avoid the additional non-zero entries, some authors [9, 1] neglect the nonlocal interaction. However, the resulting local tangent matrix is no longer consistent, and quadratic convergence is lost.

These three basic choices are summarized in Table 3.

Table 3. Properties of stiffness matrices

Matrix		Symmetry	Increased bandwidth	Nonlocal interaction	Convergence
$\mathbf{K}_{\text{sec}}$	Secant	Yes	No	No	Linear
$\mathbf{K}_{\text{sec}} + \mathbf{K}_{\text{local}}$	Local tangent	No	No	No	Linear
$\mathbf{K}_{\text{tan}}$	Consistent tangent	No	Yes	Yes	Quadratic

Quadratic Convergence Without Fill-in

If the consistent tangent matrix is chosen, equilibrium iterations read

$$\mathbf{K}^{i}_{\text{tan}}\delta\mathbf{u}^{i+1} = -\mathbf{r}^{i}\,, \tag{16}$$

where i is the iteration counter, $\mathbf{r}^{i}$ is the residual and $\delta\mathbf{u}^{i+1}$ is the iterative correction in displacements.

Due to the increased bandwidth of $\mathbf{K}^{i}_{\text{tan}}$, fill-in during the factorization is considerably larger than for a local (tangent or secant) stiffness matrix. If this additional fill-in is a critical factor, it can be avoided by accounting for the nonlocal interaction in the right-hand-side vector. The consistent tangent matrix can be expressed as

$$\mathbf{K}_{\text{tan}} = \mathbf{K}_{\text{sec}} + \mathbf{K}_{\text{local}} + \mathbf{K}_{\text{nonlocal}} \tag{17}$$

where $\mathbf{K}_{\text{sec}}$ is the secant matrix and $\mathbf{K}_{\text{local}} + \mathbf{K}_{\text{nonlocal}}$ is the non-secant matrix which accounts for the variation of the damage parameter with strains, $\partial D/\partial \varepsilon$. This matrix has a local contribution $\mathbf{K}_{\text{local}}$, which can be assembled from elementary matrices computed from the usual loop on the elements' Gauss points, and a nonlocal contribution $\mathbf{K}_{\text{nonlocal}}$, which reflects nonlocal interaction between Gauss points and is responsible for increased bandwidth. The local tangent matrix of Table 3 is $\mathbf{K}_{\text{sec}} + \mathbf{K}_{\text{local}}$.

Equations (16) and (17) can be combined into the system of equations

$$\left(\mathbf{K}^{i}_{\text{sec}} + \mathbf{K}^{i}_{\text{local}}\right)\delta\mathbf{u}^{i+1} = -\mathbf{r}^{i} - \mathbf{K}^{i}_{\text{nonlocal}}\delta\mathbf{u}^{i+1} \tag{18}$$

which can be solved with an inner iterative scheme,

$$\left(\mathbf{K}^{i}_{\text{sec}} + \mathbf{K}^{i}_{\text{local}}\right)\delta\mathbf{u}^{i+1}_{k+1} = -\mathbf{r}^{i} - \mathbf{K}^{i}_{\text{nonlocal}}\delta\mathbf{u}^{i+1}_{k} \tag{19}$$

where k is the counter for the inner iterations. Note that extra fill-in is indeed precluded, because the matrix in system (19) is local. Moreover, once $\mathbf{K}^{i}_{\text{sec}} + \mathbf{K}^{i}_{\text{local}}$ is factorized, the inner iterations have a relatively modest computational cost. Linear convergence is expected for these inner iterations k, but – and this is the key issue –, quadratic convergence without increased fill-in will be achieved for the expensive, outer equilibrium equations i. Moreover, the tolerance of the inner k loop is usually not taken as a constant, but dependent on outer iteration i to increase the efficiency. This is standard in numerical algebra [18]. In fact at the initial iterations of the Newton-Raphson method a larger tolerance is prescribed and, as i increases, the tolerance of the inner loop approaches the standard value.

If factorizing the non-symmetric matrix $\mathbf{K}^i_{\text{sec}} + \mathbf{K}^i_{\text{local}}$ is inconvenient with the finite element code at hand, matrix $\mathbf{K}^i_{\text{local}}$ can be taken to the right-hand-side. The inner loop then reads

$$\mathbf{K}^i_{\text{sec}} \delta \mathbf{u}^{i+1}_{k+1} = -\mathbf{r}^i - \left(\mathbf{K}^i_{\text{local}} + \mathbf{K}^i_{\text{nonlocal}}\right) \delta \mathbf{u}^{i+1}_{k} , \tag{20}$$

and the usual Cholesky factorization applies. More (but cheaper) inner iterations should be expected.

3.2 Reliability

Even for nonlinear problems, simply *obtaining* a finite element solution is nowadays not enough. One must also ensure the quality of the numerical results in an objective, quantifiable manner. With that purpose, we present here an adaptive strategy based on error estimation [29, 28]. The two key ingredients are a residual-type error estimator for nonlinear problems [14] and h-remeshing [31].

The issue of reliability is relevant in all fields of computational mechanics. In failure modelling of quasi-brittle materials, it is critical. Due to brittleness, the particular choice of a constitutive equation or a set of material parameters can have a very significant influence (not only quantitative but even qualitative) on the failure mechanism.

Of course, the finite element mesh also affects the numerical solution. For this reason, it is essential to keep FE errors under control when assessing the effect of material modelling. If models or sets of parameters are compared with a given mesh (deemed "sufficiently fine" but with no objective measure of its quality), the effect of FE discretization errors could be erroneously attributed to the different material models.

The key ingredient of the adaptive strategy is the error estimator. We use a residual-type error estimator first developed for linear problems in continua [12] and later extended to other problems, such as local nonlinear models, e.g. plasticity or visco-plasticity [14, 10], nonlocal nonlinear models (nonlocal damage [28]), or (linear and nonlinear) shells [13].

The focus here is in the two main issues of the problem under consideration: nonlinearity and nonlocality. More details about the error estimator can be found in the references just cited.

Error Estimation: Nonlinearity

Finite element discretization of the governing partial differential equation renders the algebraic nonlinear equilibrium equation

$$\mathbf{f}_H^{\text{int}}(\mathbf{u}_H) = \mathbf{f}_H^{\text{ext}}, \tag{21}$$

where the unknown is the nodal displacement vector $\mathbf{u}_H$, $\mathbf{f}_H^{\text{int}}(\mathbf{u}_H)$ is the vector of nodal internal forces associated with $\mathbf{u}_H$ and $\mathbf{f}_H^{\text{ext}}$ is the discretized external force term. Subscript H denotes that the working mesh has characteristic size H.

The exact error of $\mathbf{u}_H$ is defined as $\mathbf{e}_\mathbf{u}^{\text{exact}} := \mathbf{u} - \mathbf{u}_H$, where $\mathbf{u}$ is the exact solution. Of course, $\mathbf{e}_\mathbf{u}^{\text{exact}}$ cannot be computed because $\mathbf{u}$ is not available. Instead, the error $\mathbf{e}_\mathbf{u}^{\text{exact}}$ is approximated by the reference error $\mathbf{e}_\mathbf{u} := \mathbf{u}_h - \mathbf{u}_H$, where $\mathbf{u}_h$ is the finite element solution obtained with a finer mesh ($h \ll H$, the approach considered here) or a higher-order interpolation:

$$\mathbf{f}_h^{\text{int}}(\mathbf{u}_h) = \mathbf{f}_h^{\text{ext}}. \tag{22}$$

Note that (22) can also be expressed as

$$\mathbf{f}_h^{\text{int}}(\mathbf{u}_H + \mathbf{e}_\mathbf{u}) = \mathbf{f}_h^{\text{ext}}. \tag{23}$$

Computing $\mathbf{e}_\mathbf{u}$ (or, equivalently, $\mathbf{u}_h$) is computationally much more expensive than computing $\mathbf{u}_H$, because it involves solving a much larger nonlinear system of equations over a finer mesh, (22) or (23). For this reason, the basic idea of residual-type error estimators is to approximate $\mathbf{e}_\mathbf{u}$ by low-cost local computations over subdomains.

The natural subdomains for local computations are the finite elements. For this reason, the first phase of the error estimator consists on solving the nonlinear system (23) locally inside each finite element of the working mesh (interior estimate, see Table 4). To do so, each element Ω_k of size H is meshed into 4×4 elements of size h, see Fig. 1. That it, the fine mesh h is nested into the working mesh H, with $H < h/4$.

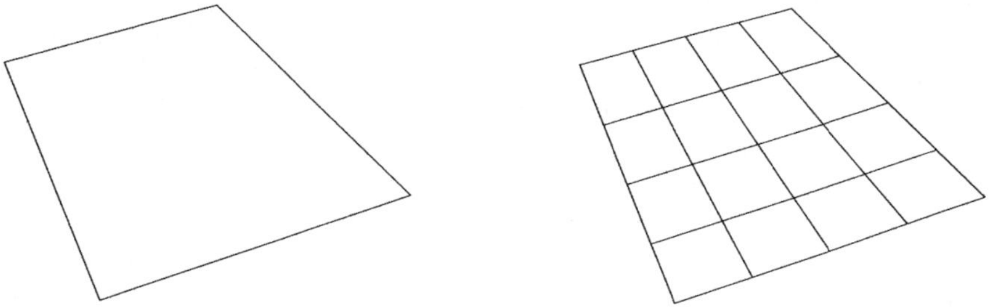

Fig. 1: Finite element of mesh H subdivided into 4×4 elements of mesh h

To avoid the expensive flux-splitting procedures of other residual-type error estimators (required to prescribe Neumann boundary conditions for each local problem), homogeneous Dirichlet boundary conditions for the error are prescribed on the element boundary $\partial\Omega_k$ (that is, $\mathbf{u}_h = \mathbf{u}_H$ on $\partial\Omega_k$). This equality is also set, over all the element Ω_k, as the initial approximation. Once error $\mathbf{e}^k_{\text{elem}}$ is obtained, its squared energy norm (based on the SPD secant stiffness matrix) is computed and added up into the global error estimate.

Table 4. Pseudocode for the first phase of the error estimation procedure. The interior estimate E is stored both locally ($\text{E}(\Omega_k)$ for $k = 1, 2, \ldots$) and globally ($\text{E}(\Omega)$ for the whole domain Ω)

`loop on elements` $k = 1, 2 \ldots$

- `Build up local refined mesh for element` Ω_k
- `Set trivial Dirichlet b.c.` $\mathbf{e}^k_{\text{elem}} = \mathbf{0}$ `on` $\partial\Omega_k$
- `Set initial approximation` $\mathbf{e}^k_{\text{elem}} = \mathbf{0}$ `in` Ω_k
- `Solve iteratively local nonlinear problem`

  $$\mathbf{r}(\mathbf{e}^k_{\text{elem}}) := \mathbf{f}^{\text{int}}_h(\mathbf{u}_H + \mathbf{e}^k_{\text{elem}})|_{\Omega_k} - \mathbf{f}^{\text{ext}}_h|_{\Omega_k} = \mathbf{0}$$

- `Compute squared local norm`

  $$\text{E}(\Omega_k) = (\mathbf{e}^k_{\text{elem}})^T \mathbf{K}^k_{\text{sec},h} \mathbf{e}^k_{\text{elem}}$$

- `Store error function:` $\mathbf{e}_{\text{elem}} \leftarrow \mathbf{e}_{\text{elem}} + \mathbf{e}^k_{\text{elem}}$
- `Upgrade global estimate:` $\text{E}(\Omega) \leftarrow \text{E}(\Omega) + \text{E}(\Omega_k)$

`end loop`

`Interior estimate`

Of course, the error $\mathbf{e_u}$ is not really zero along all element edges, as assumed during the interior estimate. For this reason, a second set of local problems is solved, over a different partition of the computational domain into subdomains. A natural choice is to associate these subdomains, called patches, to the nodes of the working mesh (patch estimate, see Table 5). If four-noded quadrilateral elements are used, a patch consists of one-fourth of each element sharing the node, see Fig. 2.

To compute the estimate $\mathbf{e}^\ell_{\text{patch}}$ for patch Λ_ℓ, the same ideas discussed for elements apply. Again, the boundary conditions and the initial approximation for the local nonlinear problem over the patch consist in setting $\boldsymbol{\eta}^\ell$ to zero over $\partial\Lambda_\ell$ and Λ_ℓ respectively.

The only difference is that orthogonality between patch estimate $\boldsymbol{\eta}^\ell$ and interior estimate $\boldsymbol{\varepsilon}$ must be imposed, to avoid accounting for the same error contribution twice. If the Lagrange multiplier is used for boundary conditions, then the orthogonality restriction can be prescribed as an additional "boundary" condition is a simple manner [14].

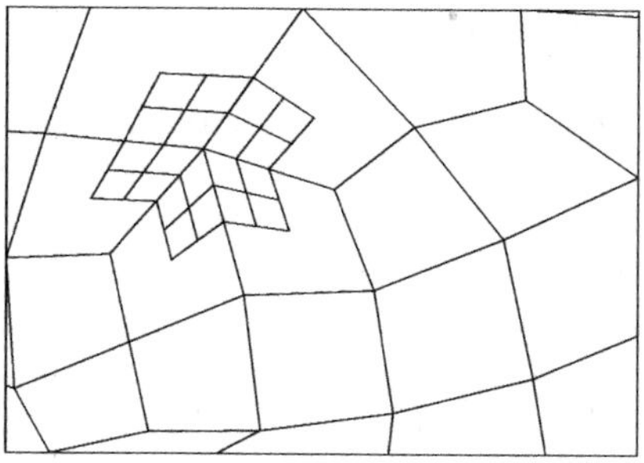

Fig. 2: Patch associated to a node of mesh H subdivided into 4×4 elements of mesh h

The proposed two-phase approach for error estimation is summarized in Tables 4 and 5.

Table 5. Pseudocode for the second phase of the error estimation procedure. The patch estimate is used to improve the estimate both locally ($\mathrm{E}(\Omega_k)$ for $k = 1, 2, \ldots$) and globally ($\mathrm{E}(\Omega)$ for the whole domain Ω)

`loop on patches` $\ell = 1, 2 \ldots$

- `Build up local refined mesh for patch` Λ_ℓ
- `Set trivial Dirichlet b.c.` $\mathbf{e}^\ell_{\text{patch}} = \mathbf{0}$ `on` $\partial\Lambda_\ell$

  `+ orthogonality to` $\mathbf{e}_{\text{elem}}$
- `Set initial approximation` $\mathbf{e}^\ell_{\text{patch}} = \mathbf{0}$ `in` Λ_ℓ
- `Solve iteratively local nonlinear problem`

  $\mathbf{r}(\mathbf{e}^\ell_{\text{patch}}) := \mathbf{f}^{\text{int}}_h(\mathbf{u}_H + \mathbf{e}^\ell_{\text{patch}})|_{\Lambda_\ell} - \mathbf{f}^{\text{ext}}_h|_{\Lambda_\ell} = \mathbf{0}$
- `Compute squared local norm`

  $\mathrm{E}(\Lambda^\ell) = (\mathbf{e}^\ell_{\text{patch}})^T \mathbf{K}^\ell_{\text{sec},h} \mathbf{e}^\ell_{\text{patch}}$
- `Upgrade global estimate:` $\mathrm{E}(\Omega) \leftarrow \mathrm{E}(\Omega) + \mathrm{E}(\Lambda_\ell)$
- `Upgrade local (element) estimate:`

  $n^\ell_{\text{over}} =$ `number of elements overlapping` Λ_ℓ
  `for` k `such that` $\Omega_k \cap \Lambda_\ell \neq \emptyset$
  $\mathrm{E}(\Omega_k) \leftarrow \mathrm{E}(\Omega_k) + \mathrm{E}(\Lambda_\ell)/n^\ell_{\text{over}}$

`end loop`

Patch estimate

Error Estimation: Nonlocality

The proposed two-phase approach for error estimation consists basically in solving two sequences of *local* nonlinear problems over subdomains (elements and patches), see Tables 4 and 5. The material model, however, is nonlocal. As a consequence, the internal forces $\mathbf{f}_{\text{int}}$ must be carefully computed in order to account for the nonlocal nature of the damage model [28].

The "natural" approach would be, see central column of Table 6, to obtain the error in strains $\mathbf{e}_{\varepsilon}$ from the error in displacements $\mathbf{e}_{\mathbf{u}}$ (in the corresponding element or patch), compute the refined strains $\boldsymbol{\varepsilon}_h$ and the local state variable Y_h. The nonlocal average over the subdomain (element k or patch ℓ), NLA_{sub}, then yields the nonlocal state variable $\widetilde{Y}_h$, which drives the damage parameter, D_h. Finally, refined stresses $\boldsymbol{\sigma}_h$ are computed.

Table 6. Standard nonlocal damage model and model modified for error estimation. Note the difference in the computation of the nonlocal state variable, steps 4 and 5

	Standard model	**Model for error estimation**
1. Error in displ.	$\mathbf{e}_{\mathrm{u}}$	$\mathbf{e}_{\mathrm{u}}$
2. Error in strains	$\mathbf{e}_{\varepsilon} = \nabla^s(\mathbf{e}_{\mathrm{u}})$	$\mathbf{e}_{\varepsilon} = \nabla^s(\mathbf{e}_{\mathrm{u}})$
3. Strain	$\boldsymbol{\varepsilon}_h = \boldsymbol{\varepsilon}_H + \mathbf{e}_{\varepsilon}$	$\boldsymbol{\varepsilon}_h = \boldsymbol{\varepsilon}_H + \mathbf{e}_{\varepsilon}$
4. Local state var.	$Y_h = Y(\boldsymbol{\varepsilon}_h)$	$Y_h \approx Y_H + \underbrace{\frac{\mathrm{d}Y}{\mathrm{d}\boldsymbol{\varepsilon}}(\boldsymbol{\varepsilon}_H)\mathbf{e}_{\varepsilon}}_{e_Y:\ \text{Error in } Y}$
5. Nonlocal state var.	$\widetilde{Y}_h = \text{NLA}_{\text{sub}}(Y_h)$	$e_{\widetilde{Y}} = \text{NLA}_{\text{sub}}(e_Y)$; $\widetilde{Y}_h = \widetilde{Y}_H + e_{\widetilde{Y}}$
6. Damage	$D_h = D(\widetilde{Y}_h)$	$D_h = D(\widetilde{Y}_h)$
7. Stresses	$\boldsymbol{\sigma}_h = (1 - D_h)\mathbf{C} : \boldsymbol{\varepsilon}_h$	$\boldsymbol{\sigma}_h = (1 - D_h)\mathbf{C} : \boldsymbol{\varepsilon}_h$

Note that the nonlocal average that transforms Y_h into $\widetilde{Y}_h$ is over a local support. This fact leads to non-physical responses, especially in zones of large damage gradients. Assume, for instance, that the error in strains is small and $\boldsymbol{\varepsilon}_h \approx \boldsymbol{\varepsilon}_H$. A small variation in $\widetilde{Y}$ is also expected ($\widetilde{Y}_h \approx \widetilde{Y}_H$). However, it may happen that $\widetilde{Y}_h \ll \widetilde{Y}_H$, because $\widetilde{Y}_h$ contains no information about nearby zones.

This point is illustrated in Fig. 3, which depicts the local state variable, the nonlocal state variable and the damage parameter for a given time increment in a zone of the working mesh H with large gradients. The circled element has a very small local state variable Y_H, see Fig. 3(a), below the threshold Y_0. However, since the elements to the right have large values of Y_H, it has a relatively large (above Y_0) nonlocal state variable $\widetilde{Y}_H$, see Fig. 3(b), which leads to damage, see Fig. 3(c). If the standard model is used to solve the local problem on the circled element during error estimation, a small error in strains leads to a small variation in the local state variable which, after nonlocal averaging over the element, results in a low value of the nonlocal state variable (that is, $\widetilde{Y}_h \ll \widetilde{Y}_H$). As a consequence, damage cannot increase in the circled element during error estimation. When estimating the error for the circled element, the nonlocal state variable $\widetilde{Y}_H$, rather than the local state variable Y_H, is representative of its mechanical properties.

For this reason, the nonlocal damage model is slightly modified for error estimation, see right column in Table 6. The difference resides in the way the nonlocal state variable $\widetilde{Y}_h$ is computed. By means of a first-order Taylor expansion, the local state variable Y_h is expressed as Y_H plus an error term e_Y. Note that the derivatives $\mathrm{d}Y/\mathrm{d}\varepsilon$ needed for computing e_Y are also required for the computation of the consistent tangent matrix, so they do not represent a computational overhead of the modified model.

The error term e_Y is averaged over the element/patch into $e_{\widetilde{Y}}$. As a consequence, $\widetilde{Y}_h$ is computed as the addition of a reference value $\widetilde{Y}_H$, which describes the real damaged stiffness, and an error term $e_{\widetilde{Y}}$.

With this modified model, a small variation in strains does result in a small variation in the nonlocal state variable (that is, $\widetilde{Y}_h \approx \widetilde{Y}_H$). Going back to figure 3, this means that the damage level of the circled element may either remain constant (for $\widetilde{Y}_h < \widetilde{Y}_H$) or increase (for $\widetilde{Y}_h > \widetilde{Y}_H$) during error estimation.

To sum up: the standard model is not capable of capturing the spread of the damaged zone associated to error estimation.

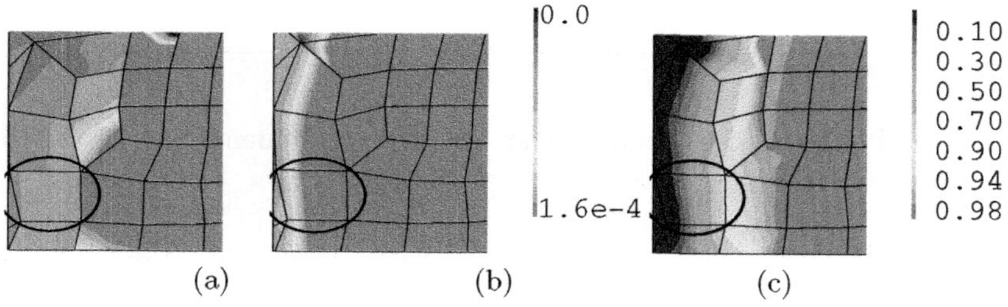

Fig. 3: Fields in a zone of large gradients: (**a**) local state variable Y; (**b**) nonlocal state variable $\widetilde{Y}$; (**c**) damage. The damage threshold is $Y_0 = 1.5 \times 10^{-4}$

4 Applications

In this section, various representative numerical examples are described in detail. The goal is to discuss, among others, the following issues: the performance of the proposed adaptive strategy based on error estimation; the influence of the material parameters on the structural response; the ability of the proposed approach to capture complex failure mechanisms; the comparative performance of the various stiffness matrices; the effect of exploiting problem symmetry on the failure pattern.

Three experimental tests will be modelled: the three-point bending test, the single-edge notched beam test, and the Brazilian (cylinder-splitting) test. These are standard tests to assess the mechanical behaviour of quasi-brittle materials.

4.1 Three-Point Bending Test

A notched beam is subjected to three-point bending, see Fig. 4. This test is modelled with the Mazars model [20] and the material parameters obtained in [4], see Table 7, by fitting the results of various experiments. Plane strain conditions are assumed.

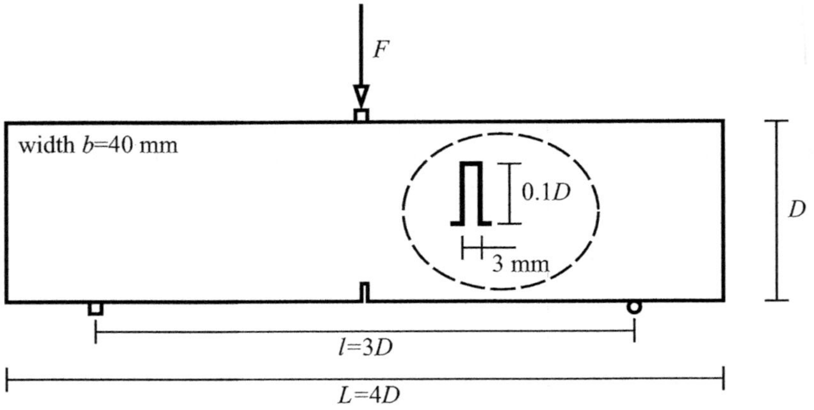

Fig. 4: Three-point bending test. Problem statement (Size D=320 mm)

If a displacement control strategy (with vertical deflection of the point under the load as the control variable) is used, the structural response of Fig. 5 is obtained. Note the abrupt softening behaviour after the peak load, typical of quasi-brittle materials.

Both in the force-deflection and the force-COD (crack-opening displacement) curves, Figs. 5(a) and 5(b), the softening branch exhibits, at its beginning a "suspicious" straight segment. To check its meaning, the numeri-

Table 7. Three-point bending test. Material parameters for the Mazars model [4]

Meaning	Symbol	Value
Young's modulus	E	38 500 MPa
Poisson's coefficient	ν	0.24
Characteristic length	l_c	40 mm
Damage threshold	Y_0	3×10^{-5}
Parameter A in compression	A_c	1.25
Parameter B in compression	B_c	1000
Parameter A in tension	A_t	0.95
Parameter B in tension	B_t	9000

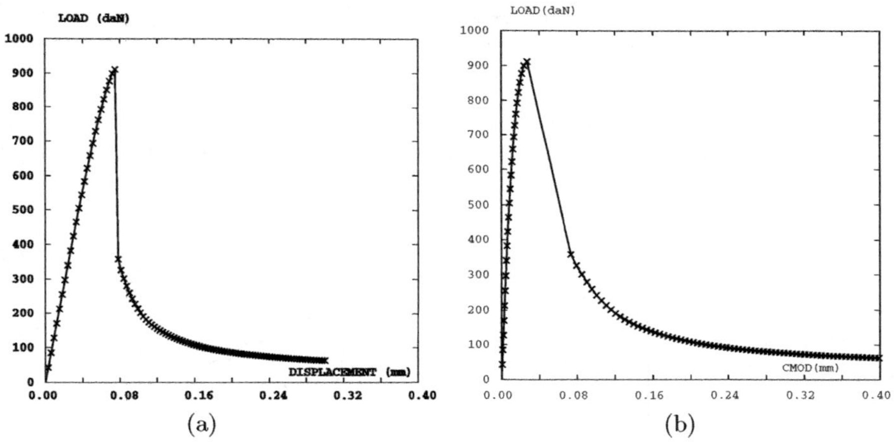

Fig. 5: Three-point bending test. Structural response obtained with displacement control: (**a**) force-vertical deflection; (**b**) force-COD (crack opening displacement)

cal test is repeated with a different control strategy: arc-length control with $\Delta s = |\Delta\mathrm{COD}|$. The results are depicted in Fig. 6.

Figure 6(a) clearly shows that the force-deflection curve exhibits a certain amount of snap-back behaviour, which cannot be captured with a displacement control strategy, Fig. 5, which assumes that the control variable (i.e. the deflection in this case) increases monotonically.

Since the snap-back is moderate, there is not much difference between Figs. 5 and 6. For other problems, however, the appropriate combination of control strategy and control variable is critical in tracking the nonlinear response [32].

The convergence behaviour of the various nonlinear solvers is summarized is Fig. 7. Note that quadratic convergence is achieved with the consistent tangent matrix, while only linear convergence is obtained with the secant matrix.

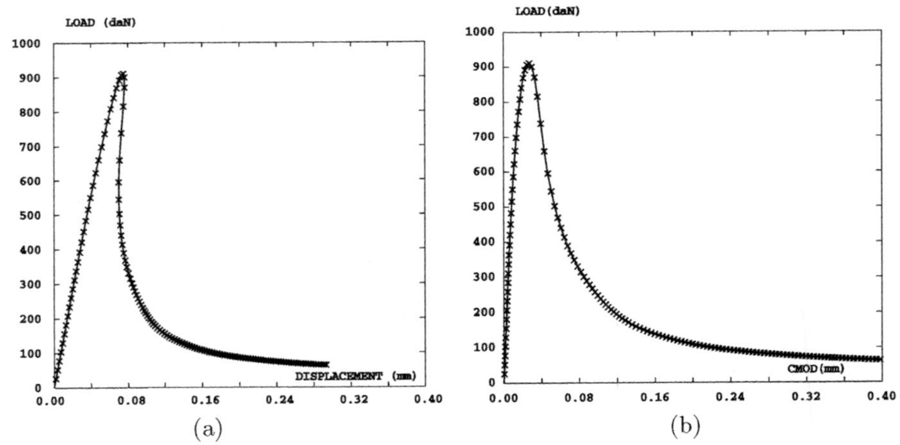

Fig. 6: Three-point bending test. Structural response obtained with arc-length control: (**a**) force-vertical deflection; (**b**) force-COD (crack opening displacement)

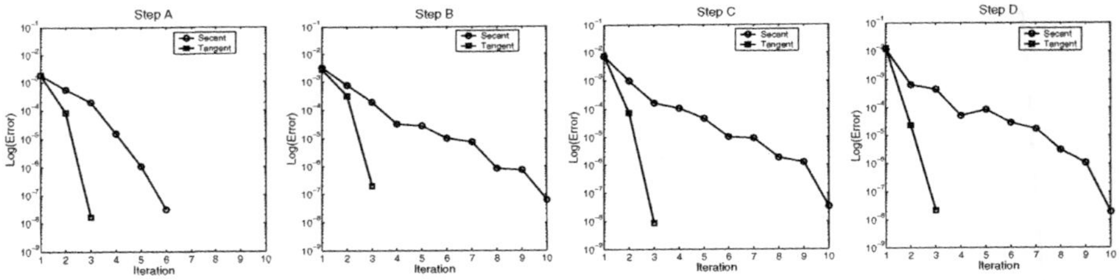

Fig. 7: Three-point bending test. Convergence history for representative load-steps with the secant matrix and the consistent tangent matrix

4.2 Single-Edge Notched Beam Test

A single-edge notched beam (SENB) is subjected to anti-symmetrical four-point bending [6]. The geometry, loads and supports are shown in Fig. 8. A plane stress analysis is performed. The concrete beam is modelled with the modified von Mises nonlocal damage model [9] with exponential damage evolution, while the steel loading platens are assumed to be elastic.

Two sets of material parameters are used for concrete [28], see Table 8. For material 1, there is a significant post-peak softening in the stress-strain law for concrete. For material 2, on the contrary, the softening is very slight, so the residual strength almost coincides with the peak strength [23]. For steel, a Poisson's coefficient $\nu = 0.2$ and a Young's modulus 10 times that of concrete are used.

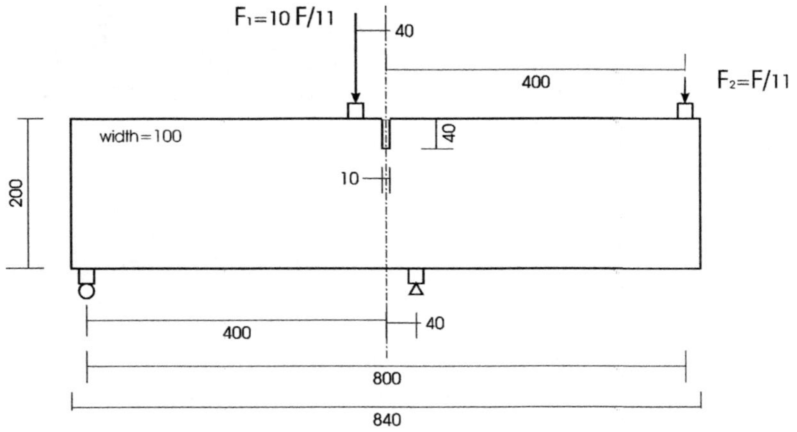

Fig. 8: Single-edge notched beam. Problem statement (all distances in mm)

Table 8. SENB test. The two sets of material parameters: large softening (material 1) and very slight softening (material 2)

		Value	
Meaning	**Symbol**	**Material 1**	**Material 2**
Young's modulus	E	28 000 MPa	35 000 MPa
Poisson's coefficient	ν	0.1	0.2
Characteristic length	l_c	10 mm	10 mm
Damage threshold	Y_0	1.5×10^{-4}	6.0×10^{-5}
Parameter A	A	0.8	0.08
Parameter B	B	9 000	8 200

SENB Test with Material 1

The results with material 1 are shown in Figs. 9–11. The initial mesh is shown in Fig. 9(a). Note that this mesh is relatively coarse, with only one element in the notch width. The final damage distribution and deformed mesh (amplified 300 times), corresponding to a CMSD of 0.08 mm, is depicted in Fig. 9(b). The curved crack pattern observed in experiments [6] is clearly captured. The error estimation procedure discussed in Sect. 3.2 is employed to compute the error field of Fig. 9(d). The error is larger in the damaged zone and near the loading platens. The global relative error (i.e. energy norm of the error in displacements over the energy norm of displacements) is 3.96%, above a threshold set a priori of 2%, so adaptivity is required.

The error field of Fig. 9(d) is translated into the mesh of Fig. 10(a). Note the element concentration in the crack and the central supports. This finer mesh leads to a better definition of the damaged zone, see Fig. 10(b). The error estimator now detects that the largest errors are associated to the *edges* of the cracked zone, see Fig. 10(d). The global relative error of 2.11% is still

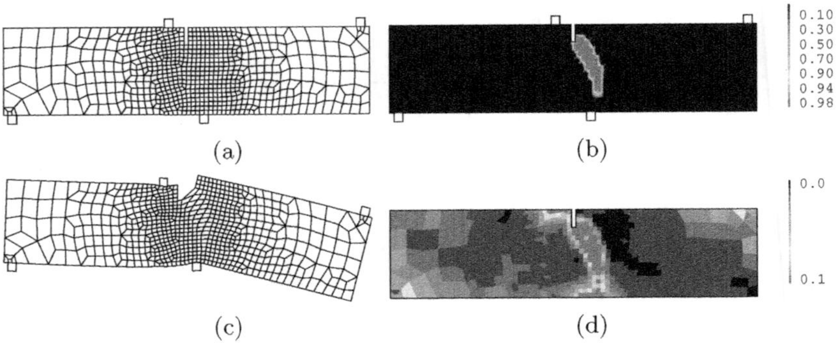

Fig. 9: SENB test. Material 1, initial approx. (**a**) Mesh 0: 659 elem., 719 nodes; (**b**) damage; (**c**) deformed mesh (×300); (**d**) error field. Global relative error: 3.96%

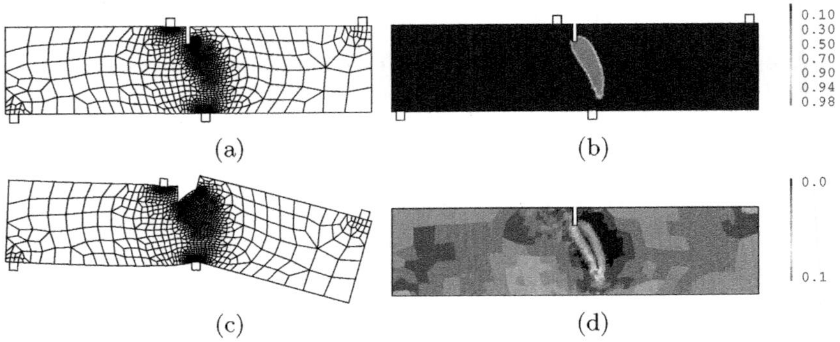

Fig. 10: SENB test. Material 1, iteration 1. (**a**) Mesh 1: 1155 elem., 1228 nodes; (**b**) damage; (**c**) deformed mesh (×300); (**d**) error field. Global relative error: 2.11%

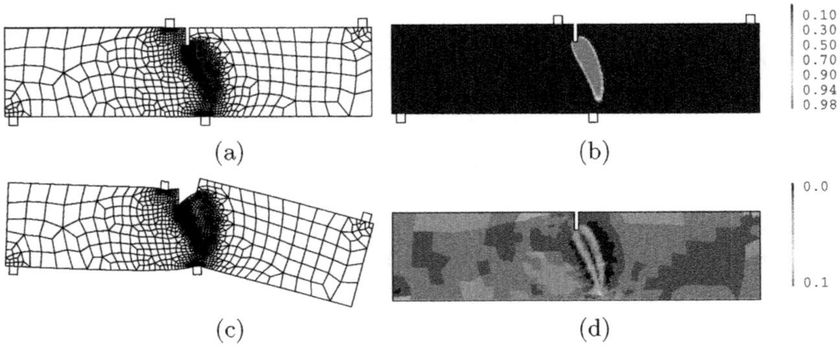

Fig. 11: SENB test. Material 1, iteration 2. (**a**) Mesh 2: 1389 elem., 1469 nodes; (**b**) damage; (**c**) deformed mesh (×300); (**d**) error field. Global relative error: 1.77%

slightly above the error goal, so another adaptive iteration is performed. The outcome of this second iteration is shown in Fig. 11. The qualitative results of iteration 1 are confirmed: (1) small elements are needed to control the error in the damaged zones and close to the loading platens and (2) error is larger in the edges than in the centre of the crack. The global relative error of 1.77% is below the threshold of 2%, so the adaptive iterative process stops.

SENB Test with Material 2

The SENB test is now reproduced with material 2, which has a stress-strain law with almost no softening [28]. A very similar law has been employed to simulate the SENB test with gradient-enhanced damage models [23].

The results are shown in Figs. 12–14. The initial mesh is the same as before, see Fig. 12(a). The change in the material parameters leads to a completely different failure pattern, dominated by bending of opposite sign in the two halves of the beam, see Figs. 12(b) and 12(c). A crack at the notch tip is also initiated, but it is only a secondary mechanism. The error estimation procedure has no difficulties in reflecting the change in the failure mode, see Fig. 12(d). The global relative error is 3.66%, so adaptivity is required.

Figures 13 and 14 illustrate the adaptive process. Note that meshes 1 and 2 are quite different from the ones obtained with material 1. The global relative errors are 2.46% and 2.13%. This value is still slightly above the threshold of 2%. However, an additional iteration is considered not necessary for the illustrative purpose of this test.

A final comparison between the two sets of material parameters is offered by Fig. 15, where the total load is plotted versus the CMSD for meshes 0 and 2. The results obtained with material 1 – a peak load of around 60 kN (with mesh 2) and post-peak structural softening, see Fig. 15(a) – are in good agreement with the experiments [6]. With material 2, on the other hand, the peak load is quite higher and no softening is observed, see Fig. 15(b).

As a final remark, note that Fig. 15 also shows a significant quantitative difference between the solutions with meshes 0 and 2. The result with the initial mesh clearly overestimates the peak load. The adaptive strategy based on error estimation enables an accurate prediction of the structural response.

4.3 Brazilian Test

The Brazilian (or cylinder-splitting) test provides an indirect measure of tensile strength of quasi-brittle materials. A cylindrical specimen is loaded along a diametral plane, see Fig. 16.

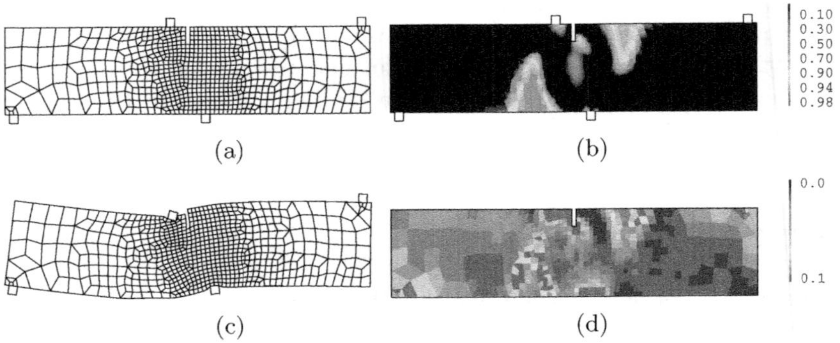

Fig. 12: SENB test. Material 2, initial approx. (**a**) Mesh 0: 659 elem., 719 nodes; (**b**) damage; (**c**) deformed mesh (×300); (**d**) error field. Global relative error: 3.66%

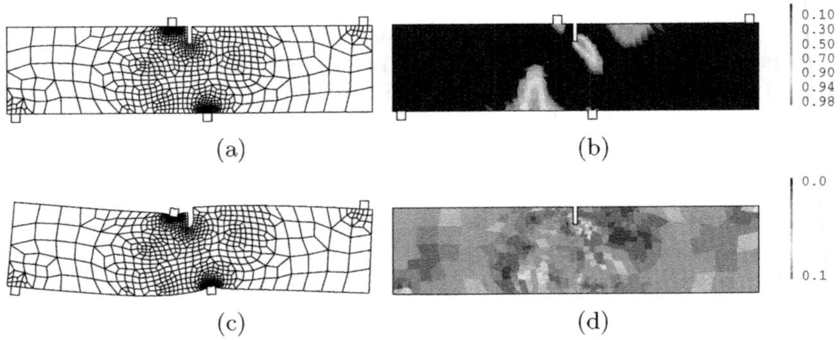

Fig. 13: SENB test. Material 2, iteration 1. (**a**) Mesh 1: 776 elem., 848 nodes; (**b**) damage; (**c**) deformed mesh (×300); (**d**) error field. Global relative error: 2.46%

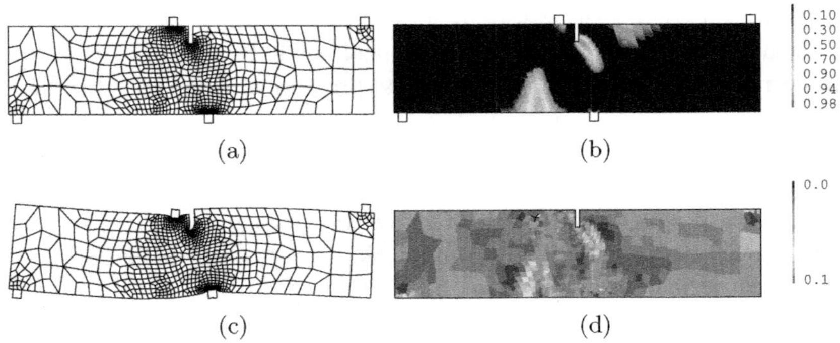

Fig. 14: SENB test. Material 2, iteration 2. (**a**) Mesh 2: 870 elem., and 954 nodes; (**b**) damage; (**c**) deformed mesh (×300); (**d**) error field. Global relative error: 2.13%

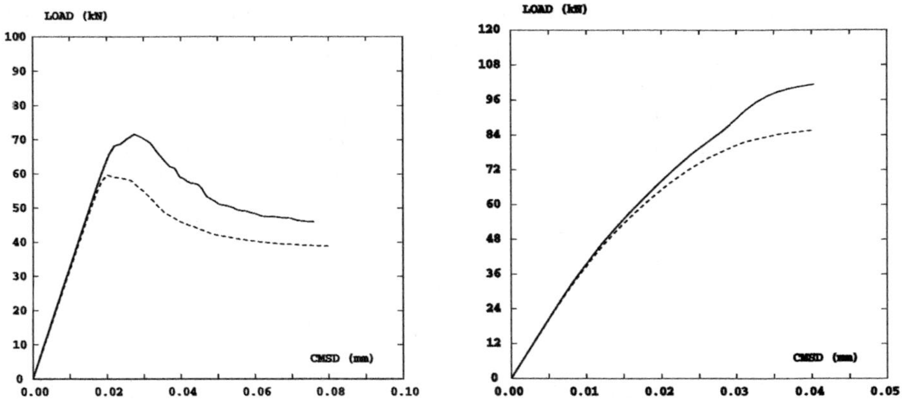

(a) (b)

Fig. 15: SENB test. Total load versus crack-mouth sliding displacement (CMSD) for meshes 0 (*solid line*) and 2 (*dashed line*): (**a**) with material 1 (large softening); (**b**) with material 2 (very slight softening)

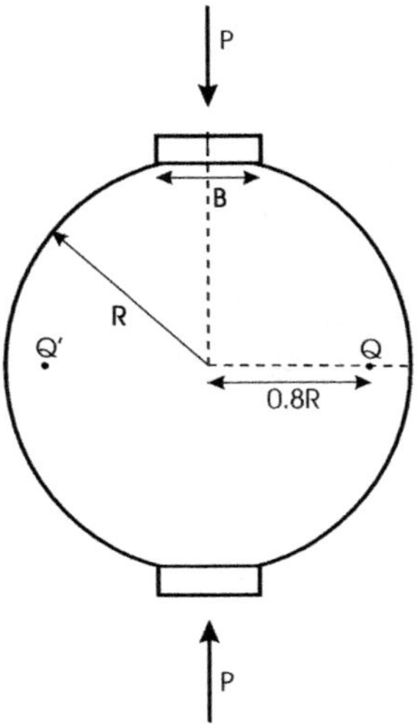

Fig. 16: Brazilian test. Problem statement: specimen of radius R and bearing strip of width B. Points Q and Q' are used to define the Crack Opening Displacement (COD)

A plane strain analysis is carried out. The Mazars model is used for the quasi-brittle specimen, while the steel loading platens are assumed to be elastic. Geometrical and material parameters are shown in Table 9.

Table 9. Brazilian test. Geometric and material parameters (Mazars model)

Meaning	Symbol	Value
Radius of specimen	R	40 mm
Width of bearing strip	B	10 mm
Young's modulus	E	37 700 MPa (specimen) 300 000 MPa (bearing strip)
Poisson coefficient	ν	0.2
Characteristic length	l_c	2 mm
Damage threshold	Y_0	10^{-4}
Parameter A in compression	A_c	1.4
Parameter B in compression	B_c	1 900
Parameter A in tension	A_t	1
Parameter B in tension	B_t	15 600

One goal of this example is to assess the influence of problem symmetry on the numerical results. For this reason, two different computational domains will be used, consisting of one-fourth and one-half of the problem domain, see Fig. 17. In both cases, adequate symmetry boundary conditions are prescribed along the symmetry axes, regarding both the displacement field and the nonlocal averaging.

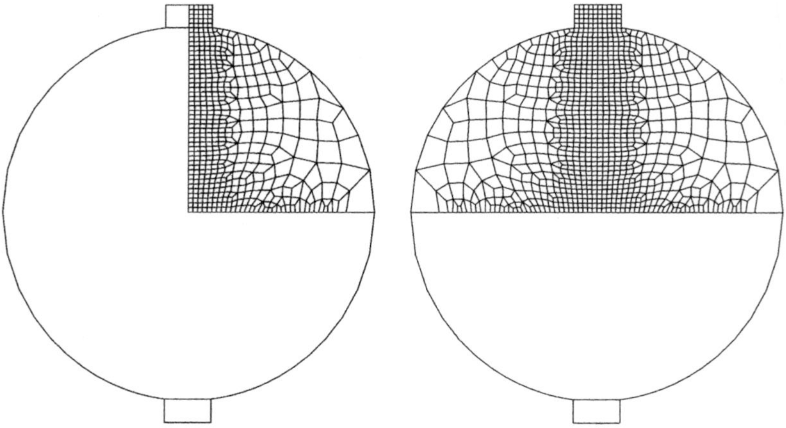

(a) (b)

Fig. 17: Brazilian test. Computational domain and finite element mesh for (**a**) one-fourth and (**b**) one-half of specimen

Figure 18 depicts the structural response. The applied vertical load is plotted versus the vertical displacement of the bearing strip and versus the crack opening displacement (defined as the separation between points Q and Q', see Fig. 16 and [27].)

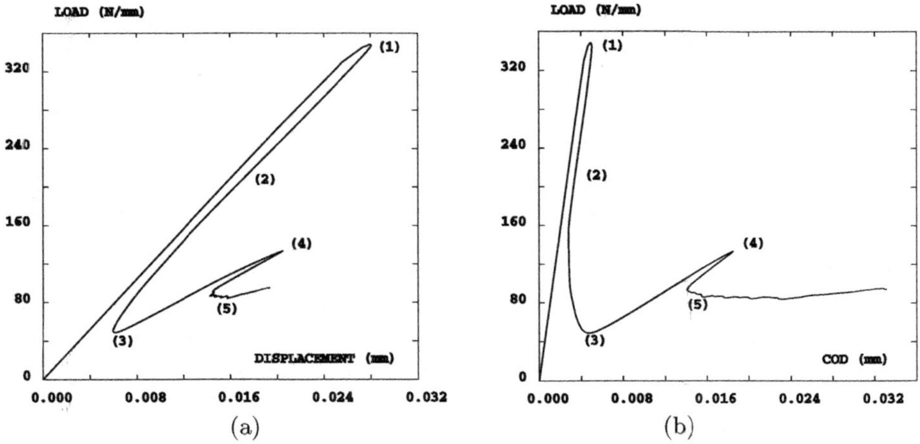

Fig. 18: Brazilian test. Response with computational domain of Fig. 17(a). Load vs. (**a**) vertical displacement; (**b**) crack-opening displacement. Damage fields at states 1–5 depicted in Fig. 19

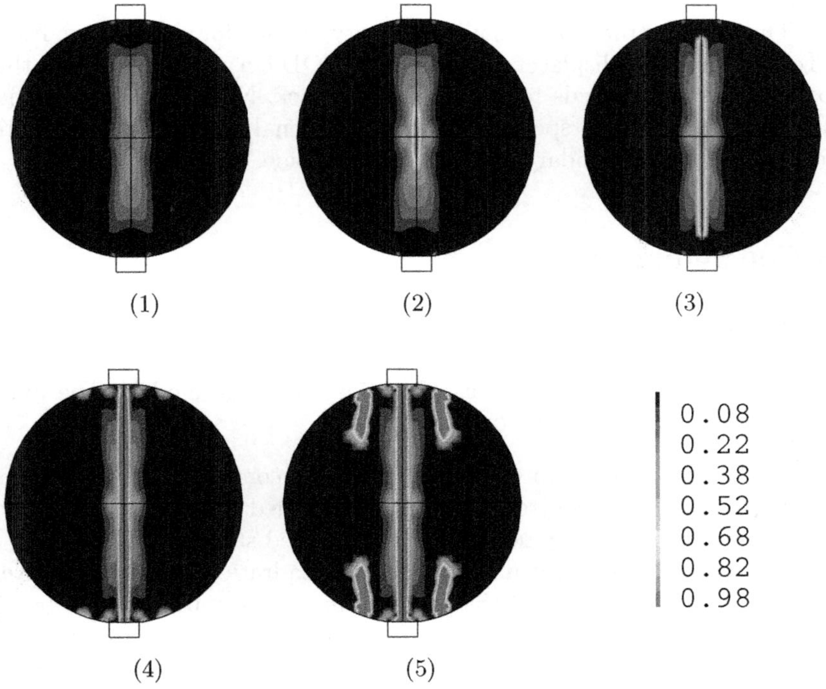

Fig. 19: Brazilian test. Crack formation with computational domain of Fig. 17(a). Primary crack starts at centre of specimen and propagates outwards. Secondary cracks caused by compression on each half specimen resisting separately

The evolution of damage is depicted in Fig. 19. Damage starts above and below the centre of the specimen (state 1) and propagates along the loading plane (state 2) until it reaches the bearing plates (state 3). When this primary crack is fully developed, the specimen is split into two half discs. Under certain conditions, each half specimen has some extra resistance under compression. After some reloading (state 4, see Fig. 18), a secondary crack system leads finally to collapse. This complex failure pattern has been experimentally observed [27] and numerically analysed in detail in [30].

Note that the double symmetry in the crack system of Fig. 19 is prescribed a priori by the choice of the computational domain. However, one would not expect such a perfectly symmetrical behaviour in the actual experiment, especially considering the quasi-brittle nature of the material.

For this reason, the numerical test is repeated with one-half of the specimen as the computational domain, see Fig. 17(b). The evolution of damage is depicted in Fig. 20. As expected, the secondary crack is no longer symmetrical with respect to the vertical axis. Rounding errors (the numerical "equivalent" of material heterogeneity in the actual experiment) induce a slight non-symmetry in the primary crack and, as a result, the secondary crack only develops in one of the half discs.

The global structural behaviour is very similar for the two analyses, see Fig. 21. The force-displacement and force-COD curves only differ in the last part, which corresponds to the secondary crack. Note the larger stiffness of the solution with half specimen as computational domain, associated to the formation of the secondary crack in only one side.

5 Concluding Remarks

Efficiency and reliability are two crucial issues in any field of computational mechanics. We have addressed them here in the context of failure analysis of quasi-brittle materials by means of nonlocal damage models. The following conclusions may be pointed out:

- Advanced control strategies (displacement control or arc-length control with appropriate control variables) are needed to account in an efficient manner for the highly nonlinear and localized structural response.
- The consistent tangent matrix provides quadratic convergence in Newton equilibrium iterations. If the extra fill-in of this matrix caused by nonlocality is a critical factor, it can be avoided with an inner iterative loop.
- An adaptive strategy driven by an error estimator is required to ensure the quality of the numerical results in an objective manner.
- The error estimator may be based on local computations – a common approach in residual-type error estimators –, provided that the damage model is slightly modified. The basic idea is that the *error* in the local state variable, rather than the variable itself, is averaged.

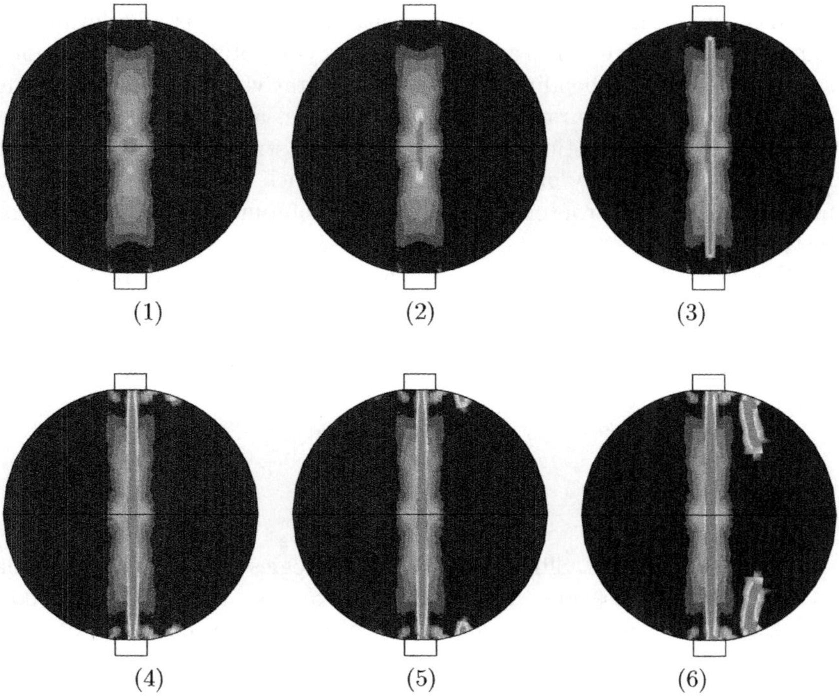

Fig. 20: Brazilian test. Crack formation computed with one-half of the specimen as computational domain, see Fig. 17(b). Note that the secondary crack system is no longer symmetrical with respect to the vertical axis, cf. Fig. 19

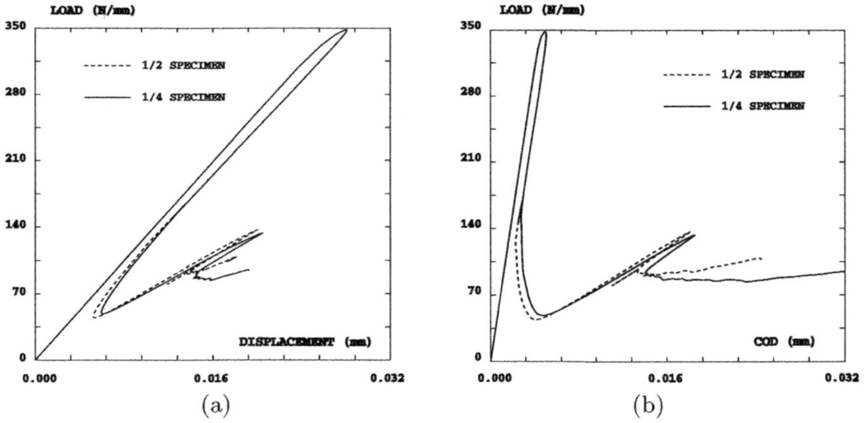

Fig. 21: Brazilian test. Effect of computational domain on mechanical behaviour. Vertical load versus (**a**) vertical displacement; (**b**) crack-opening displacement

In the numerical examples, we have attempted to cover all relevant aspects, namely: the relative performance of various control strategies and nonlinear solvers (three-point bending test); the quantitative and qualitative influence of material parameters on structural response, assessed in an reliable way thanks to the error-estimator-driven adaptive strategy (single-edge notched beam test); the ability of the proposed approach to capture complex failure mechanisms and the role of symmetry in the computational domain (Brazilian test).

We have also briefly discussed a new nonlocal damage model currently under development, based on the nonlocal average of displacements (rather than the state variable). Our preliminary results indicate that this model is sound from a physical point of view and has very attractive numerical properties.

References

[1] H. Askes and L.J. Sluys. Remeshing strategies for adaptive ALE analysis of strain localisation. *Eur. J. Mech. A-Solids*, 19(3):447–467, 2000.

[2] Z.P. Bažant and M. Jirásek. Nonlocal integral formulations of plasticity and damage: survey of progress. *J. Eng. Mech.-ASCE*, 128(11):1119–1149, 2002.

[3] Z.P. Bažant and G. Pijaudier-Cabot. Nonlocal continuum damage, localization instability and convergence. *J. Appl. Mech.-Trans. ASME*, 55(2):287–293, 1988.

[4] C. Le Bellégo. *Couplages chimie-mécanique dans les structures en béton attaquées per l'eau: étude expérimentale et analyse numérique.* PhD thesis, E.N.S. Cachan, February 2001.

[5] T. Belytschko, W.K. Liu, and B. Moran. *Nonlinear finite elements for continua and structures.* John Wiley & Sons, Chichester, 2000.

[6] A. Carpinteri, S. Valente, G. Ferrara, and G. Melchiorri. Is mode II fracture energy a real material property? *Comput. Struct.*, 48(3):397–413, 1993.

[7] M. A. Crisfield. *Non-linear finite element analysis of solids and structures. Volume 1: Essentials.* John Wiley & Sons, Chichester, 1991.

[8] R. de Borst, J. Pamin, R.H.J. Peerlings, and L.J. Sluys. On gradient-enhanced damage and plasticity models for failure in quasi-brittle and frictional materials. *Comput. Mech.*, 17(1–2):130–141, 1995.

[9] J.H.P. de Vree, W.A.M. Brekelmans, and M.A.J. van Gils. Comparison of nonlocal approaches in continuum damage mechanics. *Comput. Struct.*, 55(4):581–588, 1995.

[10] P. Díez, M. Arroyo, and A. Huerta. Adaptivity based on error estimation for viscoplastic softening materials. *Mech. Cohesive-Frict. Mater.*, 5(2):87–112, 2000.

[11] P. Díez, M. Arroyo, and A. Huerta. Adaptive simulation of the coupled chemo-mechanical concrete degradation. In H.A. Mang, F.G. Rammerstorfer, and J. Eberhardsteiner, editors, *Proceedings of the Fifth World Congress on Computational Mechanics (WCCM V)*. Vienna University of Technology, Austria, July 7-12, 2002, Vienna, Austria. http://wccm.tuwien.ac.at.

[12] P. Díez, J.J. Egozcue, and A. Huerta. A posteriori error estimation for standard finite element analysis. *Comput. Methods Appl. Mech. Eng.*, 163(1–4):141–157, 1998.

[13] P. Díez and A. Huerta. Error estimation for adaptivity in assumed-strain models for linear and nonlinear shells. *Comput. Mech.*, 2003. Submitted.

[14] A. Huerta and P. Díez. Error estimation including pollution assessment for nonlinear finite element analysis. *Comput. Methods Appl. Mech. Eng.*, 181(1–3):21–41, 2000.

[15] A. Huerta and G. Pijaudier-Cabot. Discretization influence on regularization by two localization limiters. *J. Eng. Mech.-ASCE*, 120(6):1198–1218, 1994.

[16] M. Jirásek. Nonlocal models for damage and fracture: comparison of approaches. *Int. J. Solids Struct.*, 35(31–32):4133–4145, 1998.

[17] M. Jirásek and B. Patzák. Consistent tangent stiffness for nonlocal damage models. *Comput. Struct.*, 80(14–15):1279–1293, 2002.

[18] C.T. Kelley. *Iterative methods for linear and nonlinear equations*, volume 16 of *Frontiers in Applied Mathematics*. Society for Industrial and Applied Mathematics, Philadelphia, 1995.

[19] J. Lemaitre and J.-L. Chaboche. *Mechanics of solid materials*. Cambridge University Press, Cambridge, 1990.

[20] J. Mazars. A description of micro- and macroscale damage of concrete structures. *Eng. Fract. Mech.*, 25(5–6):729–737, 1986.

[21] J. Mazars and G. Pijaudier-Cabot. Continuum damage theory – application to concrete. *J. Eng. Mech.-ASCE*, 115(2):345–365, 1989.

[22] J. Mazars, G. Pijaudier-Cabot, and C. Saouridis. Size effect and continuous damage in cementitious materials. *Int. J. Fract.*, 51(2):159–173, 1991.

[23] R.H.J. Peerlings, R. de Borst, W.A.M. Brekelmans, and M.G.D. Geers. Gradient-enhanced damage modelling of concrete fracture. *Mech. Cohesive-Frict. Mater.*, 3(4):323–342, 1998.

[24] P. Pegon and A. Anthoine. Numerical strategies for solving continuum damage problems involving softening: application to the homogenization of masonry. In *Proceedings of the Second International Conference on Computational Structures Technology*, Athens, 1994.

[25] G. Pijaudier-Cabot and Z.P. Bažant. Nonlocal damage theory. *J. Eng. Mech.-ASCE*, 118(10):1512–1533, 1987.

[26] G. Pijaudier-Cabot and A. Huerta. Finite element analysis of bifurcation in nonlocal strain softening solids. *Comput. Methods Appl. Mech. Eng.*, 90(1–3):905–919, 1991.

[27] C. Rocco, G.V. Guinea, J. Planas, and M. Elices. Mechanisms of rupture in splitting tests. *ACI Mater. J.*, 96(1):52–60, 1999.

[28] A. Rodríguez-Ferran, I. Arbós, and A. Huerta. Adaptive analysis based on error estimation for nonlocal damage models. *Revue européene des éléments finis*, 10(2–4):193–207, 2001.

[29] A. Rodríguez-Ferran and A. Huerta. Error estimation and adaptivity for nonlocal damage models. *Int. J. Solids Struct.*, 37(48–50):7501–7528, 2000.

[30] A. Rodríguez-Ferran and A. Huerta. Failure and post-failure modelling of the Brazilian test. In W.A. Wall, K.-U. Bletzinger, and K. Schweizerhof, editors, *Trends in Computational Structural Mechanics*, Barcelona, 2001. CIMNE.

[31] J. Sarrate and A. Huerta. Efficient unstructured quadrilateral mesh generation. *Int. J. Numer. Methods Eng.*, 49(10):1327–1350, 2000.

[32] A. Vila, A. Rodríguez-Ferran, and A. Huerta. A note on a numerical benchmark test: an axisymmetric shell under ring loads. *Commun. Numer. Methods Eng.*, 13(3):181–192, 1997.

Regularized formulations of strain-softening plasticity

Milan Jirásek and Simon Rolshoven

Laboratory of Structural and Continuum Mechanics, Swiss Federal Institute of Technology at Lausanne (EPFL)

Summary. This chapter provides a simple introduction into continuum-based modeling of localized inelastic processes such as cracking or formation of shear bands. The main focus is on softening plasticity models regularized by higher-gradient terms or by weighted averaging of internal variables. The chapter contains representative examples of strain-gradient plasticity, plasticity with gradients of internal variables, and strongly nonlocal plasticity models. Localization properties of such regularized formulations are analyzed in the one-dimensional setting. It is explained why these formulations act as localization limiters. Relations between different regularization techniques are elucidated.

1 Introduction

In many structures subjected to extreme loading conditions, the initially smooth distribution of strain rates abruptly changes into a highly localized one. Typically, the inelastic strain increments are concentrated in narrow zones while the major part of the structure experiences unloading. Such *strain localization* can be caused by geometrical effects (e.g., necking of metallic bars) or by material instabilities (e.g., microcracking, frictional slip, or nonassociated plastic flow). Here we concentrate on the latter case. To keep the presentation simple, we consider only the static response in the small-strain range and in the one-dimensional setting. These restrictions permit an analytical or semi-analytical treatment of the problem, while the basic characteristics of the solution remain valid in the general, multi-dimensional case.

After an illustrative example that explains why the conventional continuum theories with softening fail to provide a meaningful description of strain localization, we will give a general overview of modeling approaches that are appropriate for this purpose. Attention will then focus on softening plasticity models regularized by gradient terms or by spatial integrals. Models presented in Section 3 incorporate the gradient of strain, in the spirit of the Toupin-Mindlin strain-gradient elasticity theory. Sections 4 and 5 cover models that incorporate the second gradient of the softening variable (cumulative plastic strain), either in an explicit way or in an implicit one. Finally, Section 6 deals with models enriched by the weighted spatial average of the softening variable. When properly formulated, all of these enrichments act as localization limiters, i.e., they set the minimum width of the narrow band into which plastic flow can localize.

2 Problems with Objective Description of Strain Localization

2.1 Conventional Flow Theory of Plasticity in One Dimension

In the one-dimensional setting, standard local plasticity with linear isotropic hardening is described by the elastic stress-strain law

$$\sigma = E(\varepsilon - \varepsilon_{\rm p}) \tag{1}$$

hardening law

$$q = -H\kappa \tag{2}$$

and evolution laws

$$\dot{\varepsilon}_{\rm p} = \dot{\lambda}\frac{\partial f(\sigma, q)}{\partial \sigma} \tag{3}$$

$$\dot{\kappa} = \dot{\lambda}\frac{\partial f(\sigma, q)}{\partial q} \tag{4}$$

with loading-unloading conditions

$$\dot{\lambda} \geq 0, \quad f(\sigma, q) \leq 0, \quad \dot{\lambda}\, f(\sigma, q) = 0 \tag{5}$$

In the above relations, σ is the stress, ε is the (total) strain, $\varepsilon_{\rm p}$ is the plastic strain, E is the elastic modulus, H is the plastic modulus (positive for hardening and negative for softening), κ is the hardening variable, q is the dissipative force conjugate to κ, $\dot{\lambda}$ is the rate of the plastic multiplier, and

$$f(\sigma, q) = |\sigma| - \sigma_0 + q \tag{6}$$

is the yield function. Initially, the variables κ and q have zero values, and so the parameter σ_0 is the initial yield stress. As usual, a superimposed dot denotes differentiation with respect to time.

From the thermodynamic point of view, equations (1)–(2) are the state laws that can be derived from the free-energy potential

$$\rho\psi(\varepsilon, \varepsilon_{\rm p}, \kappa) = \tfrac{1}{2}E(\varepsilon - \varepsilon_{\rm p})^2 + \tfrac{1}{2}H\kappa^2 \tag{7}$$

where ρ is the mass density and ψ is the free energy per unit mass. Equations (3)–(5) are the complementary laws that can be derived from the dual dissipation potential $\phi^*(\sigma, q)$ defined as the indicator function of the set of plastically admissible states; for a detailed discussion, see e.g. [9], Chapter 23.

The basic equations (1)–(5) are written in a form that reveals a certain symmetry in the state laws and complementary laws. Making use of the particular definition of the yield function (6), it is possible to replace $\partial f/\partial\sigma$ in (3) by $\operatorname{sgn}\sigma$ and $\partial f/\partial q$ in (4) by 1. According to the latter equation, rewritten as $\dot{\kappa} = \dot{\lambda}$, the plastic multiplier rate can be replaced by the rate of the hardening variable and eliminated from the basic equations. From the flow rule (3), rewritten as $\dot{\varepsilon}_p = \dot{\kappa}\operatorname{sgn}\sigma$ and combined with the condition $\dot{\kappa} \geq 0$, it then follows that $\dot{\kappa} = |\dot{\varepsilon}_p|$, which gives to the hardening variable κ the physical meaning of cumulative plastic strain. To give a clear physical meaning to the variable that controls the size of the elastic domain, we can define the current yield stress $\sigma_Y = \sigma_0 - q$ and rewrite the hardening law (2) as $\sigma_Y = \sigma_0 + H\kappa$. After all these adjustments, the basic equations reduce to

$$\sigma = E(\varepsilon - \varepsilon_p) \tag{8}$$

$$\sigma_Y = \sigma_0 + H\kappa \tag{9}$$

$$\dot{\varepsilon}_p = \dot{\kappa}\operatorname{sgn}\sigma \tag{10}$$

$$\dot{\kappa} \geq 0, \quad f(\sigma, \sigma_Y) \leq 0, \quad \dot{\kappa}\, f(\sigma, \sigma_Y) = 0 \tag{11}$$

where the yield function is now given by

$$f(\sigma, \sigma_Y) = |\sigma| - \sigma_Y \tag{12}$$

From now on, we focus on the case of *softening*, characterized by a negative value of the plastic modulus H. To emphasize that, κ and H will be called the softening variable and the softening modulus, resp., and Eq. (9) will be referred to as the softening law. The linear form of the softening law cannot have an unlimited validity. The current yield stress cannot become negative, and so (9) is applicable only in the range $0 \leq \kappa \leq \varepsilon_f$ where $\varepsilon_f = -\sigma_0/H$ is the cumulative plastic strain at which the current yield stress vanishes. For $\kappa > \varepsilon_f$, σ_Y remains equal to zero.

Of course, the present softening law is highly idealized, and nonlinear softening functions would be needed to get an accurate description of real materials. However, the assumption of linear softening simplifies the analysis and the results are qualitatively the same as for more refined laws.

The stress-strain diagram under monotonic tension is shown in Fig. 2a. Note that monotonic loading means here a monotonically increasing strain, while the stress attains a peak value σ_0 at strain $\varepsilon_0 = \sigma_0/E$ and then decreases.

2.2 One-Dimensional Localization Problem

To introduce the basic concepts and to illustrate the behavior of various models, we will analyze a simple one-dimensional localization problem, which can

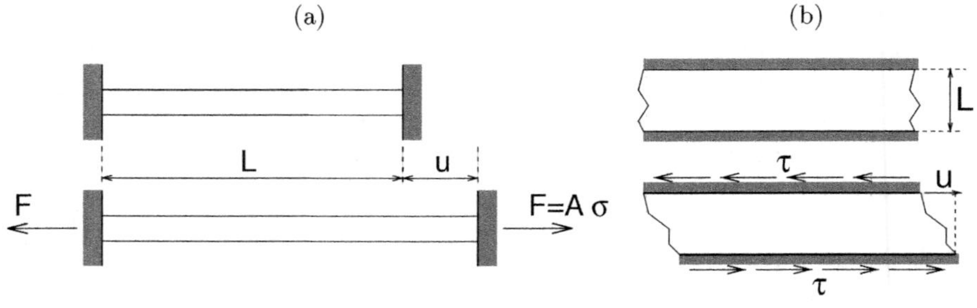

Fig. 1: (**a**) Bar under uniaxial tension, (**b**) shear layer

be interpreted either as a bar of a constant cross section A and of total length L under uniaxial tension (Fig. 1a), or as a material layer of thickness L under shear (Fig. 1b). In the examples we will discuss the tensile bar, but all the results can be reinterpreted in terms of the shear problem, if the normal stress σ is replaced by shear stress τ, normal strain ε by shear angle γ, Young's modulus E by shear modulus G, etc.

The material is assumed to obey the simple constitutive law described in the previous section, which leads to the stress-strain diagram shown in Fig. 2a. If the bar is loaded in tension by an applied displacement u at one of the supports, the response remains linear elastic up to $u_0 = L\varepsilon_0$. At this state, the force transmitted by the bar (reaction at the support) attains its maximum value, $F_0 = A\sigma_0$. After that, the resistance of the bar starts decreasing. At each cross section, stress can decrease either at increasing strain (plastic softening) or at decreasing strain (elastic unloading). For easy reference, we denote the plastic zone as $I_{\rm p}$ and the elastic zone as $I_{\rm e}$. The static equation of equilibrium implies that the stress profile must remain uniform along the bar. However, the same stress $\bar{\sigma}$ between 0 and σ_0 can be generated by strain histories with different final values of strain, and so the strain profile does not need to be uniform. For example, plastic softening can localize into an interval

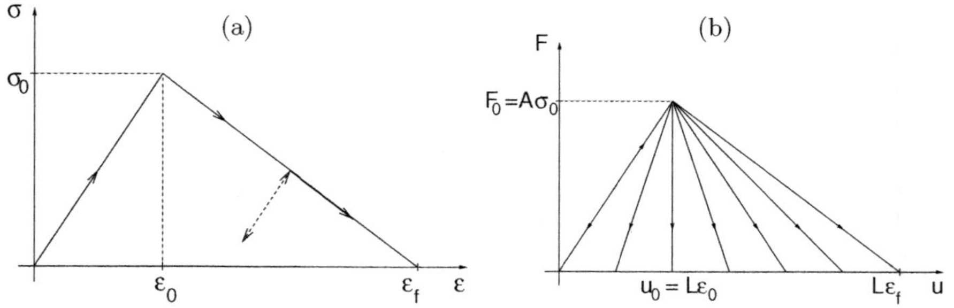

Fig. 2: (**a**) Stress-strain diagram with linear softening, (**b**) fan of possible post-peak branches of the load-displacement diagram

of length L_p while the remaining parts of the bar of total length $L - L_p$ unload elastically. When the stress is completely relaxed to zero, the strain in the softening zone is ε_f and the strain in the unloading zone vanishes; the total elongation of the bar is therefore $u = L_p \varepsilon_f$. The length L_p remains undetermined, and it can have any value between 0 and L. This means that the problem has infinitely many solutions, and the corresponding post-peak branches of the load-displacement diagram fill the fan shown in Fig. 2b.

The ambiguity is removed if imperfections are taken into account. Real material properties and sectional dimensions cannot be perfectly uniform. Suppose that the strength in a small region is slightly lower than in the remaining portion of the bar. When the applied stress attains the reduced strength, softening starts and the stress decreases. Consequently, the material outside the weaker region must unload elastically, because its strength has not been exhausted. This leads to the conclusion that the size of the softening zone cannot exceed the size of the region with minimum strength. Such a region can be arbitrarily small, and the corresponding softening branch can be arbitrarily close to the elastic branch of the load-displacement diagram. Thus the standard strain-softening continuum formulation leads to a solution that has several pathological features:

1. the softening zone is infinitely small;
2. the load-displacement diagram always exhibits snapback, independently of the structural size and of the material ductility;
3. the total amount of energy dissipated during the failure process is zero.

From the mathematical point of view, these problems are related to the loss of ellipticity of the governing differential equation. In the present one-dimensional setting, ellipticity is lost when the tangent modulus becomes zero or negative, i.e., at the peak of the stress-strain diagram. The boundary value problem then becomes ill-posed, i.e., it does not have a unique solution with continuous dependence on the given data. From the numerical point of view,

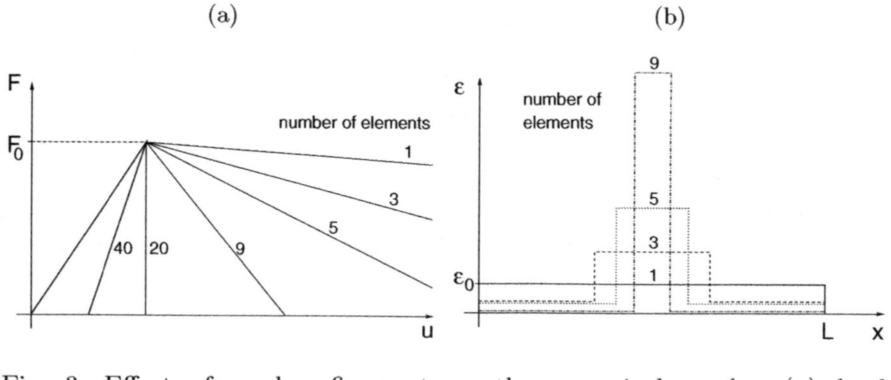

Fig. 3: Effect of mesh refinement on the numerical results: (**a**) load-displacement diagrams, (**b**) strain profiles

ill-posedness is manifested by pathological sensitivity of the results to the size of finite elements. For example, suppose that the bar is divided into N_e two-node elements with linear displacement interpolation. If the numerical algorithm properly captures the most localized solution, the softening zone extends over one element, and we have $L_{\rm p} = L/N_e$. The slope of the post-peak branch therefore strongly depends on the number of elements, and it approaches the initial elastic slope as the number of elements tends to infinity. This is documented in Fig. 3a, constructed for a stress-strain law with $\varepsilon_f/\varepsilon_0 = 20$. The strain profiles at $u = 2u_0$ for various mesh refinements are plotted in Fig. 3b (under the assumption that the weakest link is located at the center of the bar). In the limit, the profiles tend to $2\,u_0\,\delta(x - L/2)$ where δ denotes the Dirac distribution. The limit solution represents a displacement jump at the center, with zero strain everywhere else.

2.3 Classification of Models

In real materials, inelastic processes typically localize in narrow bands that initially have a small but nonzero width. Propagation and coalescence of microdefects in the localization band can eventually lead to the formation of a displacement discontinuity, e.g., of a macroscopic stress-free crack or an internal slip surface. The initial thickness of the localization band depends on the material microstructure and is usually of the same order of magnitude as the characteristic material length, determined by the size or spacing of dominant heterogeneities. Mathematically, narrow zones of highly concentrated evolving microdefects can be represented in different ways. A systematic classification can be based on the degree of regularity of the kinematic fields:

1. Models with *strong discontinuities* allow jumps in displacements across a discontinuity curve (in two dimensions) or discontinuity surface (in three dimensions). The strain field, $\boldsymbol{\varepsilon}(\boldsymbol{x})$, then consists of a regular part, obtained by standard differentiation of the displacement field, and a singular part, having the character of a multiple of the Dirac delta distribution. This is schematically shown for the one-dimensional case in Fig. 4a. In physical terms, the strong discontinuity corresponds to a sharp crack (not necessarily a stress-free one). The constitutive part of the model consists of a stress-strain law for the bulk (continuous) material and a traction-separation law for the discontinuity. Typical examples include *cohesive crack models* and *cohesive zone models*.
2. Another possible kinematic description represents the zone of localized deformation by a band of a small but finite thickness, separated from the remaining part of the body by two *weak discontinuities*, i.e., curves or surfaces across which certain strain components have a jump but the displacement field remains continuous. This is illustrated in Fig. 4b. Since the displacement is continuous, the strain components in the plane tangential to the discontinuity surface must remain continuous as well, and only the

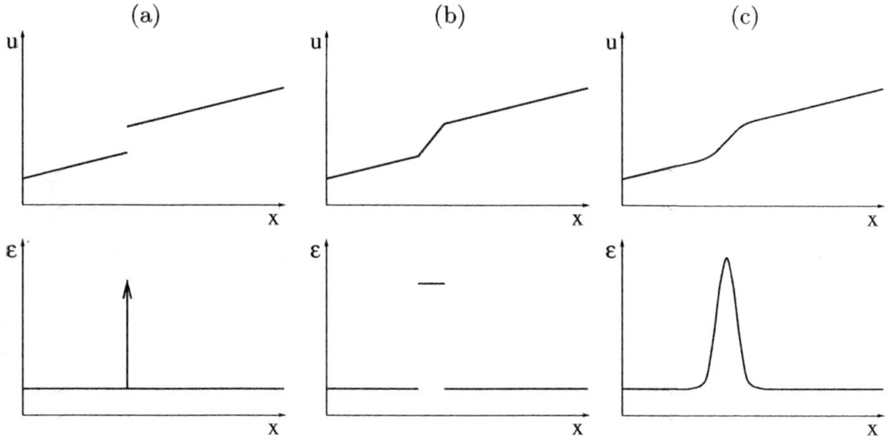

Fig. 4: Kinematic description with (**a**) one strong discontinuity, (**b**) two weak discontinuities, (**c**) no discontinuities

out-of-plane components can have a jump. In physical terms, the band between the weak discontinuities corresponds to a damage process zone with an almost constant density of microdefects. Models with localization bands bounded by weak discontinuities can be considered as simple regularizations of models with strong discontinuities. Instead of lumping all the inelastic effects into a surface, we distribute them uniformly across the width of a band of a finite thickness h. This is typically done for *smeared crack models* and their plasticity-based counterparts. To avoid pathological mesh sensitivity, certain parameters of the constitutive law that control the response after the onset of localization must be adjusted according to the thickness of the numerically resolved band, which always depends on the element size and may also depend on other properties of the discretization such as the element type, orientation of mesh lines with respect to the band, number of Gauss integration points, etc.

3. Finally, the most regular description uses a continuously differentiable displacement field, and the strain field remains *continuous*. Strain localization is manifested by high strains in a narrow band, with a continuous transition to much lower strains in the surrounding parts of the body. A typical strain profile of this type is shown in Fig. 4c. In physical terms, this corresponds to an inelastic process zone with a continuously varying concentration of defects. Solutions of this type are obtained for continuum models regularized by appropriate enrichments, which act as *localization limiters*. Various forms of such enrichments will be discussed in the subsequent sections. In particular, we will look at
   a) continua with microstructure, exemplified by strain-gradient theories;
   b) material models with gradients of internal variables; and

c) material models with nonlocal internal variables.

All enriched theories presented in this chapter abandon the assumption that the stress at a given material point is uniquely determined by the history of strain and temperature at that point. Continua with microstructure introduce additional kinematic quantities that characterize the deformation state of a material "point", e.g., the microcurvature tensor or the strain gradient. If their histories at a given point are known, the corresponding work-conjugate quantities (generalized stress measures) can be determined from the constitutive equations. So the constitutive laws are still applied locally. For other classes of enriched theories, the stress at a given point depends on the distribution of an internal variable in a neighborhood of that point. For explicit models with gradients of internal variables (Section 4), this influence is limited to an infinitesimal neighborhood, and therefore they are classified as weakly nonlocal. For implicit models with gradients of internal variables (Section 5) and models with nonlocal internal variables (Section 6), the dependence involves a finite neighborhood or even the entire body, and they are called strongly nonlocal.

A common feature of all these enriched theories is that they set a certain length scale that can be related to the material microstructure. The length scale is controlled by one or more material parameters with the dimension of length, which are either explicitly present in the governing equations, or can be deduced from a combination of material parameters with different physical dimensions (e.g., of the standard stiffness and the higher-order stiffness in strain-gradient elasticity). The presence of an intrinsic length scale means that the body cannot be decomposed into arbitrarily small material "particles" that still keep all the properties of the basic material. From this point of view, all of the enriched models mentioned here, including strain-gradient theories, can be considered as nonlocal. Due to the presence of a length scale, the enriched models are capable of describing size effects of the transitional type, more general than the simple power-type scaling resulting from models with no characteristic length; see [3] for more details.

3 Strain-Gradient Theories

3.1 Strain-Gradient Elasticity

Strain-gradient theories are extensions of the standard continuum theory that take into account the influence of the strain gradient (which is related to the second gradient of the displacement field) and, in their more general forms, even of second- and higher-order gradients of strain. For the linear elastic case, the foundations of the strain-gradient theory were laid down by Toupin and Mindlin [22, 13]. The constitutive equations can be derived from a quadratic free-energy potential that depends not only on strain but also on the strain gradient. In one dimension, such a potential is given by

$$\rho\psi(\varepsilon, \eta) = \tfrac{1}{2}E\varepsilon^2 + \tfrac{1}{2}El_e^2\eta^2 \tag{13}$$

where $\eta = \varepsilon'$ is the spatial derivative of strain and l_e is a material parameter with the dimension of length. Since the free energy density depends on two observable state variables that are *locally* independent, the state laws obtained by differentiating the potential (13),

$$\sigma = \rho\frac{\partial\psi}{\partial\varepsilon} = E\varepsilon \tag{14}$$

$$\chi = \rho\frac{\partial\psi}{\partial\eta} = El_e^2\eta \tag{15}$$

define two thermodynamic forces: the Cauchy stress σ, work-conjugate with the strain, and the so-called double stress χ, work-conjugate with the strain gradient.

The state variables ε and η are linked to the displacement field by the kinematic equations

$$\varepsilon = u' \tag{16}$$

$$\eta = u'' \tag{17}$$

The dual equilibrium equation

$$\sigma' - \chi'' + \bar{b} = 0 \tag{18}$$

follows from the principle of virtual work, extended by a term that corresponds to the work of the double stress on virtual strain gradients. The principle provides also the static (natural) boundary conditions

$$n(\sigma - \chi') = \bar{t} \tag{19}$$

$$n\chi = 0 \tag{20}$$

Here, $\bar{b}$ is the prescribed body force, $\bar{t}$ is the prescribed traction, and n is the "unit vector normal to the boundary", in one dimension reducing to a scalar equal to -1 on the left bar end and to 1 on the right bar end. The corresponding kinematic (essential) boundary conditions are

$$u = \bar{u} \tag{21}$$

$$u' = \bar{\varepsilon} \tag{22}$$

where $\bar{u}$ is the prescribed displacement and $\bar{\varepsilon}$ is the prescribed strain. Condition (21) characterizes a fixed end, while condition (19) characterizes a free end with an applied traction (which can have a zero value). Of course, at each end section we enforce only one of these conditions. In addition to that, we must enforce at each end one of conditions (20) and (22). Condition (22) would fix the strain in the direction normal to the boundary, which is not very natural. Therefore, we will use condition (20), which prescribes the double stress. In general, the right-hand side of this condition could be a given non-zero higher-order traction, but since the physical meaning of such a nonstandard boundary force is not very clear, it is usually set to zero.

Let us now look at the uniaxial tensile test, depicted in Fig. 1a. Displacements at the end sections are prescribed as $u(0) = 0$ and $u(L) = \bar{u}$ = monotonically increasing parameter that controls the loading process, and the double tractions on the boundary are assumed to vanish, i.e., $\chi(0) = 0$ and $\chi(L) = 0$. The fundamental equation governing the problem could be derived by writing the equilibrium condition (18) and the boundary conditions in terms of the unknown displacement field, using the kinematic equations (16)–(17) and the constitutive equations (14)–(15). This would lead to a fourth-order differential equation

$$Eu'' - El_e^2 u^{IV} + \bar{b} = 0 \tag{23}$$

However, owing to the special character of the one-dimensional formulation, it is sufficient to construct a second-order differential equation for strains. To this end, we first integrate the equilibrium equation (18). In the absence of body forces, the result is

$$\sigma - \chi' = t \tag{24}$$

where t is an integration constant with the physical meaning of the reaction generated by the imposed displacement at the right end section. Then we express σ and χ in terms of ε, using the constitutive equations (14)–(15) with η replaced by ε'. The resulting equation

$$E(\varepsilon - l_e^2 \varepsilon'') = t \tag{25}$$

must be solved with boundary conditions $\varepsilon'(0) = 0$ and $\varepsilon'(L) = 0$. Finally, the integration constant t is determined from the compatibility condition

$$\int_0^L \varepsilon(x)\,\mathrm{d}x = u(L) - u(0) = \bar{u} \tag{26}$$

The solution of (25) is unique and, for the boundary conditions considered here, it is simply $\varepsilon(x) = t/E$ = const. Therefore, the strain gradient and double stress vanish, and the solution is exactly the same as for standard elasticity. This could have been guessed right away, but the solution procedure outlined here will be useful for the localization analysis in the subsequent section.

3.2 Strain-Gradient Plasticity Model of Fleck and Hutchinson

Extensions of the Toupin-Mindlin theory to the inelastic range were proposed by a number of authors. Here we present the strain-gradient version of the flow theory of plasticity, formulated by Fleck and Hutchinson [7]. The structure of the constitutive equations is very similar to the standard flow theory. The strain and strain gradient are decomposed into the elastic and plastic parts. The yield condition is based on the critical value of the equivalent stress (generalization of von Mises stress), which is a scalar measure of the combined effect of Cauchy stress and double stress (more specifically, of their deviatoric parts). The dependence of the yield function on the double stress is then reflected in the associated flow rule for the plastic part of the strain gradient.

In the one-dimensional setting, the state laws derived from the free-energy potential

$$\rho\psi(\varepsilon,\eta,\varepsilon_{\rm p},\eta_{\rm p},\kappa) = \tfrac{1}{2}E(\varepsilon-\varepsilon_{\rm p})^2 + \tfrac{1}{2}El_{\rm e}^2(\eta-\eta_{\rm p})^2 + \tfrac{1}{2}H\kappa^2 \tag{27}$$

read

$$\sigma = E\,(\varepsilon-\varepsilon_{\rm p}) \tag{28}$$

$$\chi = El_{\rm e}^2\,(\eta-\eta_{\rm p}) \tag{29}$$

Here, $\varepsilon_{\rm p}$ is the usual plastic strain and $\eta_{\rm p}$ is the plastic part of the strain gradient, considered as an independent internal variable. Note that $\eta = \varepsilon'$ is the strain gradient, but $\eta_{\rm p}$ has no direct relation to the derivative of the plastic strain, $\varepsilon_{\rm p}'$.

The yield function is defined as

$$f(\sigma,\chi;\kappa) = \Sigma(\sigma,\chi) - \sigma_{\rm Y}(\kappa) \tag{30}$$

where

$$\Sigma(\sigma,\chi) = \sqrt{\sigma^2 + \frac{\chi^2}{l_{\rm p}^2}} \tag{31}$$

is the equivalent stress (called also the overall effective stress). Since σ and χ have different physical dimensions, the definition of Σ must contain a scaling parameter related to the internal structure of the material. The plastic characteristic length, $l_{\rm p}$, is in general different from the elastic characteristic length $l_{\rm e}$, defined as the square root of the ratio between the higher-order elastic stiffness and Young's modulus.

Within an associated framework, the evolution laws for plastic strain $\varepsilon_{\rm p}$, plastic strain gradient $\eta_{\rm p}$ and hardening variable κ are

$$\dot\varepsilon_{\rm p} = \dot\lambda\frac{\partial f}{\partial\sigma} = \dot\lambda\frac{\sigma}{\Sigma} \tag{32}$$

$$\dot\eta_{\rm p} = \dot\lambda\frac{\partial f}{\partial\chi} = \dot\lambda\frac{\chi}{l_{\rm p}^2\Sigma} \tag{33}$$

$$\dot\kappa = \dot\lambda\frac{\partial f}{\partial(-\sigma_{\rm Y})} = \dot\lambda \tag{34}$$

Let us go back to the one-dimensional tensile test. In the elastic range, the solution is unique and both stress and strain remain uniformly distributed, while the strain gradient and double stress vanish. Even after the onset of yielding, there exists a solution with a uniform distribution of strain, plastic strain and stress, and with vanishing strain gradient, plastic part of strain gradient and double stress. However, such a homogeneous solution is not always unique. Let us check whether a bifurcation into a nonhomogeneous solution is possible.

At the first bifurcation from a homogeneous state, the stress is still uniform and the double stress vanishes, but their rates become nonuniform. Since $\chi = 0$ and $\Sigma = \sigma$, the flow rule (32)–(33) gives $\dot{\varepsilon}_\mathrm{p} = \dot{\lambda}$ and $\dot{\eta}_\mathrm{p} = 0$, and the rate of the effective stress can be expressed as

$$\dot{\Sigma} = \frac{\partial \Sigma}{\partial \sigma}\dot{\sigma} + \frac{\partial \Sigma}{\partial \chi}\dot{\chi} = \frac{\sigma}{\Sigma}\dot{\sigma} + \frac{\chi}{l_\mathrm{p}^2 \Sigma}\dot{\chi} = \dot{\sigma} = E(\dot{\varepsilon} - \dot{\varepsilon}_\mathrm{p}) = E(\dot{\varepsilon} - \dot{\lambda}) \qquad (35)$$

Under certain circumstances (to be specified later), plastic yielding localizes into a zone I_p, in which the yield function remains equal to zero and so its rate $\dot{f}$ must vanish. After substitution of (35) and of $\dot{\sigma}_\mathrm{Y} = H\dot{\kappa} = H\dot{\lambda}$ into the consistency condition $\dot{f} = \dot{\Sigma} - \dot{\sigma}_\mathrm{Y} = 0$, we get exactly the same result for the rate of the plastic multiplier as in standard elastoplasticity, namely $\dot{\lambda} = E\dot{\varepsilon}/(E + H)$. The solution is admissible only if $\dot{\lambda}$ is nonnegative. In the zone of elastic unloading, I_e, we have $\dot{\lambda} = 0$. Since the unloading process starts from a plastic state with $f = 0$, the solution is admissible only if $\dot{f} \le 0$.

If $E+H > 0$, the loading condition is equivalent to $\dot{\varepsilon} \ge 0$ and the unloading condition to $\dot{\varepsilon} \le 0$, which is very natural. On the other hand, if $E+H < 0$, then there would be two possible solutions for $\dot{\varepsilon} < 0$ and no solution for $\dot{\varepsilon} > 0$. This pathological case corresponds to local snapback (on the level of the material point) and is excluded from future considerations. So the softening modulus must be in the range $-E < H < 0$.

In summary, the rate of the plastic multiplier in both I_e and I_p can be expressed by a single unified formula,

$$\dot{\lambda} = \frac{E\langle\dot{\varepsilon}\rangle}{E + H} \qquad (36)$$

where the Macauley brackets $\langle \ldots \rangle$ denote the positive part. The stress rate is

$$\dot{\sigma} = E(\dot{\varepsilon} - \dot{\varepsilon}_\mathrm{p}) = E(\dot{\varepsilon} - \dot{\lambda}) = E_\mathrm{t}\,\dot{\varepsilon} \qquad (37)$$

where E_t is the tangent modulus, equal to Young's modulus E in the elastic zone and to the elastoplastic modulus

$$E_\mathrm{ep} = \frac{EH}{E + H} \qquad (38)$$

in the plastic zone. The rate of the double stress at incipient localization is given by the same expression as in the elastic theory, i.e., $\dot{\chi} = El_\mathrm{e}^2\dot{\eta} = El_\mathrm{e}^2\dot{\varepsilon}'$, because $\dot{\eta}_\mathrm{p} = 0$. So the governing differential equation replacing (25) reads

$$E\dot{\varepsilon} - El_e^2\dot{\varepsilon}'' = \dot{t} \qquad \text{for } x \in I_e \tag{39}$$

$$E_{ep}\,\dot{\varepsilon} - El_e^2\dot{\varepsilon}'' = \dot{t} \qquad \text{for } x \in I_p \tag{40}$$

Note that $E > 0$ and $E_{ep} < 0$. The above differential equation has the general solution

$$\dot{\varepsilon}(x) = \frac{\dot{t}}{E} + C_1 \cosh\frac{x}{l_e} + C_2 \sinh\frac{x}{l_e} \qquad \text{for } x \in I_e \tag{41}$$

$$\dot{\varepsilon}(x) = \frac{\dot{t}}{E_{ep}} + C_3 \cos\frac{\alpha x}{l_e} + C_4 \sin\frac{\alpha x}{l_e} \qquad \text{for } x \in I_p \tag{42}$$

where C_1, C_2, C_3 and C_4 are integration constants, and

$$\alpha = \sqrt{-\frac{E_{ep}}{E}} = \sqrt{-\frac{H}{E+H}} \tag{43}$$

is a positive parameter that depends only on the ratio between the softening modulus and the modulus of elasticity. If the elastic zone or plastic zone consists of several disjoint intervals, the integration constants are of course different in each contiguous part.

To obtain a valid solution, we need to find the domains I_e and I_p and determine the integration constants such that the solution satisfies the boundary conditions at both end sections of the bar and the appropriate continuity conditions at the internal elastoplastic boundaries. To construct a strong solution, we enforce continuity of u, u', χ and $\sigma - \chi'$. In the present simple case, continuity of $\sigma - \chi'$ is satisfied automatically, since $\sigma - \chi' = t = \text{const.}$, and continuity of u would be used to obtain the displacement field by integration of the strain field but does not need to be considered when solving for the strains only. So it is sufficient to enforce continuity of ε and ε' (equivalent to continuity of u' and χ). On the external boundaries, the double stress must vanish, which gives $\varepsilon' = 0$. If the problem is formulated in the rate form, analogous conditions are imposed for the rates $\dot{\varepsilon}$ and $\dot{\varepsilon}'$.

The size and location of the elastic and plastic zones are not given in advance; they must be determined such that the solution is admissible. Recall that, according to the loading-unloading conditions, the strain rate is nonnegative in I_p and nonpositive in I_e. Since, for the present model, the strain must be continuous, these conditions imply that each internal elastoplastic boundary is in the regime of neutral loading, with a vanishing strain rate. This provides a sufficient number of additional conditions that make it possible to determine the exact location of the elastoplastic boundaries. However, the number of elastic and plastic intervals must be selected first. This is a typical situation in one-dimensional localization analysis. The choice is not unique, but the most interesting solutions are those for which the plastic strain localizes into one single interval, which can form either around one bar end, or in the interior of the bar. The number of elastic intervals is then 1 or 2, resp.

To illustrate the fundamental localization properties of the model, we look at the case of localization in the interior of the bar. Localization at the boundary will be discussed for this and other models in [11]. By imposing two boundary conditions, four internal continuity conditions and two neutral loading conditions, it is possible to determine six integration constants and the coordinates of two internal elasto-plastic boundaries. It turns out that the solution must be symmetric with respect to the middle section. The resulting distribution of strain rates, shown in Fig. 5a, is described by

$$\dot{\varepsilon}(x) = \begin{cases} \dfrac{\dot{t}}{E}\left(1 - \dfrac{\cosh \dfrac{x}{l_\mathrm{e}}}{\cosh \dfrac{L - L_\mathrm{p}}{2l_\mathrm{e}}}\right) & \text{for } x \in I_{\mathrm{e}1} \equiv [0, (L - L_\mathrm{p})/2] \\[3ex] \dfrac{\dot{t}}{E_\mathrm{ep}}\left(1 - \dfrac{\cos \dfrac{\alpha(2x - L)}{2l_\mathrm{e}}}{\cos \dfrac{\alpha L_\mathrm{p}}{2l_\mathrm{e}}}\right) & \text{for } x \in I_\mathrm{p} \equiv \left[\dfrac{L - L_\mathrm{p}}{2}, \dfrac{L + L_\mathrm{p}}{2}\right] \\[3ex] \dfrac{\dot{t}}{E}\left(1 - \dfrac{\cosh \dfrac{L - x}{l_\mathrm{e}}}{\cosh \dfrac{L - L_\mathrm{p}}{2l_\mathrm{e}}}\right) & \text{for } x \in I_{\mathrm{e}2} \equiv [(L + L_\mathrm{p})/2, L] \end{cases} \tag{44}$$

where the length of the plastic zone at incipient localization, L_p, is determined as the smallest positive solution of the transcendental equation

$$\tan \frac{\alpha L_\mathrm{p}}{2l_\mathrm{e}} + \alpha \tanh \frac{L - L_\mathrm{p}}{2l_\mathrm{e}} = 0 \tag{45}$$

Interestingly, L_p does not depend at all on the plastic characteristic length, l_p.

The graph of the normalized plastic zone size $L_\mathrm{p}/l_\mathrm{e}$ as a function of α for several values of the normalized bar length L/l_e is plotted in Fig. 5b. For fixed material parameters l_e and α, L_p is a decreasing function of the bar length and for $L \to \infty$ it approaches its minimum possible value,

$$L_{\mathrm{p},\infty} = \frac{2l_\mathrm{e}}{\alpha}(\pi - \arctan \alpha) \tag{46}$$

This minimum size is proportional to the elastic characteristic length l_e, but it also depends on the ratio H/E. For very steep softening, when H is close to its minimum admissible value, $-E$, parameter α is large and the plastic zone becomes very small. For mild softening, when the absolute value of H is small, parameter α is close to zero and the plastic zone becomes very large. In a finite bar, the size of the plastic zone is increasing with decreasing bar length and, if the theoretical size of the plastic zone solved from (45) is larger than the actual bar length, localization is impossible. The critical bar length below which bifurcation into a symmetric mode cannot take place,

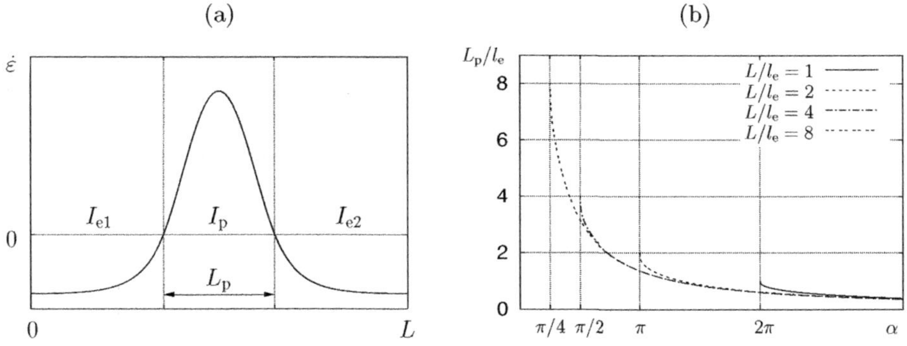

Fig. 5: (**a**) Distribution of strain rates at incipient localization, (**b**) dependence of normalized plastic zone size on parameter α

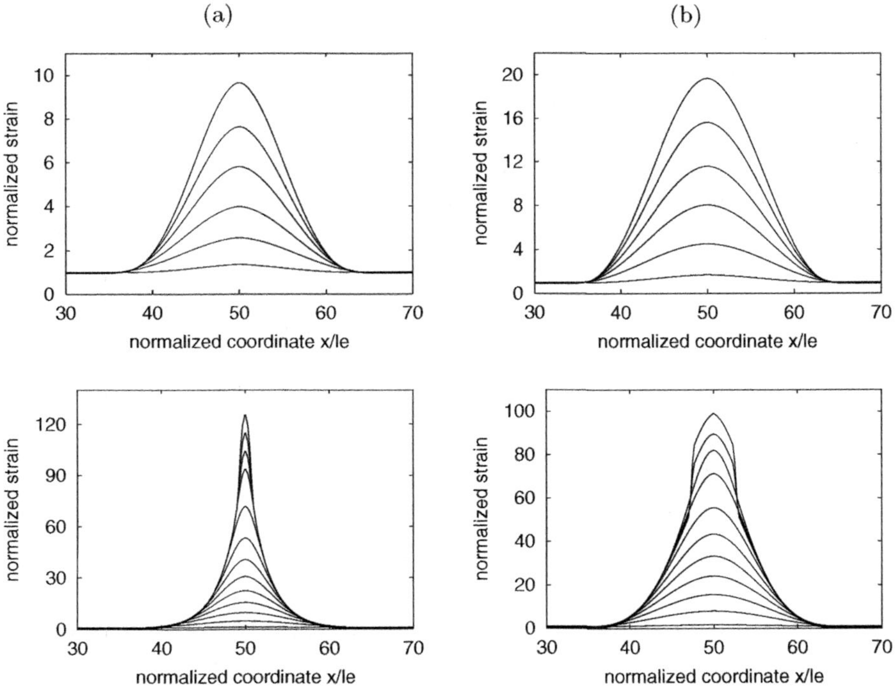

Fig. 6: Evolution of the strain profile for (**a**) $l_p = 2l_e$, (**b**) $l_p = 5l_e$

$$L_{\min} = \frac{2\pi l_e}{\alpha} \tag{47}$$

can be obtained from (45) by setting $L_p = L = L_{\min}$ and looking for the smallest positive solution.

Let us emphasize that the foregoing solution is valid only at incipient localization. As soon as a nonuniform strain profile develops, double stresses build up and the rate of the plastic strain gradient becomes nonzero. The subsequent evolution is not amenable to an analytical solution but it can be studied numerically. Fig. 6 shows the evolution of the strain profile for the same value of l_e but two different values of l_p, namely $l_p = 2l_e$ and $l_p = 5l_e$. Even though the initial size of the plastic zone is in both cases the same and the strain profiles during the initial stage of localization are very similar (Fig. 6 top), the plastic zone at late stages of the softening process is wider for larger l_p (Fig. 6 bottom).

4 Explicit Models with Gradients of Internal Variables

In contrast to strain-gradient theories, models with gradients of internal variables enrich only the constitutive equations (state laws or evolution laws) while the kinematic and equilibrium equations remain standard. A typical example is provided by plasticity models with the current yield stress dependent not only on the cumulative plastic strain but also on its Laplacean, which were pioneered by Aifantis [1, 2]. A thermodynamically based formulation of such models can be derived from a free-energy potential that depends on the gradient of the cumulative plastic strain, in addition to the usual dependence on the total strain, plastic strain, and the cumulative plastic strain.

In the one-dimensional setting, one can start from a simple quadratic potential

$$\rho\psi(\varepsilon, \varepsilon_p, \kappa, \kappa') = \tfrac{1}{2}E(\varepsilon - \varepsilon_p)^2 + \tfrac{1}{2}H\left[\kappa^2 - (l\kappa')^2\right] \tag{48}$$

where l is an internal length parameter. Note that the plastic modulus H is assumed to be negative, and so the contribution of the gradient term to the free energy is positive. The approach originally proposed in a somewhat different context by Valanis [23] and developed by several other authors [21, 18, 12] leads to the hardening law in the form

$$\sigma_Y = \sigma_0 + H[\kappa + l^2\kappa''] \tag{49}$$

and to the boundary condition $\kappa' = 0$ enforced at both end sections of the bar. To keep the second derivative bounded, the field $\kappa(x)$ must be C_1-continuous. The rate $\dot{\kappa}(x)$ must be C_1-continuous (in space) as well, and since it vanishes in the entire elastic zone, the boundary condition $\dot{\kappa}' = 0$ on the physical boundary can be replaced by the continuity condition $\dot{\kappa}' = 0$ on the boundary of the plastic zone (i.e., on the elasto-plastic interface and on the part of the physical boundary that is yielding). Here we tacitly assume that the problem has a localized solution in the strong sense. This assumption will later be confirmed. If the given data are not sufficiently smooth, we could relax the continuity conditions and enforce them in the weak sense. Even then, a solution localized into a set of zero measure would not exist.

In contrast to the strain-gradient theory, the equilibrium equation is standard and reads $\sigma' + \bar{b} = 0$. Consequently, in the absence of body forces, the stress must be uniform. This is now true even during plastic yielding, while for the strain-gradient model it was true only in the elastic range. It is therefore not necessary to make a distinction between the stress and the traction applied at the right end, and the role of t from the previous section will now be played by σ itself.

The enrichment of the softening law by the second derivative of the cumulative plastic strain regularizes the problem and prevents localization of plastic strain into an arbitrarily small region. Inside the plastic zone $I_{\rm p}$, the consistency condition $\dot{f} = \dot{\sigma} - \dot{\sigma}_Y = 0$ with $\dot{\sigma}_Y$ evaluated from (49) leads to the differential equation

$$\dot{\kappa} + l^2\dot{\kappa}'' = \frac{\dot{\sigma}}{H} \qquad \text{in } I_{\rm p} \tag{50}$$

for the unknown rate of the softening variable, $\dot{\kappa}(x)$. As already mentioned, the stress rate is uniform along the entire bar, and so the right-hand side of (50) is constant. The general solution of this differential equation reads

$$\dot{\kappa}(x) = \frac{\dot{\sigma}}{H} + C\cos\frac{x - x_0}{l} \tag{51}$$

where C and x_0 are integration constants. Formula (51) is valid inside the plastic zone $I_{\rm p} = (x_l, x_r)$, characterized by a positive value of the rate of cumulative plastic strain. Since $\dot{\kappa}$ must be continuously differentiable and must vanish outside the plastic zone, both $\dot{\kappa}$ and $\dot{\kappa}'$ must vanish at the elasto-plastic interface. Interface conditions $\dot{\kappa}(x_l) = 0$, $\dot{\kappa}(x_r) = 0$, $\dot{\kappa}'(x_l) = 0$ and $\dot{\kappa}'(x_r) = 0$ provide four equations for four unknown parameters—the integration constants C and x_0 and the coordinates x_l and x_r. It turns out that, in the idealized case of a bar with perfectly uniform properties, these conditions are not independent, which is related to the fact that the plastic zone can be located anywhere along the bar (its actual position depends on random imperfections). If we fix the value of x_0, the other constants are uniquely determined and the solution reads

$$\dot{\kappa}(x) = \frac{\dot{\sigma}}{H}\left(1 + \cos\frac{x - x_0}{l}\right) \qquad \text{for } x \in I_{\rm p} \equiv [x_0 - \pi l, x_0 + \pi l] \tag{52}$$

The plastic zone is the interval of length $L_{\rm p} = 2\pi l$ centered at x_0. It is interesting to note that the size of the localized plastic zone is directly

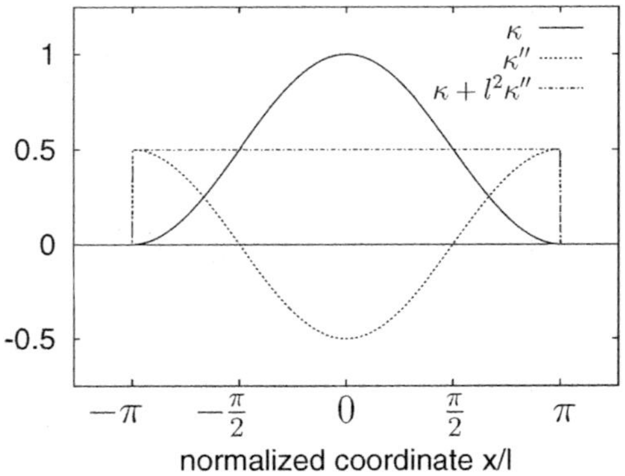

Fig. 7: Plastic strain distribution for the explicit gradient plasticity model; κ and κ'' are such that $\kappa + l^2\kappa''$ is constant in the plastic zone

proportional to the internal length parameter l and independent of all the other material parameters, including the plastic modulus H. Formula (52) is thus valid not only at the initial bifurcation from a uniform strain state, but also at later stages of the deformation process. Consequently, it can be integrated with respect to time, to get the distribution of plastic strain

$$\varepsilon_p(x) = \kappa(x) = \frac{\sigma - \sigma_0}{H}\left(1 + \cos\frac{x - x_0}{l}\right) \tag{53}$$

corresponding to the stress level σ attained in the softening regime. This plastic strain profile is graphically illustrated in Fig. 7, along with the distribution of $l^2\kappa''$ and the constant sum $\kappa + l^2\kappa''$.

It is instructive to discuss how the gradient term limits localization. Around the point that experiences the largest strain, the curvature of the plastic strain profile is negative, and due to the second-gradient term in (49) the current yield stress is higher than it would be for a standard local model. If the softening zone were too narrow, the negative curvature of the strain profile around its peak would have a large magnitude, and the current yield stress would be higher than at other points. Since the applied stress is constant along the bar, the sections around the center of the localization zone could not yield, and the zone would have to be extended in order to reduce the magnitude of the negative curvature and thus restore a constant value of the current yield stress.

5 Implicit Gradient Models

The enriched softening law (49) used in the previous section can be interpreted as the standard softening law (9) with the internal variable κ replaced by a certain "transformed internal variable", $\bar{\kappa} = \kappa + l^2\kappa''$. If the distribution of κ is known, κ'' can be directly evaluated, and this is why the model is called "explicit". The numerical implementation of such models is not easy. It is often necessary to use C_1-continuous finite elements, or mixed elements with an independent interpolation of one or more primary unknown fields in addition to the usual displacement interpolation [14, 5]. There are also problems related to the enforcement of the interface conditions at the evolving elasto-plastic boundary. New developments in this research area indicate that a more robust implementation can be achieved with implicit gradient models. Such formulations, first proposed for gradient damage [15], have recently been adapted for gradient plasticity [6, 8].

The main idea is that the transformed internal variable $\bar{\kappa}$ is defined indirectly, as the solution of a Helmholtz type differential equation with homogeneous Neumann boundary conditions. In a one-dimensional setting, the equation to be solved reads

$$\bar{\kappa} - l^2\bar{\kappa}'' = \kappa \tag{54}$$

where l is again a certain internal length parameter. The solution must satisfy the boundary conditions $\bar{\kappa}' = 0$ imposed at both end sections of the bar. The initial motivation for (54) was that the transformed variable given by the explicit definition $\bar{\kappa} = \kappa + l^2\kappa''$ satisfies (54) up to the fourth-order term $l^4\kappa^{IV}$, which might seem to be negligible if l is "small". However, this is true only if κ varies very slowly on length scales comparable to l. After the onset of localization, the characteristic wave length of the spatial distribution of κ becomes comparable to l, as is clear from the analytical solution of the one-dimensional localization problem presented in the previous section. In fact, for the plastic strain profile given by (53) the fourth-order term $l^4\kappa^{IV}$ has exactly the same magnitude as the second-order term $l^2\kappa''$, and so the above argument is not valid. As we will see, the localization properties of the implicit formulation are indeed quite different from the explicit one.

Let us suppose that, by analogy to (49), the softening law reads

$$\sigma_Y = \sigma_0 + H\bar{\kappa} \tag{55}$$

Same as in the previous section, the equilibrium equation remains standard, which means that the stress distribution in a bar of a constant cross section is uniform (same as before, we consider a state of static equilibrium with vanishing body forces). According to (55), the rate of the yield stress is $\dot{\sigma}_Y = H\dot{\bar{\kappa}}$. Combined with the consistency condition $\dot{\sigma} = \dot{\sigma}_Y$, this means that $\dot{\bar{\kappa}}$ must be constant along the entire plastic zone. Outside the plastic zone, the rate $\dot{\kappa}$ vanishes and $\dot{\bar{\kappa}}$ must satisfy the homogeneous differential equation

$$\dot{\bar{\kappa}} - l^2\dot{\bar{\kappa}}'' = 0 \tag{56}$$

which has a general solution

$$\dot{\bar{\kappa}}(x) = C_1 \cosh \frac{x}{l} + C_2 \sinh \frac{x}{l} \tag{57}$$

with arbitrary integration constants C_1 and C_2. Imposing the boundary condition $\dot{\bar{\kappa}}'(0) = 0$, we get $C_2 = 0$. Consequently, the rate of the transformed internal variable $\bar{\kappa}$ in an elastically unloading zone between the left end of the bar and the plastic zone is a multiple of the hyperbolic cosine, and it has zero slope only at $x = 0$, i.e., at the external, physical boundary. Since $\dot{\bar{\kappa}}$ is uniformly distributed in the plastic zone, there must always be a change of slope at the internal elastoplastic boundary. This means that no localized solution can satisfy (54) in the strong sense, and the only strong solution is the uniform one, with the plastic zone extending over the entire bar.

If the boundary value problem defining the transformed internal variable is rewritten in the weak form, localized solutions can be constructed. Consider a fully localized plastic strain rate $\dot{\kappa}(x) = \dot{w}\,\delta(x - x_0)$ where δ is the Dirac distribution, x_0 is the coordinate of the yielding cross section, and the scaling factor $\dot{w}$ has the physical meaning of the rate of displacement jump at the yielding section. It is easy to verify that the corresponding rate of the transformed internal variable is

$$\dot{\bar{\kappa}}(x) = \begin{cases} \dfrac{\dot{w}}{l\,t_0}\left(\cosh \dfrac{x_0}{l}\right)^{-1} \cosh \dfrac{x}{l} & \text{for } x \in [0, x_0] \\[2ex] \dfrac{\dot{w}}{l\,t_0}\left(\cosh \dfrac{L - x_0}{l}\right)^{-1} \cosh \dfrac{L - x}{l} & \text{for } x \in [x_0, L] \end{cases} \tag{58}$$

where

$$t_0 = \tanh \frac{x_0}{l} + \tanh \frac{L - x_0}{l} \tag{59}$$

is a parameter introduced for convenience.

An example of the normalized profile of $\dot{\bar{\kappa}}(x)$ for $L/l = 10$ and $x_0/l = 5$ is plotted in Fig. 8. Since the maximum rate of the transformed internal variable is attained at $x = x_0$, the softening process at all points outside the plastic zone is slower than in that zone and the degradation of yield stress is slower than the decrease of the actual stress. Therefore, the unloading condition $\dot{f} < 0$ is satisfied at all points outside $I_{\mathrm{p}} = \{x_0\}$, and the solution is admissible.

From the consistency condition $\dot{\sigma} - H\dot{\bar{\kappa}}(x_0) = 0$ we can determine the rate of the displacement jump

$$\dot{w} = \frac{l\,t_0}{H}\dot{\sigma} \tag{60}$$

It is interesting to note that even though the plastic strain rate localizes into a set of zero measure, the dissipation rate does not vanish. In fact, equation (60) represents the rate form of a cohesive law that links the traction across a discontinuity to the displacement jump. So the implicit gradient model based on softening law (55) is equivalent to a cohesive zone model. The cohesive

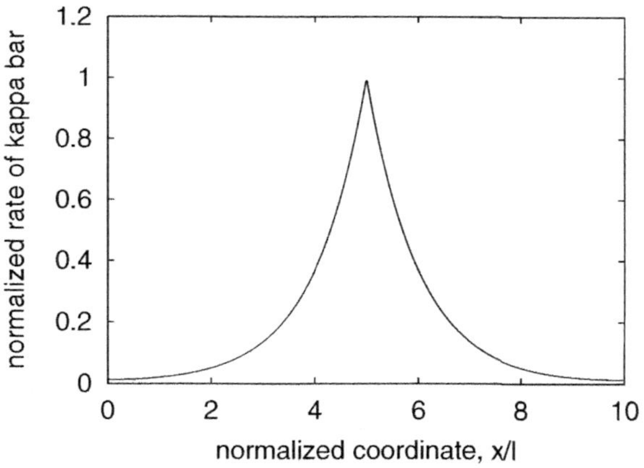

Fig. 8: Normalized profile of $\dot{\bar{\kappa}}(x)$ corresponding to plastic strain rate localized at $x_0 = 5l$

law (60) could be written as $\dot{\sigma} = H_d\,\dot{w}$ where $H_d = H/(l\,t_0)$ is the discrete softening modulus. If the plastic section is sufficiently far from both bar ends, hyperbolic tangent of x_0/l and of $(L - x_0)/l$ tends to 1 and we have $H_d = H/(2l)$.

As demonstrated, the implicit gradient model with the simplest type of softening law (55) enforces nonzero dissipation but the plastic zone collapses into a single cross section. The same is true if the linear softening function on the right-hand side of (55) is replaced by a nonlinear function of $\bar{\kappa}$ only. This is not exactly what one would expect from a localization limiter in the true sense of the word. But the model *can* act as a genuine localization limiter after a small modification. It is sufficient to make the current yield stress depend not only on the transformed internal variable, $\bar{\kappa}$, but also on the original internal variable, κ. Let us explore a generalized hardening law in the form

$$\sigma_{\mathrm{Y}} = h(\bar{\kappa}, \kappa) \tag{61}$$

where $h(0,0) = \sigma_0$ and the value of $h(\kappa, \kappa)$ tends to zero as κ tends to infinity. The rate of the current yield stress can be expressed as

$$\dot{\sigma}_{\mathrm{Y}} = H_{\mathrm{NL}}(\bar{\kappa}, \kappa)\,\dot{\bar{\kappa}} + H_{\mathrm{L}}(\bar{\kappa}, \kappa)\,\dot{\kappa} \tag{62}$$

where $H_{\mathrm{NL}} = \partial h/\partial \bar{\kappa}$ and $H_{\mathrm{L}} = \partial h/\partial \kappa$ are the (possibly variable) nonlocal and local plastic moduli.

We analyze again a possible bifurcation from a homogeneous state at which $\kappa(x) = \bar{\kappa}(x) = \kappa_c$. If $H_{\mathrm{L}}(\kappa_c, \kappa_c)$ is negative, the plastic strain rate can localize into one single cross section and the corresponding dissipation vanishes, so the model behavior gets even worse than for the previously analyzed softening law (55), which is a special case of the present law with $H_{\mathrm{L}} = 0$ and $H_{\mathrm{NL}} = H$. It will therefore be assumed that the local hardening modulus H_{L} is positive.

Substituting (62) into the consistency condition $\dot{f} \equiv \dot{\sigma} - \dot{\sigma}_Y = 0$ and replacing $\dot{\kappa}$ by $\dot{\bar{\kappa}} - l^2\dot{\bar{\kappa}}''$ according to the rate form of (54), we obtain the second-order differential equation

$$(H_L + H_{NL})\dot{\bar{\kappa}} - H_L l^2 \dot{\bar{\kappa}}'' = \dot{\sigma} \tag{63}$$

that must be satisfied by the rate $\dot{\bar{\kappa}}$ in the plastic zone. In the elastic zone, $\dot{\kappa}$ vanishes and the governing differential equation for $\dot{\bar{\kappa}}$ is (56), as already discussed.

Constructing the general solution of (63) in I_p and of (56) in I_{e1} and I_{e2}, enforcing the boundary conditions at the physical boundary, the continuity conditions and the neutral loading conditions at the elastoplastic interfaces, and checking the loading-unloading conditions, we obtain the distribution of rates $\dot{\kappa}$ and $\dot{\bar{\kappa}}$ at bifurcation from a homogeneous state. It turns out that the problem is formally equivalent to the bifurcation problem for the strain-gradient plasticity model analyzed in Section 3.2, with $\dot{\varepsilon}$ replaced by $\dot{\sigma}/H_{NL} - \dot{\bar{\kappa}}$, E by $-H_{NL}$, H by $H_L + H_{NL}$, l_e by l, and $\dot{t}$ by $\dot{\sigma}$. Parameter α is now given by

$$\alpha = \sqrt{-1 - \frac{H_{NL}}{H_L}} \tag{64}$$

The discussion of the dependence of the plastic zone size on the bar length and of the minimum bar length needed for bifurcation remains valid. Localization takes place only if α is real and positive, i.e., if $H_{NL}/H_L < -1$. Since H_L is assumed to be positive, this condition is satisfied if $H_L + H_{NL} < 0$, which is quite natural, because $H_L + H_{NL}$ is the overall plastic modulus valid for uniform strain evolution.

An implicit gradient model for metal plasticity has been developed by Geers and coworkers [6, 8]. They postulate the hardening-softening law in the form

$$\sigma_Y = [1 - \omega_p(\bar{\kappa})](\sigma_0 + H_0\kappa) \tag{65}$$

with a positive parameter H_0 and a nondecreasing function ω_p. This particular form is motivated by the concept of ductile damage, caused by the initiation and propagation of voids in a hardening matrix. The term $\sigma_0 + H_0\kappa$ represents linear increase of the effective yield stress in the matrix, while the multiplicative factor $1 - \omega_p$ reflects the reduction of the nominal yield stress due to the presence of voids. The hardening process in the matrix is driven by the local plastic strain, while the softening process is driven by the transformed plastic strain that depends not only on the local value of the plastic strain but also on its distribution in the whole body. This is why the model can be classified as strongly nonlocal. Plastic yielding in a certain small region leads to the degradation of the yield stress in the whole body, but this effect decays with increasing distance between the "source" and the "receiver".

For hardening-softening law (65), the plastic moduli are given by $H_L = [1 - \omega_p(\bar{\kappa})]H_0$ and $H_{NL} = -(\sigma_0 + H_0\kappa)\omega_p'(\bar{\kappa})$ where ω_p' is the derivative of ω_p

with respect to its argument. For comparison with other models, we consider a specific type of damage function

$$\omega_{\rm p}(\bar\kappa) = \frac{(H_0 - H)\bar\kappa}{\sigma_0 + H_0\bar\kappa} \tag{66}$$

that would lead to linear softening if the plastic strain remained uniform. Parameter $H < 0$ is the overall softening modulus for this fictitious case. In the actual test, localization takes place if the length of the plastic zone $L_{\rm p}$ determined from (45) is smaller than the bar length L, which is the case if parameter

$$\alpha = \sqrt{-1 + \frac{(\sigma_0 + H_0\kappa)\,\omega_{\rm p}'}{(1-\omega_{\rm p})H_0}} = \sqrt{-\frac{\kappa + \sigma_0/H_0}{\kappa + \sigma_0/H}} \tag{67}$$

exceeds the critical value $\alpha_b = 2\pi l/L$. Since $H_0 > 0$ and $H < 0$, parameter α is real and positive right at the onset of yielding, i.e., at $\kappa = 0$. For increasing κ, the value of α increases and can become arbitrarily large as κ tends to $\kappa_c = -\sigma_0/H$, which is the value at which the yield stress would vanish if the response remained uniform. The localization condition $\alpha > \alpha_b$ is satisfied when the plastic strain exceeds the critical value

$$\kappa_b = \frac{-\sigma_0 H_0^{-1} - \sigma_0\alpha_b^2 H^{-1}}{1+\alpha_b^2} = -\frac{\sigma_0}{HH_0}\,\frac{HL^2 + 4\pi^2 H_0 l^2}{L^2 + 4\pi^2 l^2} \tag{68}$$

This can happen right at the onset of yielding, or later. The former case occurs if $\alpha_0 = \sqrt{-H/H_0}$ is larger than α_b, i.e., if $-H/H_0 > 4\pi^2 l^2/L^2$. This is true for sufficiently steep softening and long bars. The latter case occurs for mild softening and short bars.

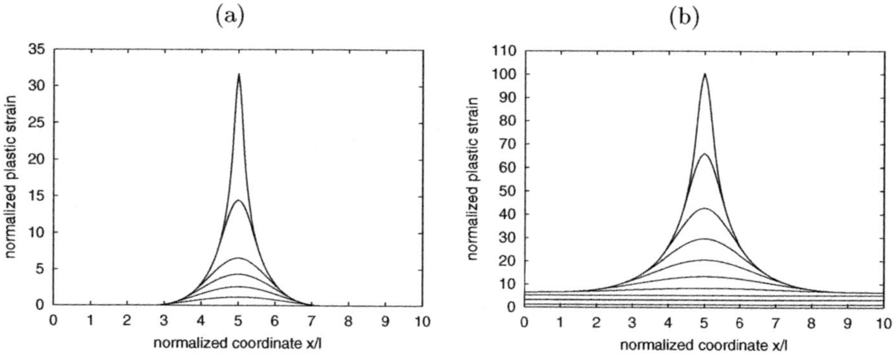

Fig. 9: Plastic strain evolution for the implicit gradient plasticity model: (**a**) steep softening, localization right at the onset of yielding, (**b**) mild softening, delayed localization

The two possible scenarios are documented in Fig. 9. The results have been computed with parameter ratios $H_0/E = 0.1$ and $L/l = 10$. The spatial coordinate is normalized by l and the plastic strain by the limit elastic strain, $\varepsilon_0 = \sigma_0/E$. Fig. 9a shows the evolution of the plastic strain profile for $H/E = -0.1$. In this case, $\alpha_0 = \sqrt{-H/H_0} = 1$ is larger than $\alpha_b = 2\pi l/L = \pi/5 = 0.628$, and the plastic strain immediately localizes into an interval of length $L_p = 4.72\, l$ (solved from (45)). For milder softening with $H/E = -0.02$, parameter $\alpha_0 = \sqrt{-H/H_0} = 0.447$ is smaller than α_b, and the plastic strain profile initially remains uniform until the plastic strain attains the critical value $\kappa_b = 6.98\,\varepsilon_0$, as shown in Fig. 9b. After that, the plastic strain increments progressively localize. In both cases, the active part of the plastic zone keeps shrinking as the applied displacement is increased. Let us emphasize that only the symmetric solution is studied here, and localization at one boundary is artificially surpressed.

6 Nonlocal Integral Models

In the previous section, the transformed internal variable $\bar{\kappa}$ was defined as the solution of the differential equation (54) with homogeneous Neumann boundary conditions. We have also seen that if the right-hand side of (54) is a multiple of the Dirac distribution, the (weak) solution of the boundary value problem can be written in a closed form; see (58). For an arbitrary (sufficiently regular) right-hand side, the solution of the boundary value problem can formally be expressed as

$$\bar{\kappa}(x) = \int_0^L G(x,\xi)\kappa(\xi)\,\mathrm{d}\xi \tag{69}$$

where $G(x,\xi)$ is the Green function of the problem, describing $\bar{\kappa}(x)$ for the right-hand side of (54) given by a Dirac distribution centered at ξ. In view of (58), the Green function is

$$G(x,\xi) = \frac{1}{l\,t(\xi)}\frac{\cosh(\tilde{x}/l)}{\cosh(\tilde{\xi}/l)}, \tag{70}$$

where $\tilde{x} = x$ and $\tilde{\xi} = \xi$ for $0 \le x \le \xi \le L$, $\tilde{x} = L - x$ and $\tilde{\xi} = L - \xi$ for $0 \le \xi \le x \le L$, and

$$t(\xi) = \tanh\frac{\xi}{l} + \tanh\frac{L-\xi}{l} \tag{71}$$

Formula (69) shows that the transformed internal variable $\bar{\kappa}$ at a receiver point x is obtained as a weighted average of the original internal variable κ sampled at source points ξ all over the bar, using the weight function $G(x,\xi)$ that decays with increasing distance of the source from the receiver. In this context, $\bar{\kappa}$ is called the nonlocal internal variable, because its value at a given point depends on the distribution of the local internal variable κ over the whole bar. This is in contrast to the transformed internal variable $\bar{\kappa} = \kappa + l^2\kappa''$ used by the explicit gradient model, which depends only on the distribution of κ in an infinitesimal neighborhood of the point of interest. Therefore, the explicit gradient model from Section 4 is classified as weakly nonlocal, while the implicit gradient model from Section 5 is said to be strongly nonlocal. Equation (69) represents a definition of the transformed internal variable by a nonlocal averaging integral, which is fully equivalent to the implicit definition of $\bar{\kappa}$ as the solution of (54) with appropriate boundary conditions.

Motivated by the above considerations, one can think of a wider class of strongly nonlocal models that define the transformed internal variable by weighted nonlocal averaging with a suitably chosen weight function, not necessarily equal to the Green function of a simple boundary value problem. Equation (69) is rewritten as

$$\bar{\kappa}(x) = \int_0^L \alpha(x,\xi)\kappa(\xi)\,\mathrm{d}\xi \tag{72}$$

where the weight function α is relatively arbitrary but should satisfy some basic requirements. First, it is reasonable to assume that α is nonnegative, so that a monotonic increase of the local variable κ never leads to a decrease of the nonlocal variable $\bar{\kappa}$ at any point. Second, in an infinite, macroscopically homogeneous bar, the weight $\alpha(x,\xi)$ should depend only on the distance between points x and ξ. Therefore, one may set $\alpha(x,\xi) = \alpha_\infty(x-\xi)$ where $\alpha_\infty(r)$ is an even function that attains its maximum at $r=0$, is nonincreasing for $r>0$, and tends to zero for $r\to\infty$.

All these conditions are satisfied by the Green function of equation (54) constructed on an infinite domain (with boundary conditions replaced by the condition that the solution should remain bounded for $|x|\to\infty$), which is given by

$$G_\infty(r) = \frac{1}{2l}\exp\left(-\frac{|r|}{l}\right) \tag{73}$$

They are also satisfied by the Gaussian function

$$\alpha_\infty^{\mathrm{Gauss}}(r) = \frac{1}{l\sqrt{2\pi}}\exp\left(-\frac{r^2}{2l^2}\right) \tag{74}$$

or by the truncated polynomial function

$$\alpha_\infty^{\mathrm{poly}}(r) = \frac{15}{16R}\left\langle 1-\frac{r^2}{R^2}\right\rangle^2 \tag{75}$$

that are often used as weight functions by nonlocal integral models. All these functions are plotted in Fig. 10. Parameters l and R have the dimension of

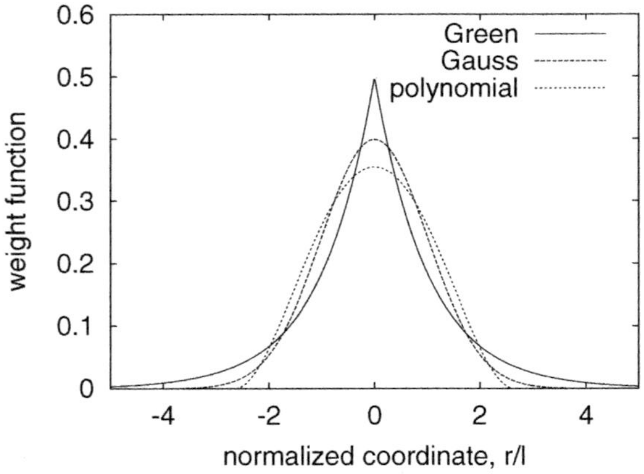

Fig. 10: Functions that can serve as weight functions in the nonlocal averaging integral

length and they set the internal length scale of the model. Parameter R in (75) is called the radius of nonlocal interaction, because the weight $\alpha_\infty^{\mathrm{poly}}$ vanishes at distances greater than or equal to R. The scaling factors $1/(l\sqrt{2\pi})$ and $15/(16R)$ in (74) and (75) are determined from the normalizing condition

$$\int_{-\infty}^{\infty} \alpha_\infty(r)\,\mathrm{d}r = 1 \tag{76}$$

which ensures that the nonlocal counterpart of a uniform field $\kappa(x) = \kappa_0$ is $\bar{\kappa}(x) = \kappa_0$, in other words, that a uniform field is not modified by the averaging operator. This holds also for the Green function (73), as can be verified directly by integration, or indirectly by the argument that among all solutions of differential equation (54) with $\kappa(x) = \kappa_0 = \text{const.}$ on the right-hand side, $\bar{\kappa}(x) = \kappa_0$ is the only one that remains bounded for $|x| \to \infty$.

An infinitely long bar is of course only an idealized concept that represents a sufficiently long bar with the plastic zone localized far from both ends. If the bar is short (compared to the internal length parameter), or if localization takes place in the proximity of a boundary, the averaging operator may need some adjustment. For the nonlocal averaging derived from the implicit gradient model, this adjustment consists in using the Green function constructed on a finite interval $[0, L]$. It is important to realize that in this case the Green function depends on the choice of boundary conditions. The implicit gradient model typically uses the homogeneous Neumann boundary conditions, which are loosely motivated by the requirement that the transformed field $\bar{\kappa}$ corresponding to a uniform distribution of κ should also be uniform. This is a natural requirement in an infinite domain occupied by a macroscopically homogeneous material, but it becomes somewhat questionable in a finite domain. With a nonlocal operator preserving a uniform field, one can hope to exclude spurious boundary effects, but in reality it turns out that even for operators with this property, the boundary acts as an attractor for localization, at least in the one-dimensional setting; see [10] for more details.

Despite the foregoing criticism, the requirement that nonlocal averaging should preserve a uniform field is widely used by common integral-type nonlocal models, and it leads to the normalizing condition

$$\int_0^L \alpha(x,\xi)\,\mathrm{d}\xi = 1 \qquad \forall x \in [0,L] \tag{77}$$

There is more than one way of satisfying this condition. The most widely used technique simply scales $\alpha_\infty(x-\xi)$ by the factor $1/V_r(x)$ where

$$V_r(x) = \int_0^L \alpha_\infty(x-\xi)\,\mathrm{d}\xi \tag{78}$$

is the area under the graph of $\alpha_\infty(x-\xi)$ taken for a fixed x as a function of ξ and considered only for $\xi \in [0,L]$. Clearly, $V_r(x)$ is equal (or very close) to 1 for x sufficiently far from the boundary and decreases to the minimum value 0.5 as x approaches the boundary. The scaled weight function, defined as

$$\alpha(x,\xi) = \frac{1}{V_r(x)}\alpha_\infty(x-\xi) \tag{79}$$

is not symmetric with respect to x and ξ but satisfies the normalizing condition (77). Fig. 11 shows the scaled weight function derived from the truncated polynomial function (75) in the proximity of the boundary at $x = 0$. In Fig. 11a, $\alpha(x,\xi)$ is plotted as a function of x for fixed values of $\xi = 0$, 0.5, 1, and 1.5. In Fig. 11b, $\alpha(x,\xi)$ is plotted as a function of ξ for fixed values of $x = 0$, 0.5, 1, and 1.5.

By inspection of (70) we find that the Green function is symmetric with respect to x and ξ, even on a finite domain. This is related to the fact that

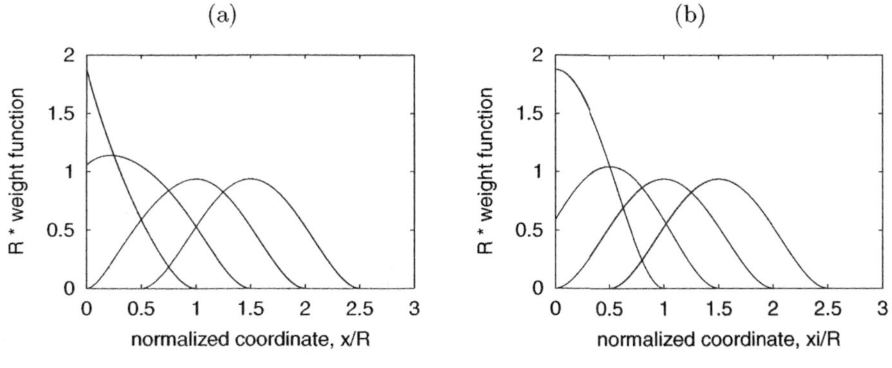

Fig. 11: Scaled weight function $\alpha(x,\xi)$ as a function of (**a**) x, (**b**) ξ

the differential operator used by the implicit gradient model is self-adjoint. A definition that provides a symmetric and normalized weight function "derived" from an arbitrary weight function $\alpha_\infty(x-\xi)$ has recently been proposed in [4, 17]. The idea is to add to α_∞ a multiple of the Dirac distribution according to the formula

$$\alpha(x,\xi) = \alpha_\infty(x-\xi) + [1 - V_r(x)]\,\delta(x-\xi) \tag{80}$$

Far from the boundary, the factor $1 - V_r(x)$ vanishes and the original weight function $\alpha_\infty(x-\xi)$ is recovered.

Once the choice of the weight function is made, an integral-type nonlocal plasticity model can be constructed along the same lines as the implicit gradient plasticity model described in the previous section. All the basic equations remain valid, with the transformed internal variable $\bar{\kappa}$ interpreted as the nonlocal average defined in (72). Same as for the implicit gradient model, the formulation using the simple softening law (55) turns out to be equivalent to a cohesive zone model and does not act as a true localization limiter. With a more sophisticated softening law in the form (61), localization of plastic strain into a single cross section is excluded, provided that the "local" hardening modulus $H_{\rm L}$ is positive. Plastic strain localizes into a finite zone if the overall modulus $H_{\rm L} + H_{\rm NL}$ reaches a critical negative value that depends on the bar length and for very long bars is close to zero.

An integral nonlocal model of the above type was proposed in [24] and later analyzed in [16, 20, 19]. In the case of linear softening, the softening law (61) reads

$$\sigma_{\rm Y} = \sigma_0 + H\,[m\bar{\kappa} + (1-m)\kappa] \tag{81}$$

where $H < 0$ is the overall softening modulus valid for uniform yielding and m is a fixed dimensionless parameter. Identifying $H_{\rm L} = (1-m)H$ and $H_{\rm NL} = mH$, we find that the condition $H_{\rm L} > 0$ is satisfied if $m > 1$. For this range of parameters m, the model acts as a true localization limiter. The distribution of plastic strain rates at bifurcation from a homogeneous state can be found from the consistency condition combined with the equilibrium condition, which finally leads to

$$m\int_0^L \alpha(x,\xi)\dot{\kappa}(\xi)\,{\rm d}\xi + (1-m)\dot{\kappa}(x) = \frac{\dot{\sigma}}{H} \tag{82}$$

This is a Fredholm integral equation of the second kind. It should be enforced only in the plastic zone $I_{\rm p}$ while outside this zone the rate $\dot{\kappa}(x)$ vanishes. The equation can be solved for different assumed plastic intervals $I_{\rm p} \subset [0, L]$, but the solution is admissible only if it satisfies the loading condition $\dot{\kappa}(x) \geq 0$ for all $x \in I_{\rm p}$. So the length of the interval $I_{\rm p}$ must be adjusted iteratively until an admissible solution is found. The resulting profiles of $\dot{\kappa}(x)$ are shown in Fig. 12 for different values of m. For $m \to 1^+$, the size of the plastic zone

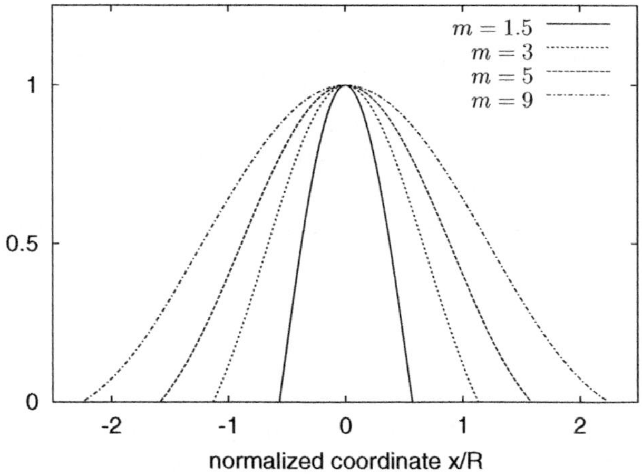

Fig. 12: Localized plastic strain profiles for different values of parameter m

L_p tends to zero, and for $m \to \infty$ it tends to infinity. For a fixed value of m, it is proportional to the internal length imposed by the nonlocal averaging operator through the length parameter that appears in the definition of the weight function, e.g., l in (74) or R in (75). Of course, if the weight function is set equal to the Green function (70), the solution is exactly the same as that obtained in Section 5 for equivalent parameters $H_L = (1 - m)H$ and $H_{NL} = mH$.

One particular property of the present model is that the size of the plastic interval and the shape of the plastic strain profile do not change during the softening process. This is because equation (82) remains valid for finite increments of plastic strain if the stress rate on the right-hand side is replaced by the finite stress increment. Even if the linear softening law (81) is replaced by a nonlinear one in the form

$$\sigma_Y = h_0 \left(m\bar{\kappa} + (1 - m)\kappa \right) \tag{83}$$

where h_0 is a given function describing the evolution of the yield stress in the case of uniform yielding, the distribution of plastic strain can be solved from (82) with $\dot{\kappa}$ replaced by κ and the right-hand side replaced by the value κ_0 for which $h_0(\kappa_0) = \sigma =$ current stress level.

7 Concluding Remarks

We have presented four selected formulations of softening plasticity that prevent the boundary value problem from becoming ill-posed. From the physical points of view, they act as localization limiters that control the size of the localized plastic zone and enforce nonzero dissipation. Three of these formulations are based on gradient-type enrichments and the fourth on an integral-type enrichment of the governing equations. They are meant as representative examples of enriched continuum theories that regularize problems with strain localization. By analyzing the localization problem in the one-dimensional setting, we have been able to demonstrate some basic properties using analytical or semi-analytical solutions. Generalization to multiple dimensions is relatively straightforward but of course requires more powerful numerical techniques.

References

[1] E. C. Aifantis. On the microstructural origin of certain inelastic models. *Journal of Engineering Materials and Technology, ASME*, 106:326–330, 1984.

[2] E. C. Aifantis. The physics of plastic deformation. *International Journal of Plasticity*, 3:211–247, 1987.

[3] Z. P. Bažant. *Scaling of Structural Strength.* Hermes-Penton, London, 2002.

[4] G. Borino, B. Failla, and F. Parrinello. Nonlocal damage theory with self-adjoint integral kernel. In *Proceedings of the 6th National Congress SIMAI*, Chia Laguna, Italy, May 2002. CD-ROM.

[5] C. Comi and U. Perego. A generalized variable formulation for gradient dependent softening plasticity. *International Journal for Numerical Methods in Engineering*, 39:3731–3755, 1996.

[6] R. A. B. Engelen, M. G. D. Geers, and F. P. T. Baaijens. Nonlocal implicit gradient-enhanced elasto-plasticity for the modelling of softening behaviour. *International Journal of Plasticity*, 19:403–433, 2003.

[7] N. A. Fleck and J. W. Hutchinson. Strain gradient plasticity. In J. W. Hutchinson and T. Y. Wu, editors, *Advances in Applied Mechanics*, volume 33, pages 295–361. Academic Press, New York, 1997.

[8] M. G. D. Geers, R. A. B. Engelen, and R. J. M. Ubachs. On the numerical modelling of ductile damage with an implicit gradient-enhanced formulation. *Revue européenne des éléments finis*, 10:173–191, 2001.

[9] M. Jirásek and Z. P. Bažant. *Inelastic Analysis of Structures.* John Wiley and Sons, Chichester, 2002.

[10] M. Jirásek and S. Rolshoven. Comparison of integral-type nonlocal plasticity models for strain-softening materials. *International Journal of Engineering Science*, 2003. In press.

[11] M. Jirásek and S. Rolshoven. On gradient plasticity models for strain-softening materials. *International Journal of Solids and Structures*, 2003. Submitted for review.

[12] T. Liebe and P. Steinmann. Theory and numerics of a thermodynamically consistent framework for geometrically linear gradient plasticity. *Inter-*

national Journal for Numerical Methods in Engineering, 51:1437–1467, 2001.

[13] R. D. Mindlin. Second gradient of strain and surface tension in linear elasticity. *International Journal of Solids and Structures*, 1:417–438, 1965.

[14] J. Pamin. *Gradient-dependent plasticity in numerical simulation of localization phenomena.* PhD thesis, Delft University of Technology, Delft, The Netherlands, 1994.

[15] R. H. J. Peerlings, R. de Borst, W. A. M. Brekelmans, and J. H. P. de Vree. Gradient-enhanced damage for quasi-brittle materials. *International Journal for Numerical Methods in Engineering*, 39:3391–3403, 1996.

[16] J. Planas, G. V. Guinea, and M. Elices. Basic issues on nonlocal models: uniaxial modeling. Technical Report 96-jp03, Departamento de Ciencia de Materiales, ETS de Ingenieros de Caminos, Universidad Politécnica de Madrid, Ciudad Universitaria sn., 28040 Madrid, Spain, 1996.

[17] C. Polizzotto. Remarks on some aspects of nonlocal theories in solid mechanics. In *Proceedings of the 6th National Congress SIMAI*, Chia Laguna, Italy, May 2002. CD-ROM.

[18] C. Polizzotto and G. Borino. A thermodynamics-based formulation of gradient-dependent plasticity. *European Journal of Mechanics/A: Solids*, 17:741–761, 1998.

[19] S. Rolshoven and M. Jirásek. On regularized plasticity models for strain-softening materials. In R. de Borst, J. Mazars, G. Pijaudier-Cabot, and J. G. M. van Mier, editors, *Fracture Mechanics of Concrete Structures*, pages 617–624, Lisse, 2001. Balkema.

[20] L. Strömberg and M. Ristinmaa. FE-formulation of a nonlocal plasticity theory. *Computer Methods in Applied Mechanics and Engineering*, 136:127–144, 1996.

[21] T. Svedberg and K. Runesson. A thermodynamically consistent theory of gradient-regularized plasticity coupled to damage. *International Journal of Plasticity*, 13:669–696, 1997.

[22] R. A. Toupin. Elastic materials with couple-stresses. *Archive for Rational Mechanics and Analysis*, 11:385–414, 1962.

[23] K. C. Valanis. A gradient theory of internal variables. *Acta Mechanica*, 116:1–14, 1996.

[24] P. A. Vermeer and R. B. J. Brinkgreve. A new effective non-local strain measure for softening plasticity. In R. Chambon, J. Desrues, and I. Vardoulakis, editors, *Localisation and Bifurcation Theory for Soils and Rocks*, pages 89–100, Rotterdam, 1994. Balkema.

Fuzzy data in geotechnical models

Michael Oberguggenberger

Institut für Technische Mathematik, Geometrie und Bauinformatik
Universität Innsbruck, michael@mat1.uibk.ac.at

1 Introduction

This paper addresses the question of data uncertainty in geotechnical models. An adequate understanding of the influence of input parameter variability on the output of geotechnical computations requires that the uncertainty itself is captured in mathematical terms. Our goal is to present some of the methods that have been developed for this task and now usually are subsumed under the heading "imprecise probability" (e.g. [5, 41]). Two essential features are that, first, the appropriate model of the data uncertainty should reflect and incorporate the level of information available on the data and, second, must be able to transmit it through numerical computations and deliver an output whose uncertainty is formulated in the same terms. These issues are important for calibrating, verifying and validating models in geotechnical sciences, but even more so in practical design and safety analysis.

To specify more clearly the role of uncertainty analysis, let us have a short look at scientific modelling in engineering. On the one side, there is reality (with soils, materials etc.). On the other side, there is the model of reality (formulated in mathematical terms and containing physical laws and constitutive equations) plus correspondence rules (prescribing how to translate one into the other). The physical model defines state variables and parameters. The values of the parameters have to be determined from information extracted from the real world and serve as input in the physical model. This plus the design of the structure enters in numerical computations that result in the proper dimensions, construction procedures, assessment of safety measures. The actual implementation usually undergoes several transitional phases of construction - each requiring different modelling - until the final structure is completed. Ideally, the model output should provide a design that works, reliable guidelines for action, and aids for decision making (as emphasized e.g. in [8]). Here the question of accuracy, or at least robustness, becomes decisive. Predicted data are valuable only to the extent that their uncertainty and possible fluctuations have been assessed as well.

Major uncertainties in the procedure just outlined are of course the choice of soil model (continuum or granular), the adequate selection of state variables and parameters (constant or not), the proper limit state function (failure due to bounds exceeded by average values or due to localized disturbances?), the proper incorporation of transitional states of construction, and parameter variability. This latter point is the issue to be addressed in the sequel.

Parameter variability can be attributed to a large number of causes. There are random fluctuations, lack of information, random measurement errors, but also systematic measurement errors (deriving e.g. from uncontrollable changes of the properties of the material caused by its extraction in bore holes). There are fluctuations due to spatial inhomogeneity, errors made by assigning parameter status to state variables; and variability arising from the fact that parameters have to carry the burden of model insufficiency. The available information on data uncertainty may range from frequency distributions obtained from large samples, values from small samples or single measurements, interval bounds, to experts' point estimates and educated guesses from experience.

We believe that different mathematical approaches should be used, adapted to the type of information available. In Section 2 we shall present a spectrum ranging from sensitivity analysis, interval analysis, fuzzy sets, possibility theory, random sets to probability theory. All these theories come with appropriate correspondence rules (from model to reality and vice versa, often referred to as semantics) and can be correctly and consistently entered in numerical computations. In Section 3 we focus on computations with fuzzy sets and exhibit some applicative examples (vibrations of bell towers subject to an elastic soil coupling and a finite element simulation with fuzzy parameters in tunnelling).

In conclusion, we believe that formulating, modelling and evaluating the uncertainty in input and output are necessary requirements in engineering computations. Probability theory is not the only means of doing so; the last decades have seen the development of a number of theories with correspondence rules that allow to formalize different levels of information more accurately. Nevertheless, whatever modelling approach is chosen, uncertainty is there and persists. A possible strategy in the face of this fact could be a shift from quantitative prediction to qualitative description of scenarios and to explanation. A better understanding of possible system responses can be gained through uncertainty analysis. In this sense one should understand the role of practical engineering not so much as the production of factual knowledge but rather as decision making with the help of scientific tools.

For additional theoretical background on the concepts discussed here we refer the interested reader to the following sources: interval analysis [26], fuzzy sets [2, 7, 19, 20, 42, 43], possibility theory [7, 45], random sets [37, 42]. As a selection of references related to the more specific applications envisaged here we mention: fuzzy sets in engineering [12, 15, 18, 22, 24, 25, 30], especially in geotechnics [6, 10, 11] [13, 14, 21, 31], random sets and possibility theory in engineering [16, 17, 33, 38, 39], probability in geotechnics [1, 8, 10, 35], comparative studies [9, 20, 23, 32, 40]; see also the unifying concept of *clouds*, recently proposed by [28].

2 Models of parameter variability

This section serves to discuss a variety of mathematical descriptions of uncertainty. We shall consider an input parameter which we call (upper case) A, while the specific values it may take (realizations) will be denoted by lower case letters. Further, the computational engineering model produces an output as a function $F(A)$ of the input A.

a. Deterministic values: The simplest approach is what in engineering terminology usually is called *deterministic description*, that is, the parameter A is described by a single value a, see Figure 1.

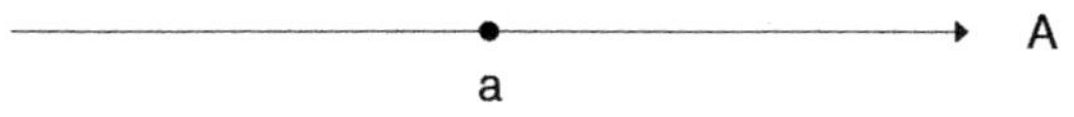

Fig. 1: Deterministic value a of parameter A

The semantics is simply that a is an expert estimate (or educated guess). However, even in the deterministic approach, much more can actually be said about the influence of input variability on the output, namely, one can perform a sensitivity analysis. In its simplest form, it might consist in the computation of the outputs $F(a_1), \ldots, F(a_n)$ corresponding to several input values. In more advanced form, it might involve Taylor expansion of the function F at the point a and information about the gradient or higher derivatives of F at a.

b. Intervals: The next level in modelling uncertainty is interval analysis. The uncertainty of the input A is described by an interval $[a_L, a_R]$, signifying bounds in terms of a worst/best case assumption. This way, the total variability is captured, but no detailed information on the uncertainty is provided (Figure 2). The functional evaluation $F(A)$ results in a set of values (an interval, if F is continuous) and subsumes again the full range of the possible output without further fine structure.

Fig. 2: Interval bounds for parameter A

c. Fuzzy sets: In a first interpretation, a fuzzy set A (over the real numbers) is a family of parametrized intervals. The parametrization is done in terms of levels $\alpha, 0 \leq \alpha \leq 1$. Each level α has a corresponding interval A^α so that $A^\beta \subset A^\alpha$ if $\alpha \leq \beta$. Thus the intervals are stacked and can be depicted by their left/right contour functions, see Figure 3. More generally, one could allow the A^α to be arbitrary, stacked subsets of a given set of objects under investigation (complex numbers, vectors, matrices, functions or the like).

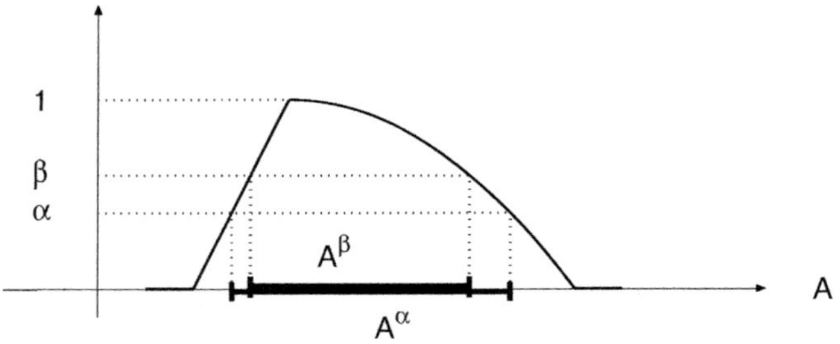

Fig. 3: Fuzzy set as family of intervals

In a second interpretation, the contour function is taken as the primary object, and a fuzzy set A (over the real numbers) is just a map from the real line to the interval $[0,1]$, assigning to each real number a a value $\pi_A(a) \in [0,1]$. This value may be interpreted as the membership degree to which a belongs to the fuzzy set A, or in the language of parameters, as the degree of possibility that the parameter A takes the value a, see Figure 4. In classical set theory, the membership degree is either 0 or 1; fuzzy set theory allows gradual membership as well. The intervals from the first interpretation are now the level sets $A^\alpha = \{a : \pi_A(a) \geq \alpha\}$.

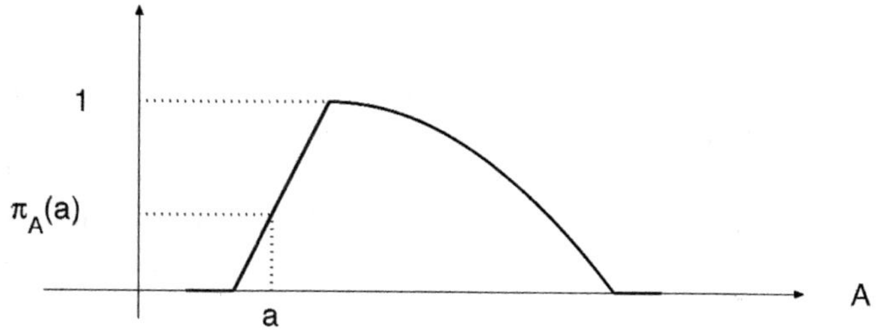

Fig. 4: Fuzzy set as degree of possibility

The evaluation of a function F of fuzzy parameters will be discussed in detail in the next section.

d. Possibility theory: The notion of *possibility* provides an interpretation for a fuzzy set and an operational method of constructing it. We start by interpreting the scale $0 \leq \alpha \leq 1$ as signifying risk levels (*risk* meant in a colloquial sense) that the given parameter fluctuates in the interval A^α. The scale serves to encode subjective risk assessments and has to be defined verbally by the designing engineer in advance, but then remains fixed during the whole modelling process. By default, level $\alpha = 1$ signifies the standard value of parameter A, $\alpha = 2/3, 1/3, 0$ might indicate high, medium, and low risk, respectively, see Figure 5. The level α is termed *degree of possibility.*

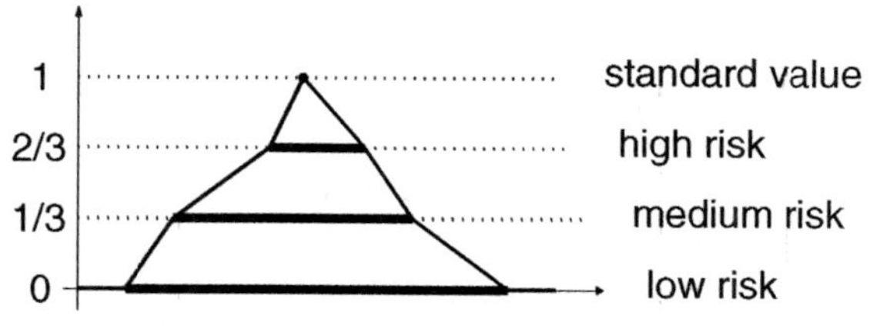

Fig. 5: Fuzzy set as intervals with risk value

It is envisaged that the risks leading to parameter fluctuations at the corresponding possibility level are established in an analysis of scenarios. We refer to [22] where the procedure has been demonstrated in the case of network analysis of a civil engineering project. An alternative way of establishing the semantics of *possibility* is to start from the notion of *potential surprise* and define possibility as its complementary notion [36] or as a transformed quantity [27] thereof.

Similarly to probability theory, one can introduce a *possibility distribution* on the underlying set, defining a degree of possibility for each subset by $\pi_A(S) = \sup\{\pi_A(a) : a \in S\}$, giving the degree of possibility that the parameter A takes a value in S. The possibility distribution is monotonic, i.e., $\pi_A(S) \le \pi_A(T)$ if $S \subset T$, but not additive (as opposed to a probability distribution).

e. Random sets: A *random set*, sometimes also referred to as a *Dempster-Shafer structure*, is given by finitely many subsets $A_i, i = 1, \dots, n$ of a given set, called the *focal sets*, each of which comes with a *probability weight* $m(A_i), \sum m(A_i) = 1$. In a frequentist interpretation, this might correspond to a sample of size n of interval data for a parameter A, the probability weight being approximated by the relative frequency. The difference to a histogram then is that the focal sets A_i may overlap. In a subjectivist interpretation, the focal sets A_i may be (possibly conflicting) estimates given by different experts and the weights might correspond to each expert's relative credibility. In any case, they can be interpreted as describing supporting sets for unknown probability distributions, see Figure 6.

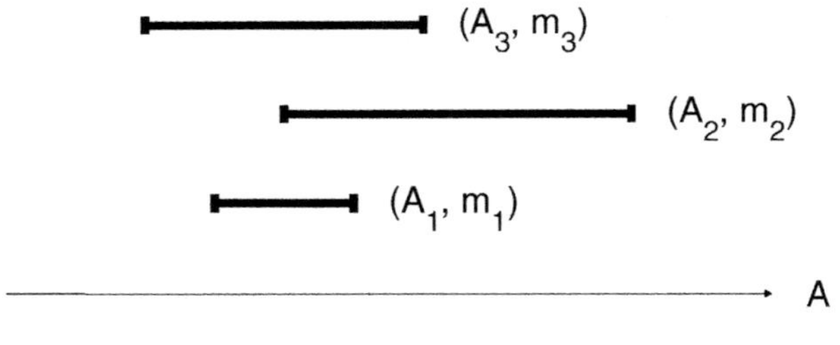

Fig. 6: A random set

In this way one may derive lower and upper probabilities for any set S by

$$\underline{P}(S) = \sum_{A_i \subset S} m(A_i)\,, \quad \overline{P}(S) = \sum_{A_i \cap S \neq \emptyset} m(A_i)\,,$$

which alternatively are interpreted as the degree of belief and the degree of plausibility, respectively, that the realizations of parameter A lie in S. The *contour function* of a random set is the function $a \to \overline{P}(\{a\})$, assigning each singleton a its plausibility. This may be viewed as the membership degree defining a fuzzy set and thus in turn provides a third interpretation for fuzzy sets (Figure 7).

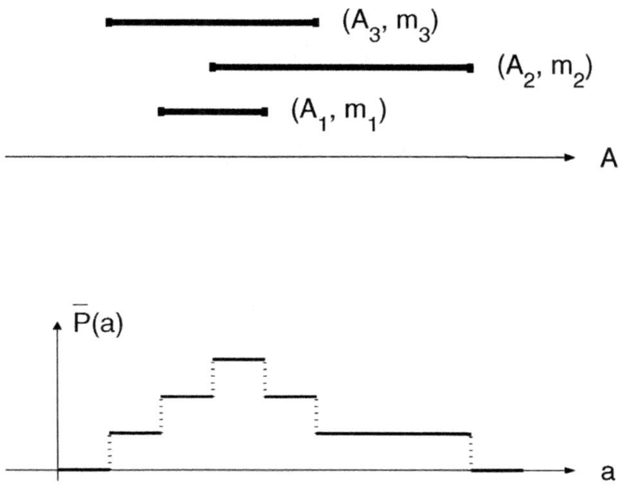

Fig. 7: Contour function of a random set

f. Probability: The most informative, but also most stringent description of the uncertainty of a parameter A is by means of *probability*. If the probability distribution is given by a density $p_\lambda(a)$, the probability that the realizations of the parameter A lie in a set S is

$$P(A \in S) = \int_S p_\lambda(a)\, \mathrm{d}a\,,$$

see Figure 8.

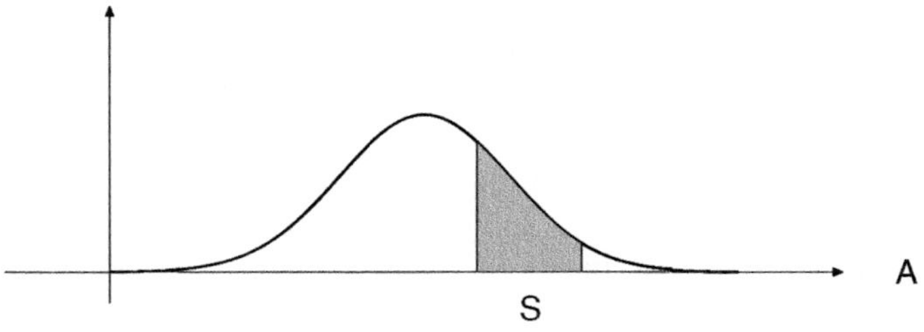

Fig. 8: Probability distribution

The notation p_λ indicates that, usually, the probability distributions arise as members of a class of distributions which in turn are parametrized by parameters λ. For example, the class of Gaussian normal distributions $\mathcal{N}(\mu, \sigma^2)$

is given by the Gaussian densities $p_\lambda(a)$ with parameters $\lambda = (\mu, \sigma)$. Thus the correct specification of a probability distribution requires determination of the type it belongs to as well as the values of its parameters.

3 Computing with fuzzy sets

In this section we take up the question of propagating the uncertainty of a fuzzy input A through a function F. If the input is a fuzzy set with membership function $\pi_A(a)$, say, the output will also be a fuzzy set, described by a membership function $\pi_{F(A)}(b)$. According to the *Zadeh extension principle* [44], it is given by

$$\pi_{F(A)}(b) = \sup\big\{\pi_A(a) : F(a) = b\big\}\,.$$

In case the input consists of a vector of parameters $A = (A_1, \ldots, A_m)$, the extension principle takes the form

$$\pi_{F(A)}(b) = \sup\big\{\min\big(\pi_{A_1}(a_1), \ldots, \pi_{A_m}(a_m)\big) : F(a_1, \ldots, a_m) = b\big\}\,.$$

Note that this comes naturally from the possibility interpretation: to compute the degree of possibility that $F(A)$ takes the value b, one has to look for all combinations $a_1, \ldots, a_m$ producing the value b; each single combination gets the smallest possibility among its participants, while b gets the supremum of all possibility degrees that can be obtained in this way. In case F is continuous and the level sets of $A_1, \ldots, A_m$ are compact $(0 < \alpha \le 1)$, this corresponds exactly to computing the range of the function F on each level set,

$$F(A)^\alpha = F(A^\alpha)\,, \quad \text{respectively,} \quad F(A_1, \ldots, A_m)^\alpha = F(A_1^\alpha, \ldots, A_m^\alpha)\,.$$

When the A_j are intervals, in addition, the set $F(A_1, \ldots, A_m)$ is an interval as well. The computation of its boundaries is then a task of global optimization: finding the minimum and maximum value of F on the set A^α. In any case, the procedure is consistent: if the input data are structured as stacked intervals, so is the output. For further reading on using the extension principle in differential equations, see [3, 4, 34]; in finite element computations, see [12, 14, 23].

We now turn to presenting practical examples related to geotechnical questions. The first example concerns bell-induced vibrations of bell-towers. In a simplified model [29], the situation is described by two coupled pendulums obeying the system of ordinary differential equations

$$\begin{aligned}
&\psi'' L_1^2(m_1 + m_2) + m_1 L_1 L_2[\phi''(\cos\phi + \psi\sin\phi) - \phi'^{\,2}(\sin\phi - \psi\cos\phi)] \\
&\qquad\qquad + g\psi(m_1 L_1 + m_2 L_1) + k\psi = 0\,, \\
&\phi''(m_2 L_2^2 + \theta_2) + m_2 L_1 L_2[\psi''(\cos\phi + \psi\sin\phi) + \psi'^{\,2}(\sin\phi - \psi\cos\phi)] \\
&\qquad\qquad + m_2 L_2 g\sin\phi = 0\,,
\end{aligned}$$

where ψ_j denote the angles of inclination of the tower and the bell, L_j their lengths, m_j their masses, g the acceleration due to gravity and k a stiffness constant describing the coupling of the soil/tower (assumed elastic), see Figure 9.

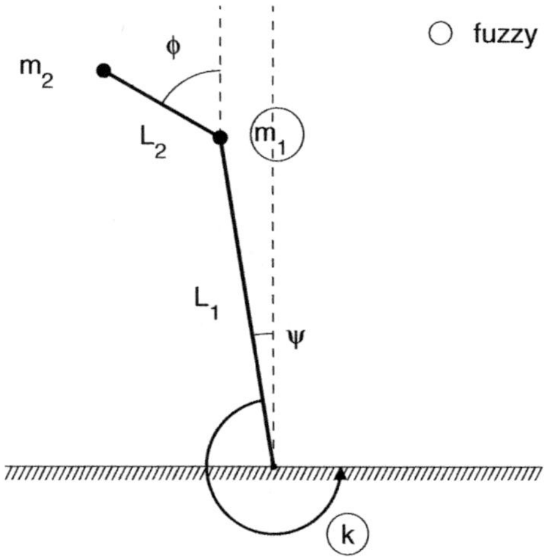

Fig. 9: Bell-tower and bell as double pendulum

If the investigation concerns an existing tower, all data can be measured sufficiently precisely except for the mass m_1 of the tower and the stiffness constant k. Thus we model these parameters by fuzzy numbers; for the purpose of our example, we choose them as depicted in Figure 10.

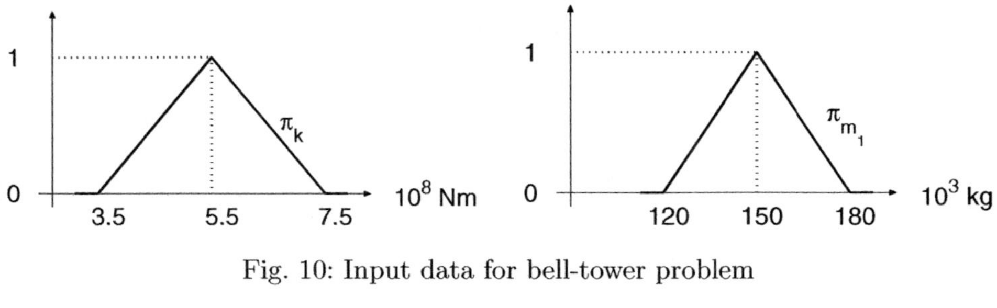

Fig. 10: Input data for bell-tower problem

The remaining parameters have been assigned deterministic values (see [30] for more details on this problem). We then compute the output $\phi(t), \psi(t)$

of angles of inclination at time t, depending on the fuzzy parameters m_1, k according to the extension principle. We may depict the results as fuzzy numbers depending on time t either by showing the level sets (Figure 11) or by plotting the time-depending membership degree as a three-dimensional plot (Figure 12).

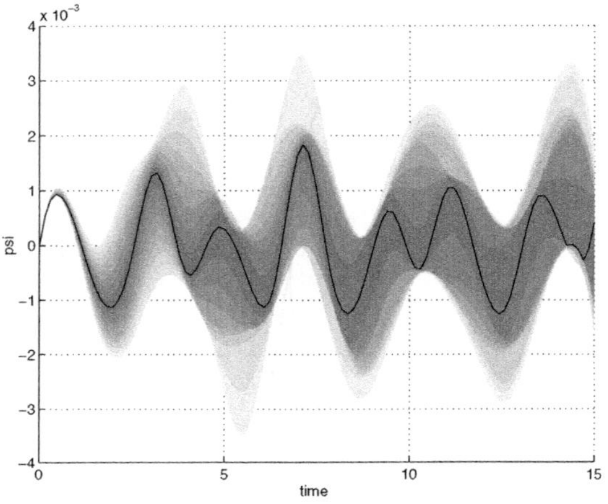

Fig. 11: Level sets for angle of inclination $\psi(t)$

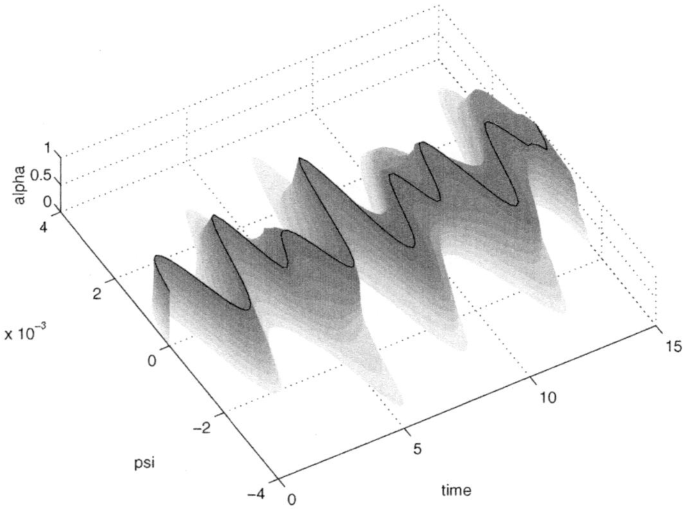

Fig. 12: Fuzzy angle of inclination $\psi(t)$

Further functions of the output of interest, like the dominant frequency, can be computed from there, making use of the extension principle again.

The second example deals with a finite element model of a tunnel excavation according to the New Austrian Tunnelling Method. The model is two dimensional, the tunnel is assumed to be circular, the soil is described by a Drucker-Prager model. As is common modelling practice, the finite element computation proceeds in stages: first the cavity is formed and the stresses at fixed tunnel surface are computed. Second, the stresses are reduced by a certain factor A and the corresponding displacements at the tunnel surface are computed. Third, the lining is put in place. The final output consists of variables of interest, like the normal stress or the bending moment in the lining. For the sake of demonstration, we assume that the modulus of elasticity E of the soil and the reduction factor A are fuzzy numbers as given in Figure 13. As a sample output, the fuzzy bending moments in the lining are shown in Figure 14, the shading corresponding to the level sets as indicated.

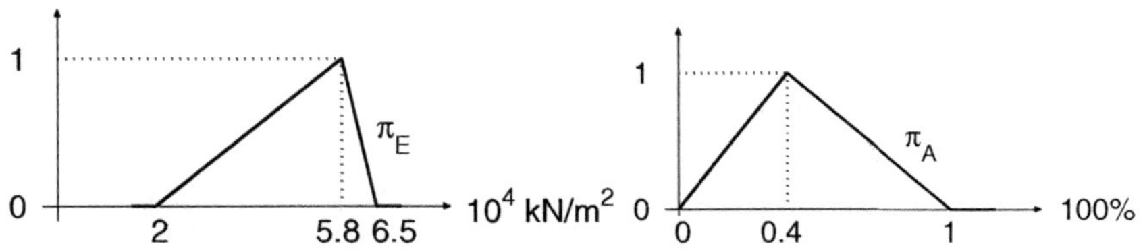

Fig. 13: Fuzzy modulus of elasticity and reduction factor

Conclusion: In the paper we argued that, for a complete geotechnical analysis, it is mandatory to address and model data uncertainty on mathematical terms. We saw that this is indeed possible, and that simple models do exist; a fuzzy approach appeared particularly robust. A useful analysis of output variability is possible, as is the implementation in numerical schemes, based on the extension principle. This is but one step, albeit an important one, in understanding the uncertainties that come along in engineering modelling. Explaining and addressing data variability not only complements theory and numerics, but actually is indispensable for producing results that can be meaningfully interpreted. Fuzzy set theory can serve as a helpful tool in this regard, either in case the available information on data uncertainty is vague, or in case one wants to get a qualitative understanding of the output range, using a description of input variability by stacked intervals.

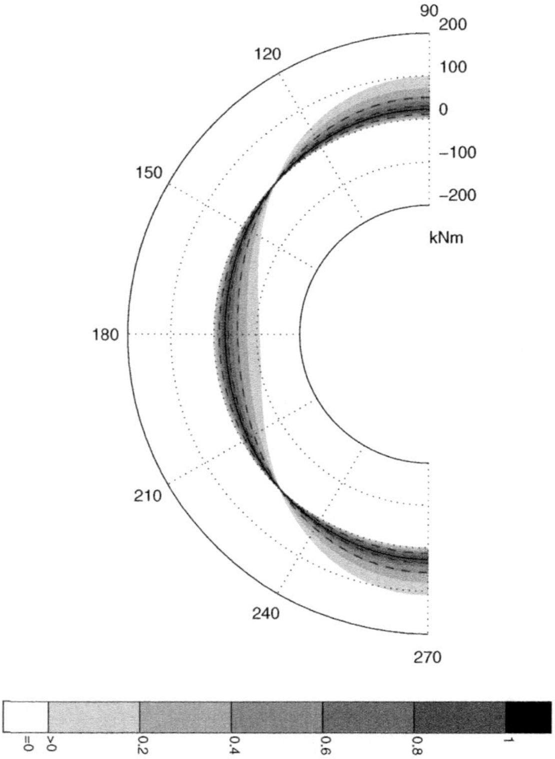

Fig. 14: Fuzzy bending moment

Acknowledgements: I am grateful to Thomas Fetz who programmed and performed the fuzzy FE-analysis of the tunnelling example and generously allowed me to reproduce his results in Figure 14. Computations for the bell-tower example from [30] have been implemented by Simon Pittschmann.

References

[1] Auvinet G, González J L (2000) Computers and Geotechnics 26:247-261
[2] Bandemer H, Näther W (1992) Fuzzy Data Analysis. Kluwer Acad Publ, Dordrecht
[3] Bonarini, A Bontempi G (1994) ACM Trans Modelling Comp Sim 4:285–313.
[4] Buckley J J (1992) Fuzzy Sets and Systems 50:1–14.

[5] de Cooman G, Fine T, Seidenfeld T (2001) ISIPTA'01, Proceedings of the Second Symposium on Imprecise Probabilities and Their Applications. Shaker Publ BV, Maastricht

[6] Dodagoudar G R, Venkatachalam G (2000) Computers and Geotechnics 27:101-115

[7] Dubois D, Prade H (1988) Possibility Theory. Plenum Press, New York

[8] Einstein H H (2001) Felsbau 19(5):72–84

[9] Elishakoff I (1999) Whys and Hows in Uncertainty Modelling. Probability, Fuzziness and Anti-Optimization. Springer-Verlag, New York

[10] Fellin W, Oberguggenberger M (2002) From probability to fuzzy sets: the struggle for meaning in geotechnical risk assessment. In: Pöttler R, Klapperich H, Schweiger H F (eds) Probabilistics in GeoTechnics: Technical and Economic Risk Estimation. United Engineering Foundation, Verlag Glückauf GmbH, Essen: 29–38

[11] Fellin W, Oberguggenberger M (2002) Das Problem der Unsicherheit der Versagenswahrscheinlichkeit und mögliche Alternativen. In: Workshop stochastische Prozesse in der Geotechnik. Weimar: to appear

[12] Fetz Th (1997) Finite element method with fuzzy parameters. In: Troch I, Breitenecker F (eds) Proceedings IMACS Symposium on Mathematical Modelling. ARGESIM Report 11:81–86 Vienna

[13] Fetz Th, Hofmeister M, Hunger G, Jäger J, Lessmann H, Oberguggenberger M, Rieser A, Stark R F (1997) Bauingenieur 72:33–40

[14] Fetz Th, Jäger J, Köll D, Krenn G, Lessmann H, Oberguggenberger M, Stark R (1999) Computer-Aided Civil and Infrastructure Engineering 14:93–106

[15] Fetz Th, Oberguggenberger M, Pittschmann S (2000) International Journal of Uncertainty, Fuzziness and Knowledge-Based Systems 8:295–309

[16] Fetz Th (2001) Sets of joint probability measures generated by weighted marginal focal sets. In: de Cooman G, Fine T, Seidenfeld T (eds) ISIPTA'01, Proceedings of the Second Symposium on Imprecise Probabilities and Their Applications. Shaker Publ BV, Maastricht: 171–178

[17] Hall J, Lawry J (2001) Imprecise probabilities of engineering system failure from random and fuzzy set reliability analysis. In: de Cooman G, Fine T, Seidenfeld T (eds) ISIPTA'01, Proceedings of the Second Symposium on Imprecise Probabilities and Their Applications. Shaker Publ BV, Maastricht: 195–204

[18] Haldar K, Reddy R K (1992) Fuzzy Sets and Systems 48:201–210.

[19] Kandel A (1986) Fuzzy Mathematical Techniques with Applications. Addison-Wesley, Reading

[20] Klir G J, Wiermann M J (1998) Uncertainty-Based Information. Elements of Generalized Information Theory. Physica-Verlag, Heidelberg.

[21] Leitgeb M (2001) Grundbruchberechnung mit unscharfen Bodenkennwerten. Dipl Ing Thesis, Leopold-Franzens-Universität, Innsbruck

[22] Lessmann H, Mühlögger J, Oberguggenberger M (1994) Der Bauingenieur 69:469–478

[23] de Lima B S L P, Ebecken N F F (2000) Finite Elements in Analysis and Design 34:211-232

[24] Möller B (1997) Bauingenieur 72:75–84

[25] Möller B (1999) Special Issue on Fuzzy Modelling. Computer Aided Civil and Infrastructure Enginering 14:2

[26] Neumaier A (1990) Interval Methods for Systems of Equations. Cambridge University Press, Cambridge

[27] Neumaier A (2001) Fuzzy modeling in terms of surprise. Preprint, Universität Wien,
http://www.mat.univie.ac.at/ neum/papers.html#surprise

[28] Neumaier A (2002) Clouds, fuzzy sets and probability intervals. Preprint, Universität Wien,
http://www.mat.univie.ac.at/ neum/papers.html#cloud

[29] Niederwanger G (1991) Über die Horizontalkraft schwingender Glocken mit horizontal verschiebbarem Auflager. Projektbericht, Universität Innsbruck

[30] Oberguggenberger M, Pittschmann S (1999) Mathematical and Computer Modelling of Dynamical Systems 5:181–202

[31] Oberguggenberger M, Russo F (2001) Fuzzy, probabilistic and stochastic modelling of an elastically bedded beam. In: de Cooman G, Fine T, Seidenfeld T (eds) ISIPTA'01, Proceedings of the Second Symposium on Imprecise Probabilities and Their Applications. Shaker Publ BV, Maastricht: 293–300

[32] Oberkampf W L, Helton J C, Sentz K (2001) Mathematical Representation of Uncertainty. American Institute of Aeronautics and Astronautics, Reston VA, AIAA 2001 - 1645 Non-Deterministic Approaches Forum, Seattle, WA

[33] Oberkampf W L, Helton J C (2002) Investigation of Evidence Theory for Engineering Applications. American Institute of Aeronautics and Astronautics, Reston VA, AIAA 2002 - 1569 Non-Deterministic Approaches Forum, Seattle, WA

[34] Otto K N, Lewis A D, Antonsson E K (1995) Fuzzy Sets and Systems 70:15–29

[35] Rackwitz R (2000) Computers and Geotechnics 26:199-223

[36] Shackle G L S (1969) Decision - order and time in human affairs. Cambridge Univ Press, Cambridge

[37] Shafer G (1976) A Mathematical Theory of Evidence. Princeton University Press, Princeton NJ

[38] Tonon F, Bernardini A, Mammino A (2000) Reliabilty Engineering & System Safety 70:241–261

[39] Tonon F, Bernardini A, Mammino A (2000) Reliabilty Engineering & System Safety 70:263–282

[40] Tonon F, Bernardini A, Elishakoff I (2001) Chaos, Solitons and Fractals 12:1403–1414

[41] Walley P (1991) Statistical Reasoning with Imprecise Probabilities. Chapman & Hall, London
[42] Wolkenhauer O (2001) Data Engineering. Fuzzy Mathematics in Systems Theory and Data Analysis. John Wiley & Sons, Inc, New York
[43] Zadeh L (1965) Information and Control 8:338–353
[44] Zadeh L (1975) Information Sci 8:199-250; 8:301-357; 9:43–80
[45] Zadeh L A (1978) Fuzzy sets and systems 1:3-28

GPSR Compliance

The European Union's (EU) General Product Safety Regulation (GPSR) is a set of rules that requires consumer products to be safe and our obligations to ensure this.

If you have any concerns about our products, you can contact us on ProductSafety@springernature.com

In case Publisher is established outside the EU, the EU authorized representative is:

Springer Nature Customer Service Center GmbH
Europaplatz 3
69115 Heidelberg, Germany

Batch number: 10755244

Printed by Printforce, the Netherlands